# A LIVING COUNTRYSIDE?

# Perspectives on Rural Policy and Planning

Series Editors:
**Andrew Gilg**
University of Exeter, UK
**Professor Keith Hoggart**
Kings College, London, UK
**Professor Henry Buller**
University of Exeter, UK
**Professor Owen Furuseth**
University of North Carolina, USA
**Professor Mark Lapping**
University of South Maine, USA

*Other titles in the series*

# A Living Countryside?
## The Politics of Sustainable Development in Rural Ireland

*Edited by*

JOHN McDONAGH
*National University of Ireland, Galway*

TONY VARLEY
*National University of Ireland, Galway*

SALLY SHORTALL
*Queen's University Belfast, Northern Ireland*

**Routledge**
Taylor & Francis Group

LONDON AND NEW YORK

First published 2009 by Ashgate Publishing

2 Park Square, Milton Park, Abingdon, Oxon OX14 4RN
711 Third Avenue, New York, NY 10017, USA

*Routledge is an imprint of the Taylor & Francis Group, an informa business*

First issued in paperback 2016

**British Library Cataloguing in Publication Data**
A living countryside? : the politics of sustainable
    development in rural Ireland. -- (Perspectives on rural
    policy and planning)
    1. Rural development--Government policy--Ireland. 2. Rural
    development--Environmental aspects--Ireland. 3. Rural
    population--Ireland. 4. Ireland--Rural conditions.
    I. Series II. McDonagh, John. III. Varley, Tony.
    IV. Shortall, S.
    333.7'615'09415-dc22

**Library of Congress Cataloging-in-Publication Data**
McDonagh, John.
    A living countryside? : the politics of sustainable development in rural Ireland / by John
McDonagh, Tony Varley and Sally Shortall.
        p. cm. -- (Perspectives on rural policy and planning)
    Includes bibliographical references and index.
    ISBN 978-0-7546-4669-3 (hardcover)
    1. Rural development--Ireland. 2. Rural development--Government policy--Ireland.
    I. Varley, Tony. II. Shortall, Sally. III. Title.

HN400.3.Z9C643 2009
307.1'41209415--dc22

                                                                    2009005310

ISBN 978-0-7546-4669-3 (hbk)
ISBN 978-1-138-26441-0 (pbk)

# Contents

# List of Figures

# List of Tables

# Notes on Contributors

**John Barry** is Reader in the School of Politics, International Studies and Philosophy at Queen's University Belfast and a founding member of the Institute for a Sustainable World at QUB. His areas of research include – green moral and political theory; normative aspects of environmental and sustainable development politics policy; environmental governance; the greening of citizenship and civic republicanism; green politics in Ireland, North and South.

**Anne Byrne** is a Sociologist working in the School of Political Science and Sociology, National University of Ireland Galway. Her research and publication interests include social identities, rurality, gender, participative group and community relations, historical sociology and research methodologies with a focus on narrative and story.

**Peter Doran** is a Lecturer in sustainable development at the School of Law, Queens University, Belfast. He has worked in the environment and sustainable development policy area for over ten years at the United Nations, in the NGO sector, and in legislatures in both Northern Ireland and the Republic of Ireland. He is a senior editor and writer on the 'Earth Negotiations Bulletin' for the International Institute for Sustainable Development at United Nations negotiations on climate change and sustainable development.

**John Feehan** is a Senior Lecturer in the Department of Environmental Resource Management at University College Dublin where much of his research and teaching has focused on environmental heritage evaluation and management. A major concern of his recent work has been the decline of the natural and cultural heritage of the rural landscape and of rural communities. His recently published *Farming in Ireland: History, Heritage and Environment* has been widely acclaimed. He is Senior Editor of *Tearmann*, the *Irish Journal of Agri-environmental Research*.

**John Fennessy** joined COFORD as Programme Research Manager/Tree Improvement and Non Wood Forest Products in April 2002 having previously been Seed Manager and Manager of Tree Improvement with Coillte, the state forestry board. John is the Irish Representative on the Management Committee of the British and Irish Hardwoods Improvement Programme (BIHIP) as well as Chairman of the BIHIP Oak Group. He is National Coordinator of the European Forest Genetic Resources Programme (EUFORGEN) and is a technical member and Past President of the Society of Irish Foresters.

**Brendan Flynn** is a Lecturer in the Department of Political Science and Sociology, at NUI, Galway. His teaching and research interests include Irish and European environmental policies, EU policies, and European politics. He is the author of the *Blame Game: Rethinking Ireland's Sustainable Development and Environmental Performance* (Dublin: Irish Academic Press, 2006).

**Seamus Grimes** is a Personal Professor in the Department of Geography at NUI, Galway. He has published widely on a range of issues related to Information and Communication Technologies and regional and rural development in Ireland. Having started his academic career in Galway, Professor Grimes carried out his postgraduate studies at the University of New South Wales in Sydney. In recent years, his research has been focused on foreign investment by technology corporations in Ireland and on the challenge that Ireland faces in developing a more sustainable knowledge-based economy.

**Trutz Haase** has been an independent Social and Economic Consultant since 1995. Previously, he worked for the Northern Ireland Economic Research Centre (Belfast), the Combat Poverty Agency (Dublin) and the Educational Research Centre at St. Patrick's College (Dublin). Mr. Haase has worked for various Irish Government departments, local authorities and non-governmental agencies. He is best known for his work on the development of an Irish Index of Relative Affluence and Deprivation which features in the current Irish Regional and Local Development Plans.

**Maria Heneghan** is a National Rural Tourism Specialist with Teagasc, the Agriculture and Food Development Authority in Ireland. With 15 years experience in rural tourism, Maria's work focuses on facilitating individuals, community groups, Government bodies, agencies and institutions in developing rural tourism policies and projects. She is a member of the National Advisory Group on Rural Tourism to Dept. of Community, Rural and Gaeltacht Affair; the National Federation for Rural Tourism; the Evaluation committee Agri Tourism funding Leader and the National Judging Panel for Rural Tourism Award.

**Colleen Hennessy** recently completed her M.A. in Community Development at NUI, Galway. She completed a field placement and research work on emerging transport needs in town and rural areas and the effect of Irish transport policy on social exclusion and the local development sector. She is currently working as a consultant to the Community and Voluntary Sector O'Leary and Associates Training and Consultancy in Kerry.

**Seosamh Mac Donnacha** is Academic Co-ordinator of Acadamh na hOllscolaíochta Gaeilge, NUI, Galway. He lectures in language planning and is interested in the study of language planning as a process. He has published widely on issues relating to language planning, including a major baseline study of Gaeltacht schools: Staid

Reatha na Scoileanna Gaeltachta (2005) and has co-authored the 'Comprehensive Linguistic Study of the Use of Irish in the Gaeltacht', published by the Department of Community, Rural and Gaeltacht Affairs (2007).

**Ruth McAreavey** is a Lecturer in Spatial and Environmental Planning in Queen's University Belfast. She teaches rural development; economic development; and research methods. Her main research interests are rural development theory, policy and practice; research methodologies; interdisciplinary research and migrant communities. She is currently conducting research for the Nuffield Foundation on migrant communities in Northern Ireland.

**John McDonagh** is a Lecturer in the Geography Department at NUI, Galway. His research interests are in rural geography, sustainable environments and the changing geography of peripheral rural regions. More recently his research has involved consideration of the emerging discourse surrounding the provision of, and access to, rural transport in Ireland. He has been involved in European funded projects and completed two previous books – *Renegotiating Rural Development in Ireland*, Ashgate (2001) and *Economy, Society and Peripherality* (ed.), Arlen House (2002).

**Joan McGinley** graduated with a Masters in Rural Development form NUI, Galway and attained her Ph.D. from UCD. As well as having a strong community activist background, Joan also has a career in environmental research with a strong research interest in marine policy. She published a book with Croughlin Press in 1991 entitled *Ireland's Fisheries Policy.*

**Brian McGrath** is a Lecturer and Programme Director of the MA in Community Development at the School of Political Science and Sociology, NUI Galway. He is also Research Fellow with the Child and Family Research Centre at NUI Galway. His research currently focuses on migration, social support, young people, intergenerational support and community development.

**David Meredith** joined Teagasc in 2001 having spent a number of years working as a consultant to the Marine Institute and the broader marine food industry in Ireland. Between 2001 – 2004 he worked on two EU funded projects. The AsPIRE project examined aspatial factors that influence regional development whilst the SUPPLIERS project assessed the implications of supply chain structures and dynamics for rural food producers. More recently David has worked on a foresight study of rural areas.

**Brian Motherway** earned his Ph.D. for research on environmental debates in Ireland, focusing on questions of power asymmetries and competing discourses. He has been involved in several large environmental sociology research projects in Ireland, on topics from environmental attitudes to participatory governance. He is currently Head of Business Programmes at Sustainable Energy Ireland.

**Gerard Mullally** is a Lecturer and Co-ordinator of Postgraduate Studies in the Department of Sociology University College Cork. His publications include chapters on the governance of sustainable electricity and corporate social responsibility reporting in Ireland. The research outlined in this book forms part of the project funded by the Environmental Protection Agency ERTDI Programme 2000–2006 Sub-measure 2, Sustainable Development conducted in Cleaner Production Promotion Unit, UCC.

**Deirdre O'Connor** is Lecturer in Resource Economics in the School of Biology and Environmental Science, University College Dublin. Her research interests include agricultural and rural policy analysis, with particular reference to the impact of EU policy on rural communities; food policy – especially the issue of food poverty in Ireland and agri-environmental policy analysis in the Irish and European contexts.

**Conchúr Ó Giollagáin,** a native of Dublin now living in An Cheathrú Rua, Co. Galway, is the Head of the Language Planning Unit in Acadamh na hOllscolaíochta Gaeilge, NUI, Galway. He is the Academic Director of the Acadamh's M.A. in Language Planning, the first academic course in this discipline offered in an Irish university. He has co-authored the Government-commissioned Gaeltacht survey: Comprehensive Linguistic Study of the Use of Irish in the Gaeltacht (2007). Alongside his interest in language planning, his published works include research on sociolinguistics, linguistic anthropology and oral biography.

**Eamon O'Shea** is a Personal Professor in the Department of Economics and Director of the Irish Centre for Social Gerontology (ICSG) at the National University of Ireland Galway. His main research interests are on the economics of ageing, rural gerontology and the economics of the welfare state.

**John Phyne** is a Professor of Sociology and Coordinator of the Interdisciplinary Studies in Aquatic Resources Program at St. Francis Xavier University, Antigonish, Nova Scotia, Canada. He has conducted research on the salmon aquaculture industries of Atlantic Canada, Chile, the Faroe Islands, Norway and the Republic of Ireland. He is planning to research the factors behind the regional variations in capital formation and coastal governance in the fledgling aquaculture industry of his native province of Newfoundland, Canada.

**Henrike Rau** is a Lecturer in Political Science and Sociology at the National University of Ireland, Galway. Her Ph.D. research work focused on the comparative study of German and Irish time perspectives and temporal practices. Her current research interests include time-related aspects of environmental change and sustainability, and the study of socio-temporal and cultural consequences of car-dependency and high mobility.

**Stephen Roper** is Professor of Business Innovation at Aston Business School, Birmingham, England. Before joining Aston in 2003, Stephen was Assistant Director of the Northern Ireland Economic Research Centre, Belfast. Stephen has teaching and research interests in innovation, entrepreneurship and local and regional development with a particular focus on Ireland and Northern Ireland. Stephen has worked extensively with public agencies in Ireland and Northern Ireland on issues relating to innovation policy, R&D, business innovation and policy evaluation.

**Yvonne Scannell** is a Professor in Irish and EU Environmental and Planning Law at the Law School, Trinity College, Dublin and practises as an environmental lawyer with Arthur Cox, Solicitors, Dublin. She is the author of six books and numerous articles on environmental and planning law.

**Mark Scott** is Senior Lecturer in Regional and Urban Planning in the School of Geography, Planning and Environmental Policy, University College Dublin. In addition to planning practice, Mark has worked in the University of Ulster and Queen's University Belfast and is a Corporate Member of the Irish Planning Institute. His primary research interests relate to spatial planning, rural housing, local governance and community planning.

**Sally Shortall** is a Reader in Sociology and was previously Director of the Gibson Institute for Land, Food and Environment in Queen's University Belfast. Her main research interests are rural development policy, social changes in farming practice and the role of women on farms and in rural development. She has published widely on these topics, and she is a co-editor of *Rural Gender Relations: Issues and Case Studies* (2006) recently published by CAB International.

**Hilary Tovey** is a Senior Lecturer in Sociology, a Fellow of Trinity College Dublin, and past President of the European Society for Rural Sociology. She was the co-ordinator (2004–2007) of the EU-funded research project CORASON (A Cognitive Approach to Rural Sustainable Development) in which 12 participating European countries completed more than 80 case studies of knowledge dynamics in local projects for rural sustainable development. Recent publications include *A Sociology of Ireland* – third edition (with Perry Share and Mary Corcoran, 2007) and *Environmentalism in Ireland: Movement and Activists* (IPA Press, 2007).

**Roy W. Tomlinson** is Senior Lecturer in Geography, Queen's University Belfast. His current research interests are biomass and soil carbon stocks in Ireland but he has a more general interest in land cover mapping and habitat conservation.

**Sharon Turner** is the Professor of Environmental Law at the School of Law, Queen's University Belfast. She is the lead author of the foremost book on Environmental Law in Northern Ireland and has written extensively on EU environmental law generally. She was seconded to the Department of the Environment in NI as its Senior Legal Adviser from 2002–2004 and continues to advise Government on the implementation of EU environmental law and the handling of EU infraction litigation, including being part of the expert panel tasked by the Department of the Environment to undertake an independent review of the arrangements for environmental governance in Northern Ireland.

**Tony Varley** is a Senior Lecturer in Political Science and Sociology at NUI, Galway. His main research interests fall within the sociology of development and agrarian politics.

# List of Abbreviations

| | |
|---|---|
| A&E | Accident and Emergency |
| ADLs | Activities of Daily Living |
| | |
| BAG | Burren Action Group |
| BIM | An Bord Iascaigh Mhara (Irish Sea Fisheries Board) |
| BMW | Border, Midland and Western region |
| | |
| CAP | Common Agricultural Policy |
| CBAIT | Cross-Border Aquaculture Initiative |
| CDB | City/County Development Board |
| CFP | Common Fisheries Policy |
| CLAMS | Coordinated Local Aquaculture Management Systems |
| CSO | Central Statistics Office |
| | |
| DAFRD | Department of Agriculture, Food and Rural Development |
| DARD | Department of Agriculture and Rural Development |
| DCMNR | Department of Communications, Marine and Natural Resources |
| DEHLG | Department of Environment, Heritage, and Local Government |
| DELG | Department of the Environment and Local Government |
| DED | District Electoral Division |
| DOE | Department of Environment |
| DOM | Department of the Marine |
| DoMNR | Department of the Marine and Natural Resources |
| DRD | Department of Regional Development |
| DSL | Digital Subscriber Line |
| | |
| EAGGF | European Agricultural Guidance and Guarantee Fund |
| EC | European Commission |
| ED | Enumerative District |
| EEA | European Economic Area |
| EEC | European Economic Community |
| EFF | European Fisheries Fund |
| EFSA | European Food Safety Authority |
| EFZs | Exclusive Fishery Zones |
| EHS | Environment and Heritage Service |
| EMAS | European Eco-Management and Audit System |
| EPA | Environmental Protection Agency |

| | |
|---|---|
| ERHA | Eastern Region Health Authority |
| ESDP | European Spatial Development Perspective |
| EU | European Union |
| | |
| FAO | Food and Agriculture Organization |
| FCILC | Foyle, Carlingford and Irish Lights Commission |
| FDI | Foreign Direct Investment |
| FFI | Family Farm Income |
| | |
| GCC | Global Commodity Chains |
| GDA | Greater Dublin Area |
| GDP | Gross Domestic Product |
| GRT | Gross Registered Tonnage |
| | |
| HBSC | Health Behaviour in School-aged Children Survey |
| | |
| IAoE/EI | Irish Academy of Engineering/Engineers Ireland |
| IBEC | Irish Business and Employers Confederation |
| ICTs | Information and Communication Technologies |
| ICZM | Integrated Coastal Zone Management |
| IFQC | International Fish Quality Certification |
| IISD | International Institute for Sustainable Development |
| IQM | Irish Quality Mussels |
| IQS | Irish Quality Salmon |
| IRDA | Irish Rural Dwellers Association |
| ISDN | Integrated Services Digital Network |
| ISPG | Irish Salmon Producers Group |
| IT | Information Technology |
| | |
| LEADER | Liaisons Entre Actions de Développement de l'Économie Rurale |
| | |
| MAGP | Multi-Annual Guidance Programme |
| MEY | Maximum Economic Yield |
| MIP | Minimum Import Price |
| MNPWP | Mourne National Park Working Party |
| MSY | Maximum Sustainable Yield |
| Muintir | Muintir na Tíre |
| | |
| NAPs | Nitrate Action Plans |
| NESF | National Economic and Social Forum |
| NDP | National Development Plan |
| NHA | National Heritage Area |
| NI | Northern Ireland |
| NIPA | Northern Ireland Peace Agreement |

NIS        Northern Ireland Seafood
NITB       Northern Ireland Tourist Board
NSS        National Spatial Strategy
NUTS       Nomenclature of Territorial Units for Statistics
NVZ        Nitrate Vulnerable Zone

OECD       Organization for Economic Cooperation and Development
OFMDFM     Office of the First Minister and Deputy First Minister

PAWS       Plantations on Ancient Woodland Sites
PoP        Points of Presence
PSTN       Public Switched Telephone Network

RAP        Rural Action Project
RCN        Rural Community Network
RDS        Regional Development Strategy
REPS       Rural Environmental Protection Scheme
RoI        Republic of Ireland
RIAI       Royal Institution of Architects Ireland
RPGs       Regional Planning Guidelines

SACs       Special Areas of Conservation
SBM        Single Bay Management
SISRG      Seafood Industry Strategy Review Group
SFI        Science Foundation Ireland
SMILE      Sustainable Mariculture in Northern Irish Lough Ecosystems
SMP        Stock Management Plans
SPAs       Special Protection Areas
SRP        Stock Recovery Plans

TSG        Tourism Sustainability Group

UAA        Utilized Agricultural Area
UFU        Ulster Farmers Union
UK         United Kingdom of Great Britain and Northern Ireland
UN         United Nations
UNWTO      World Tourism Organization

WCED       World Commission on Environment and Development
WHB        Western Health Board
WSSD       World Summit on Sustainable Development
WTO        World Trade Organization

# Chapter 1

# The Politics of Rural Sustainability

Tony Varley, John McDonagh and Sally Shortall

## Introduction

What is most striking about the concept of 'sustainable development' today is how ubiquitous it has become and consequently how various and potentially contestable its meanings have proved to be. Certainly there is no shortage of critics of the concept. Michael Redclift's (2005) review article carries the revealing title: 'Sustainable Development (1987–2005): An Oxymoron Comes of Age'. What impresses Timothy Luke (2005, p. 228) is how 'the intellectual emptiness of sustainable development has clung to it from the moment of its official articulation by the World Commission on the Environment and Development'.

Despite its complexity and potential contestability, it is nonetheless possible to find in the concept of sustainable development some useful general features (see Baker 2006, pp. 212–3). The broad normative ideal of sustainability that it inscribes may be a difficult one to live up to in practice (some would even see the task as utopian), but it can be argued that as a general ideal it is morally commendable and that it can potentially provide a standard of sorts by reference to which actors in the real world (and those who study them) can position themselves.

As soon as we move from the rather lofty general level, however, matters begin to become more complicated. At the lower altitudes the normative ideal of sustainable development inevitably encounters the question of 'whose sustainability?' or 'sustainable for whom?' Such a question implies two things: that different interests (or at least those who speak for them) must at some point decide what is 'sustainable' and 'unsustainable' for them in light of their own specific circumstances; and that what different interests take to be 'sustainable' and 'unsustainable' may throw up a number of possibilities as regards how they choose to deal with one another. They may, for instance, agree and co-operate, disagree and come into conflict or perhaps both agree and disagree in ways that mix co-operation and contention in varying proportions.

What is also evident is that the question of 'whose sustainability?' can be seen as presenting itself at different junctures: in initially deciding what is to be understood by sustainable development, in moving to give effect to what has been decided and in reflecting on and assessing the outcomes of what is ultimately achieved. In the real world these junctures may overlap, as when the process of implementation has a significant bearing on what people come to understand discursively by sustainable development. Assessing the outcomes may present other difficulties.

Even if only a moderately ambitious standard of sustainable development is taken, it might be objected that the ideal can never be realized in any final or enduring sense. The point of this objection is that sustainable development, far from being achieved once and for all, has always to be regarded as to some degree 'under construction' or even 'all to play for'. At the very least such an objection raises the issue of whether the gains of 'sustainable development' can be maintained in the longer term.

When we descend then from the higher altitudes – or the realm of general normative ideals – sustainable development quickly begins to stand out for its potential contestability. Such contestability spans different junctures: in deciding what constitutes sustainable development initially, in the manner of its pursuit and in assessing outcomes. It is this potential contestability, in all its variety, that opens the door to the 'politics of sustainability', a politics that revolves around discursive struggles over how sustainable development is to be properly understood at different junctures by those representing and defending different interests.

Two broad ideal-typical possibilities – what we will term the consensual and the contentious – can be introduced to characterize the sort of directions that this politics of sustainability, centred on sustainable development's potential contestability, might conceivably take. On the one hand, notions of what constitutes sustainable development may – where conceptions, implementations and outcomes are concerned – be widely shared. This can result in considerable consensus and co-operation between a diversity of actors willing to pull together to give effect to what they can mutually agree. On the other hand, notions of what constitute sustainable development, its pursuit and the assessment of its achievements may become the subject of considerable disagreement and contention between different interests. How such disagreement and contention take shape, and how they are handled and resolved (if at all), involves political processes every bit as much as does the construction of consensus and the organization of co-operation around some conception of sustainable development, its pursuit and the interpretation of its outcomes.

To examine the context from which the recent concept of 'sustainable development' has emerged can help us explore further the suggestion that sustainable development always implies a politics of sustainability. Here the unavoidable starting point is the World Commission on the Environment and Development (WCED), better known as the Brundtland Commission, which was convened by the UN General Assembly in 1983. Its establishment, in a global context where the world's population was experiencing unprecedented growth, reflected a keen concern with the accelerating deterioration of the human environment and depletion of non-renewable natural resources, and with the consequences of that deterioration and depletion for economic, social and political development in the near and distant future.

The publication in 1987 of the Brundtland report, *Our Common Future*, would spark a debate about the nature of 'sustainable development', and the prospects of achieving it, that still continues. Central to the conclusions of the Brundtland report was the realization that 'a new development path was required, one that sustained human progress not just in a few places for a few years, but for the entire planet into the distant future' (Dresner 2002, p. 31). For this new path to be 'sustainable' it would have to be (in the famous formulation) a form of 'development which meets the needs of the present without compromising the ability of future generations to meet their own needs' (ibid., p. 31). Such a conception of sustainable development, 'simple and vague' as it may be (ibid., p. 31), is useful in so far as it draws our attention to how critical are notions of temporality, survival and balance to the Brundtland formulation.

What can be concluded is that a principle of 'survival via balance' informs the Brundtland conception of sustainable development. It is not just a matter of balancing the needs of present and future generations. Ever since the Brundtland report's appearance an ever more pressing political challenge has been how some workable 'balance' might be struck between the competing demands of economic growth and environmental protection (ibid., p. 63). A more ambitious and complex form of dynamic balance would be required when social development (encompassing a range of social inclusion needs) is added to the equation.

Beneath the surface of the Brundtland and other conceptions of sustainable development it is possible to discover a functionalist notion of 'survival via balance'. Typically the 'survival' in question relates to some 'system' that can encompass global society in all its vastness, nationally organized societies or even smaller territorial and social formations (such as regions). Within a functionalist conceptual framework each of these territorial and social formations can be seen as constituting a system of elements in which each element contributes variously, but critically, to the survival prospects of the system as a whole. What sustainability implies here is that a necessary condition of survival is that some workable balance must be struck between the different elements that make up the system.

Of course functionalist notions of 'survival via balance' were well known in social science long before the recent advent of 'sustainable development' as a distinctive concept and political/developmental project. Early modernization perspectives began by being quite upbeat about the prospects for ultimately striking some acceptable and workable balance (however dynamic and shifting) that would deliver social, economic and political development, and thus survival in the form of a progressive and sustainable future to people in general (see So 1990). For some of those who influenced modernization theory the market was to be seen as fundamentally a self-governing mechanism for balancing the demand and supply of a myriad of commodities. For others what balance was achievable did not occur spontaneously and autonomously; it had to be consciously created and maintained, thus suggesting that the pursuit of sustainability, in the 'survival via balance' sense, could never be anything other than a political process.

There were of course always those in the social sciences who questioned whether 'survival via balance' is attainable both in general and in the longer term. The critics (as exemplified by Marxist and Dependency writers) acknowledge that system survival may be possible to some degree (certainly in the short run) while still insisting on seeing such survival as built on deeply problematic relationships of domination and exploitation between the different elements that make up the system. Imbalances that reflect power differences between classes, states and regions abound in the world as theorized by Marxist and Dependency theorists. In a world that is organized around patterns of uneven and unbalanced development, disturbances, dislocations and conflicts tend to thrive and to find a multitude of expressions (Kitching 1989). In short, according to the critics of 'survival via balance', 'survival via imbalance' provides an infinitely better description of the way the world actually works.

What we have in the social sciences then are two different and conflicting visions of sustainability. One sees the existing order (or orders) surviving or being 'sustained' on the basis of 'balance', and the other on the basis of 'imbalance'. In considering how such balance or imbalance is achieved and maintained, attention focuses variously on how market forces, states and organized economic and social interests interact with each other. Another germane issue concerns the sort of politics that the pursuit of sustainability based on balance or imbalance produces. While early modernization theory may have often emphasized consensual politics and Dependency and Marxist positions contentious politics, many real world situations can be encountered where the consensual and the contentious co-exist or are mixed together. A final issue to be addressed concerns itself with the question of 'outcomes' – who 'wins' and who 'loses', and with what consequences, when development based on balance or imbalance is being pursued?

### Rural sustainable development

Specifically in relation to rural areas, Karl Marx observed presciently in the middle of the nineteenth century how the countryside would increasingly find itself subject 'to the rule of the towns' within the emerging city-centred system of modern capitalist industrial societies (Marx and Engels [1848] 1992, p. 7). Urban domination of the countryside was certainly set to increase and to condemn many rural inhabitants to a future of uneven and imbalanced development that would have migration, dislocation and marginalization on a global scale among its litany of sharp consequences (Roberts 1995).

Of course, threatened rural populations did not always meekly accept their fate as declining groups in a system dominated by ever more powerful urban-based industrial interests and the states that took their side. In particular the rural populists (and their intellectuals) would launch a critique of the dominant urban-based industrial model that has endured (in many different guises) since the nineteenth century (Lipton 1977; Kitching 1989). Explicitly identifying with those who were losing out under the dominant version of modernity, the broad populist challenge has been to find a *modern* alternative to an increasingly dominant urban-based social formation that had capitalist industrialization as its economic centre. Fundamental to many populist political projects historically (frequently dismissed as utopian by their critics) has been the desire to deliver sustainable futures to rural populations under severely imbalanced and hostile conditions.

In a simplified world three perspectives can be identified in looking at how 'systems' achieve and maintain sufficient 'balance' or 'imbalance' to 'survive' or 'sustain' themselves. The system may be seen from the standpoint of those who 'win', those who 'lose' or from those who adopt a social science functionalist perspective that seeks to remain 'above the fray' by looking at the dynamics of system survival from a purportedly free-floating 'systems' perspective. At the core of a populist analysis of 'sustainable development' is some notion of systems surviving or sustaining themselves on the basis of 'imbalance' rather than 'balance'.

The process and underlying dynamics of survival are seen, according to this populist version of 'survival via imbalance', to be suffused with power relations. Inequality and relative powerlessness may be core features of how social, economic and political orders were sustained in the past, but sustainable futures will depend for rural populists on how well relatively powerless groups (and the powerful interests who sometimes take their side) can alter the power structures embedded in patterns of imbalanced and uneven development by re-negotiating them to their own advantage.

With the benefit of hindsight we now know that many populist challenges to the *status quo* would end in failure (Kitching 1989). We further know that, in many instances, the context for pursuing a politics of rural survival (and 'sustainability') has become steadily more adverse. Adding appreciably to this adversity are the imbalances that emanate from the challenges of globalization, the rapidly advancing commodification of the 'consumption' countryside (Marsden 1999), the increasing (and often competing) demands on rural resources and the high levels of social exclusion often found in rural areas.

One specific marker of adversity (as both cause and consequence) – that which finds expression in the scale of population decline in the remoter disadvantaged rural areas – would prove important in urging policy makers at the European Union (EU) level to embrace the ideal of rural sustainable development. Concerns that the progressively more ecologically damaging character of intensive farming in favoured agricultural areas would have to be addressed as a matter of priority proved to be another driving force. From the context then it is clear that the European idea of using the notion of rural sustainable development to counter problems such as desertification and ecological degradation – problems that in time would prompt the introduction of such measures as the LEADER area partnerships and the Rural Environmental Protection Scheme (REPS) (European Commission 1988; Kearney et al. 1994) – was substantially born out of crisis conditions.

Taking advantage of the post-Bruntland and post-Rio swing to 'sustainable development', what the European embrace of rural sustainable development signalled as well was a desire to be innovative in the policy sphere. The elements of the European intervention – in particular the notions of subsidiarity, partnership, participation and empowerment – were offered as the building blocks of a purportedly new model of development aimed at achieving balanced economic, social and environmental change. In promoting this new model 'sustainable development' (as to a lesser degree was true of its predecessor and close family relation, 'integrated rural development') has proved to be an evocative (if still often vague) concept.

An optimistic reading might suggest that the prospects for the project of rural sustainable development (and its attendant politics) are bright enough. Certainly it would appear that the relevant policy actors and organized rural interests can agree that rural policy should be founded on a normative commitment to 'sustainability' and to 'a living countryside'. To go by the commitments of the Cork Declaration of 1996, the Salzburg Conference of 2003 and the 3rd Report on Economic and Social Cohesion (2004), a strong commitment to the ideal of a living countryside has materialized within the EU. It has further been broadly accepted that pursuit of this ideal will require dynamic local actors who, in partnership with the state and the EU, can develop their capacities to develop the local economy while safeguarding the environment and promoting social inclusion.

But are such acceptances and commitments enough of themselves? For some the real challenge of striving for rural sustainable development is to be found at the level of implementation. Without significant progress at this level the whole approach can never begin to realize its promise. Nor are the challenges at this level to be underestimated. Whether the striking of some sort of workable and lasting balance between social, economic and environmental considerations can be achieved in general at the level of implementation remains substantially an open question. What we can be sure of is that questions relating to what the ideal of sustainable rural development might mean for specific rural groups, and how it might be understood in specific policy areas, become unavoidable at this level.

So far one policy area, that occupying the interface between agriculture and the environment, has featured very prominently when efforts have been made to give effect to sustainable development. Some critics have indeed suggested that efforts at pursuing sustainable rural development have so far hardly gone beyond the 'greening' of farming (Lowe and Ward 2007). While there are solid reasons why much of the current EU funding provision for rural development should go towards stimulating environmental farming, progress is likely to be limited as long as policies remain centred excessively on agriculture and therefore confined largely to one sector (Marsden 1999; Bryden 2005).

While regulation and the offering of incentives have been used to achieve environmental sustainability in rural Europe, the approach to economic and social sustainability has relied considerably on the promotion of endogenous models of rural development. The LEADER programme, launched in 1991, was thus designed to encourage 'bottom-up' development in rural areas on a 'partnership' basis. This partnership approach was aimed at making the products and services of rural areas more competitive, adding value to local production and improving the quality of life in rural areas. It was anticipated that local 'participation' would be a central element in both the design and implementation phases of the local area partnerships.

There is now a substantial literature assessing the early performance of endogenous models of rural development in delivering on economic growth, social inclusion and more integrated rural development (Curtin and Varley 1997; Edwards et al. 2000; Shucksmith 2000; Commins 2004; Scott 2004; Bryden 2005; McAreavey 2006; Shortall 2008). What generally can be said is that while some notable advances have been recorded, the potential of endogenous and partnership-based approaches still remains substantially unrealized.

Why this is so can be attributed to a number of reasons. At a broad level there can be little doubt but that sustainable rural development has been subject to a lack of clarity vis-à-vis wider EU and state policy goals (Marsden 1999; Davis 1999). In particular it has yet to become clear how the model of bottom-up development might optimally relate to regional, national, and EU policies more generally. There is some evidence to suggest that policies at these different levels have sometimes been in serious conflict with one another.

The pursuit of rural sustainability, as others have noted (Bryden 1994; Lowe and Ward 2007), is substantially influenced by the choices of political elites and, in a context of perceived incompatible ends and limited resources, its pursuit can be expected to entail the making of hard decisions about priorities for rural areas. It follows that the pursuit of rural sustainability is likely to be attended by considerable struggle, as one conception of sustainability comes to vie with another or others and as competition over incompatible ends and the distribution of scarce resources generates tensions and conflicts. When the politics of sustainability become contentious (as against consensual) in these ways, whose interests come to the fore and prevail can tell us much about the distribution of political, economic and social power within and outside the countryside.

Against such a backdrop both the opportunities available (and that can be created) to advance conceptions of sustainable rural development and the obstacles that lie in their path will be considered in this volume across a range of cases. Questions of how the opportunities to hand are being taken up and the obstacles negotiated, at the centre of the politics of rural sustainability, feature in many guises throughout this book. By way of answer some of our contributors highlight how a reluctance to make decisions unfavourable to powerful interest groups can be a major barrier to both the implementation and achievement of different sustainable development agendas.

## Sustainable development in rural Ireland

Contemporary rural Ireland has changed utterly since the peasant and patriarchal society encountered by Arensberg and Kimball (2001) in the 1930s, or even since the rapidly changing society described by Hannan and Katsiaouni (1977) in the 1970s. The contemporary countryside in Ireland, no less than elsewhere, is now being challenged as never before by agricultural restructuring, declining service provision, depopulation and counter-urbanization, communication and infrastructural deficits and the degradation of the natural environment (McDonagh 2007).

What is now an appreciably smaller agricultural sector has to contend with global food systems, and global regulation, most keenly felt in the revised Common Agricultural Policy (CAP) reforms being imposed on the EU by the World Trade Organization (WTO). Global, EU and national regulations impact significantly on the environmental, social and economic choices being made by rural actors, particularly in relation to land use. What is further evident is that for some time now rural Ireland has been functioning less and less as a purely production-centred space. In the new circumstances the consumption-type demands being made on the countryside are large and are expanding all the time. Such demands range from the supply of leisure and recreation to the provision of a living space for many urban commuters and migrants who choose to live in rural areas.

Rural areas, to the extent that they find themselves facing broadly the same challenges, share much in common. This is not to say of course, as will become apparent presently, that there are not substantial economic, social and environmental differences between rural areas in Ireland. The accelerated pace of change can make for a more differentiated countryside. It is to an overview of how some of these differences surface in individual chapters that we now turn.

In recent years environmentalism has emerged as a powerful ethical and political force (Marsden 1999). A key driver of its emergence has been the rapidly evolving regulatory environmental framework, and six chapters in all explore how this framework has impacted on the prospects for sustainable development in rural Ireland. Yvonne Scannell and Sharon Turner, in an overview of relevant EU and national law, observe that the fluid and open-ended nature of sustainable development as a general phenomenon has led 'many to argue that it is unsuited to precise legal definition'. The reluctance to give the concept precise and enforceable legal meaning is evident among both ruling politicians and the courts. Nonetheless sustainable development has acquired a number of general meanings that reflect the circumstances of its historical emergence in the 1980s; and here Scannell and Turner suggest that 'it is fairly safe to say that legislative duties to "promote" or "have regard" to the principle of sustainable development do not necessarily mean that environmental interests must always triumph over other interests'. All that is required is that such interests 'must always be taken into account and balanced with other interests'. Crucially, however, the authors opine that 'it is highly unlikely that the courts in the Republic and NI [Northern Ireland] will question whether the correct balance has been achieved'.

Both Governments on the island are shown to have moved some distance to honour their common commitments to integrate the principle of sustainable development as promoted by the EU into 'key legal and policy frameworks'. Enforcement of environmental rules, however, has often been problematic, though it is suggested that while financial penalties imposed by the EU on member states have been few in number to date, 'in reality the threat of EU fines has a very real impact'. The increase in new rural housing (a policy domain which currently lies beyond the EU's competence) is introduced as a case to test the extent to which the 'evolving policy and legal rhetoric' centred on sustainable development 'has been put into practice'. Here the judgement is that sustainable development has fallen down in view of the way 'both elected members of local authorities and many planning officers have for various reasons (good and bad) failed to halt the progressive despoliation of the countryside by one-off houses'. Planning law may lie beyond the EU's competence at present, but Scannell and Turner suggest that the proliferation of rural housing north and south of the border will in time very likely fall foul of the Water Framework Directive, the Urban Waste Water Treatment Directive and the Habitats Directive. In such an eventuality it is predicted that 'the need to contain the risk of EU infraction in relation to these key Directives may ultimately force both Governments to strengthen legislative controls and adopt formal guidance ensuring an environmentally sustainable approach to decision making concerning rural housing'.

Using the Nitrates Directive and the Birds and Habitats Directive as examples, Brendan Flynn's chapter also explores the impact of EU environmental policies on the prospects for sustainable development in Ireland. Focusing on the dimension of policy implementation, Flynn contends that good intentions are often stymied at the level of implementation. It is for this reason that the EU's impact on Irish environmental policies has been patchy at best. The EU may have 'provided a modern template of environmental laws that are basically sound', but 'poor implementation' north and south of the border has meant that 'policy failures have been largely sourced within Ireland and not in Brussels'. Furthermore, in view of the EU's basically confederal structure, the 'phenomenon of poor implementation' is seen as a structural problem that is likely to persist. In accounting for why EU environmental policy has tended to fall down at the level of implementation in Ireland, Flynn points in particular to the power of organized farming interests to delay and water down EU legislation.

If the largely legalistic approach to environmental policy adopted by the EU has fallen short, Flynn suggests that very probably integrating 'EU-level funding with EU-level legal norms more closely and in a systematic manner' will offer better prospects. In his view, 'Only such a savvy approach has the promise of unblocking the scope for powerful domestic interests to politically dilute or even derail the implementation process'. Another of Flynn's suggestions is that the establishment of an all-Ireland Environmental Protection Body, as proposed in the original Mitchell draft for a power-sharing agreement in NI, might strengthen the institutional framework required to deal more effectively with environmental issues.

The wider external and internal political and administrative context can always be expected to influence the pursuit of sustainable development administratively. Much of the external political and administrative context in Ireland has derived from the opportunities opened up by EU membership and funding as well as by acceptance of Local Agenda 21 (LA21) in the years following the Earth Summit in Rio de Janeiro in 1992. It is within this context that Gerard Mullally's and

Brian Motherway's chapter considers the institutional capacity required to advance official commitments to sustainable development *via* new governance arrangements. After a discussion of institutional design and the building of institutional capacity for sustainable development, attention shifts to 'some of the existing institutions engaged in governance *of* sustainable development' and to how these have recently fared and undergone change. Strategies for sustainable development are then reviewed with a view to tracing 'the intended approach to the creation of institutions *for* sustainable development'. All this is meant to set the scene for a review of the experience of Local Agenda 21 (LA21) as a leading means of instituting local and regional sustainable development in Ireland.

Over the last decade or so substantial policy and institutional innovation for implementing sustainable development in Ireland can be observed 'on the regional to national, and national to European levels'. On both sides of the Irish border significant strategic, integrative and participative capacities have been created. Nonetheless Mullally and Motherway feel obliged to conclude that 'we are still at the stage of developing capacities for the governance *of* sustainable development, that in effect means it will remain secondary to the established priorities of socio-economic development'. One suggested reason for this is that the internal political and administrative context within which the new governance arrangements – such as the partnerships associated with the City/County Development Boards (CDBs) in the south and the Local Strategic Partnerships in the north – has been heavily influenced by pre-existing patterns of governance in Ireland.

Another of the authors' conclusions is that the 'partnerships for sustainable development have remained the poor relation of social partnership bodies at the national level'. 'At the local level', it is suggested, 'we find less of a sense of the capacity to influence sustainable development outcomes or indeed to advance the implementation of policies'. As much as 'official discourse is increasingly couched in the language of governance *for* sustainable development', the underlying pattern suggests to Mullally and Motherway that it is 'the governance *of* sustainable development' that continues to be the prime focus. A key conclusion therefore is that if governance *for* sustainable development is what we want then the many serious challenges to building appropriate capacity building at the levels of the state and civil society will have to be taken a lot more seriously.

In response to a renewed interest in regional spatial planning in Europe and in Ireland, discernible both before and after the publication of the *European Spatial Development Perspective* in 1999, Mark Scott sets himself the task of comparing NI's Regional Development Strategy (RDS) with the Republic of Ireland's (RoI) National Spatial Strategy (NSS). As Scott sees it there is much scope for regional planning – even if less developed than urban planning in its tools and discourses – to address conflicts that arise around sustainable development or 'competing sustainabilities'. He does accept, however, that the vexed question of one-off rural housing has resisted any solution along such lines to date. As evidence of this he reminds us how 'both the RDS and NSS were careful to avoid detailed policy prescription on rural housing'.

For the potential of regional planning to come into its own, regional policy and rural development will have to be regarded as intimately related. The continuing decline of agriculture is a case in point; it adds urgency to the pursuit of balanced regional development, built around the designation of effective 'gateways' and 'hubs' (comparable to the 'growth centres' of the 1960s). Based on the experience to date, however, Scott believes that 'key questions remain in relation to the capacity of the selected gateways and hubs to effectively counterbalance the dominance of Ireland's eastern corridor and to disperse the benefits of development to rural areas'. If regional planning is to counter such imbalance and so contribute to rural development, Scott sees it as essential that the 'diversity of rural Ireland' be respected. Rural areas subject to strong and weak 'urban influence' will thus require different policy responses. The Republic's NSS may be laudable in this respect, though Scott suggests that local as well as regional difference are worthy of the close attention of planners. All this points to the necessity for 'a more interactive and collaborative style of local policy-making to enable planning officials and rural development stakeholders to explore new "storylines" of rurality to provide a common departure point for developing an area-based, integrated and holistic approach to rural sustainable development'.

Hilary Tovey's account of environmental management begins from the position that 'an environmental regulatory regime which devalues local and lay knowledges makes rural environmental "governance" almost impossible to achieve'. Typically in the contemporary world, and clearly environmental management is no exception here, local and lay knowledges coexist in a context where science has established itself as a superior form of knowledge. Historically such dominance has been deeply embedded in environmentalism. European environmentalism, for instance, can be pictured 'as a struggle by enlightened core elites against "backwardness" and "ignorance" about environmental issues among rural populations'. Yet rural environmental management can never be solely a matter for elites. It is always a pressing concern for 'rural people themselves who face problems in sustaining their livelihoods or their desired quality of life'. The question then for Tovey is how the different sorts of knowledge of environmental managers and rural actors are being brought to bear on Irish environmental management. Drawing on the ideas of Bruno Latour in the main, Tovey sets out to discover what storylines different actors use to construct or 'translate' environmental management.

In the Department of the Environment in both NI and the RoI scientists and scientific discourse occupy the commanding heights where environmental management is concerned. How such dominance works in practice is explored in the cases of the National Parks and Wildlife Service (NPWS) and the Heritage Council (HC), each of which constitutes its core advisory activities around a 'a scientific understanding of nature conservation'. While the NPWS interprets environmental management as the 'scientific management of nature', the HC interprets it as 'heritage conservation'. And while the NPWS managers address themselves overwhelmingly to fellow scientists and their funding bodies, the HC managers choose to throw the net wider, addressing themselves to 'the nation' as a whole rather than to a scientific elite. Lay actors and lay knowledges may be seen by HC managers as crucial to the project of heritage conservation, but this cannot conceal the way the HC's 'continued existence and funding depends on the recognition that it is an expert body capable of giving expert, uncontested advice to government'. All this allows the HC's approach to environmental management to be seen as a form of 'bottom-up but science-based nature conservation'.

Tovey links the interpretation of environmental management as 'sustainable development' with lately introduced governance structures among the recently reformed local authorities and LEADER committees, each of which professes to be participative, democratic and accountable. In this new context of democratic experimentation 'environmental management is valued as much for its capacity to develop "community" as to "conserve" nature'. Compared to those institutional contexts where scientific discourse monopolizes power, the interpretation of environmental management as sustainable development appears to be the 'most successful in "moving the world."' The reason for this is that its network is more popular and inclusive, something that is seen to make for greater policy effectiveness. Even here, however, there is still the danger that '"development" will be given more emphasis than "sustainability."' In other words, as Tovey tellingly puts it, 'the contradiction between economic growth and the protection of nature has not gone away, even if we now have a discourse which says that it has'.

Crucially, as well, sustainable development 'appears to be implemented in ways which prioritize scientific, professional and managerial over local and lay forms of knowledge'. All this points to a potential conflict, not only between 'development' and 'sustainability', but between elitist scientific and local democratic forms of environmental knowledge. It came to be recognized under the influence of post-Rio style democratization that 'an environmental management agenda which is unable to recognize and incorporate a diverse range of knowledges of nature is likely...to be judged both ineffective and undemocratic, and hence itself "unsustainable"'. What this implies is the need for 'a shift from "environmental management" to "environmental governance"'. Judged against this standard the Irish interpretation of environmental management as sustainable development reveals how numerous and formidable are the challenges of achieving 'cognitive justice' in which expert and non-expert knowledges can interact as something akin to equals in practice.

Rural areas are distinguished by the way they are home to primary production industries. The main questions posed in the four chapters that discuss such industries in Ireland ask how viable they are economically and environmentally and what contributions they can make to the wider rural economy and society. As a source of rural employment Irish agriculture has been a rapidly contracting form of economic activity. John Feehan and Deirdre O'Connor's chapter on agriculture and multifunctionality begins by charting 'the decline of on-farm sustainability', something that has brought a sharp drop in both the number of farms and farmers as well as an increase in the size of those agricultural holdings that remain. Our attention is drawn to an often neglected feature of this restructuring: the way that 'the ability to support one's enterprise from local resources – always a challenge, always dependent on intelligent management learned over ages – has been lost'. Since it is no longer possible for the majority of farmers to support a family by relying on agriculture alone, pluriactivity – in the forms of 'alternative land-utilizing enterprises' and off-farm employment – has become the norm. Here Feehan and O'Connor describe how forestry, on-farm tourism, and organic farming have come to function as Irish examples of multi-functional land use.

Another facet of their discussion is concerned with the move from 'production' to 'consumption' in land use. An important consequence of this move is the tensions being generated in current agricultural and rural policy and that find reflection in the way farmers are being encouraged to move in different directions: to be at once competitive as agriculturalists in an increasingly liberalized global market while gearing themselves to meet an array of post-productivist demands, all in the context of 'a complex and rapidly-changing market and policy environment'.

The approach Roy Tomlinson and John Fennessy take to forestry is to see it as a complex and dynamic resource that varies in accordance with 'the different values' it has for society. Over time economic, political and social forces have all contributed to shaping the development of forestry and of forestry policy. In the contemporary period forestry has come to be regarded as at once a renewable resource, an alternative land use, a provider of wildlife habitats, an environment for recreation, a carbon store and a source of raw material for timber-based industries. Each of these aspects of forestry is discussed, as is the manner forestry relates to the wider rural society and economy and to the environment.

On occasion the relationship between forestry and society in Ireland has been troubled. In some parts of Ireland (Leitrim especially), for instance, the extension of forestry has been seen locally as both a threat to agriculture and to the local environment. Considerable attention is paid to how the extension of forestry has the potential to have negative as well as positive environmental impacts. One prominent negative impact has taken the form of the 'enhanced acidification of soils, streams and lakes'. As 'efficient "scavengers" of acid pollutants and acid-precursors', the exotic conifers that have accounted for the main type of forest planting in the RoI have thus imposed their own environmental costs. It seems that in heavily forested counties like Wicklow 'during periods of easterly airflow (from urbanized and industrialized Great Britain) inputs of nitrates and sulphates increased stream acidity in forested catchments, probably due to "scavenging" by conifers'. Tomlinson and Fennessy suggest that 'in existing forests species mix may need to be widened and include more broadleaves to reduce the scavenging effect'. It is also to the relative advantage of broadleaves (particularly native varieties) that they 'generally have higher biodiversity than conifers'. In spite of the complexities and some negative effects, Tomlinson and John Fennessy conclude their overview with the observation that 'the balance of existing analysis suggests that objectives to increase forest cover are valid economically, socially and environmentally'.

Sustainability may for long have been a core aspiration of fisheries management systems, but this policy aim has encountered many obstacles at the level of implementation. A leading problem today is the growing imbalance that results from the way fish landings have declined with overfishing while both the consumption of fish and the fishing effort itself have significantly increased. For all the difficulties David Meredith and Joan McGinley, in their review of the impact of EC/EU Common Fisheries Policy on sea fisheries North and South, see the ideal of sustainability as important where fish stocks are concerned. At the very least it forces us to think about 'the conditions whereby available resources may be exploited in a rational manner'.

Against this backdrop the main contention of Meredith and McGinley is that the EU has historically been unable to implement an effective fisheries management system. The authors outline the limitations of early approaches to fisheries management that focused excessively on the biological aspects of fisheries and too little on the role of fishers and their possible impact on fish stocks. A major contributor to destructive fishing practices has been the political reluctance to take appropriate unpopular action, although this has begun to change in the very recent past. Experience has shown that supra-national policies applied at local level are rarely successful. Accordingly, recent revisions to the Common Fisheries Policy have led to the bestowal of greater responsibilities on national and regional authorities. This has inspired new governance structures as well as the introduction of financial and criminal penalties for non-compliance. In general Meredith and McGinley can contend that current fishery problems are primarily those of management. Their comparison of the Irish sea fisheries regime North and South highlights the importance of governments and state authorities adopting a more appropriate fisheries policy (in particular as regards EU-supported decommissioning schemes).

A distinguishing feature of aquaculture as compared to fishing is that fish stocks are farmed rather than hunted. To frame his discussion of the 'sustainability' of Irish aquaculture, John Phyne relies on insights drawn from the literatures on political economy, global commodity chains and environmental risk. Pride of place is given to global commodity chain analysis. What is different about the food industry in the post-Fordist era is the way import and export agents and giant supermarket firms have taken over from processors in exerting the decisive influence over both prices and quality. For all its explanatory force, Phyne sees global commodity chain analysis as incomplete unless supplemented by an analysis that pays attention to social relations at the point of production.

In his discussion the part capital and labour play in the social organization of Irish aquaculture is thus given prominence along with 'buyer-driven' markets and the attempts by a range of official actors to monitor and regulate the environmental impacts of fish farming. Within such a context Irish aquaculture's 'social sustainability' is seen to depend on the way industry actors can accommodate themselves to new commercial and regulatory requirements while providing the residents of Irish coastal communities with an 'inclusive consultative role – especially in matters relating to the environmental impacts of aquaculture'.

This discussion of 'primary production and sustainability' is followed by a section titled 'information technology, tourism and sustainability'. How well situated is rural Ireland to benefit from the growth of the knowledge economy? Seamus Grimes and Stephen Roper address this question by looking at a number of relevant aspects of the knowledge economy. In view of the tendency for R&D and innovation, the foundation of 'sustainable competitive advantage' in the current globalized business environment, to become more concentrated in urban centres – as the software industry in Dublin and in Belfast (to a lesser extent) well illustrate – the Irish evidence suggests that rural areas are unlikely to be able to participate directly in the development of R&D and innovation. What about the ability of rural small firms to overcome spatial disadvantages by adopting ICTs? Here the research shows how unrealistic was the early optimism that the IT revolution of itself would allow the countryside to shake off its historic spatial and other disadvantages. Rural areas lag behind in the provision of the most up-to-date communicative technology (such as broadband), though the rural North compared to the rural South has achieved significantly better coverage.

At the level of policy, Grimes and Roper explore a significant underlying tension in EU policy for the knowledge society that springs from trying simultaneously to raise competition and to promote 'greater social cohesion between regions'. In view of the historic difficulties experienced by endogenously led development, the Irish strategy has been to rely heavily on attracting inward investment and on 'global sources of knowledge and global demand to spur regional development'. Such a policy, it can readily be argued, has increasingly favoured urban areas. Apart altogether from the centripetal tendencies at work, peripheral rural areas in particular tend to lack the capacity to take advantage of the opportunities that the new information and knowledge society bring. If anything, the likely 'urban bias' of the new 'knowledge-intensive sectors' can be expected to widen 'developmental gaps between urban and more rural areas' as time passes.

Ruth McAreavey, John McDonagh and Maria Heneghan review the different ways in which rural tourism and sustainability have been linked together in Ireland and consider the challenges that rural communities involved in tourism now face. Case studies are relied on to explore whether a sustainable development approach can usefully be applied to rural tourism. By way of conclusion what emerges is an argument that stresses the need for a collaborative approach among a wide range of rural and non-rural dwellers. This is seen to be necessary if the growing demand for access to the countryside is to be adequately met.

Five chapters follow that consider the prospects for rural sustainability in the light of different forms of social differentiation. How demography impinges on the prospects of sustainable rural communities is the question Trutz Haase poses. His discussion begins with the observation that 'the general ideal of demographically-balanced, self-sustaining and economically-viable communities may be more a product of ideology than of actual historical reality'. Against an historical backdrop of demographic imbalance in rural Ireland, Haase's specific question asks: '...to what extent can poverty in rural Ireland explain weak demography or to what extent is poverty in rural Ireland the outcome of weak demography?' Such a question begs another: how is 'deprivation' to be adequately conceptualized?

Conceptions of deprivation, Haase argues, need to 'go beyond considerations of income poverty at the individual level, to relate the experience of individuals, groups and communities to the prevailing social context'. It is important that deprivation indices not 'be reduced to poverty outcomes alone, but must also include measures of the risk of poverty'. Rural and urban deprivations are seen as differing in their underlying causes, the forms they take and in the policy responses they should properly evoke.

Taking 'population loss and increased age and dependency rates' as 'the census indicators most relevant to rural deprivation', Haase finds that these phenomena, far from diminishing or disappearing once the RoI briefly joined the ranks of the tiger economies, would actually grow dramatically. Continuing depopulation raises the question of whether it is 'poor labour market conditions alone' or 'a growing disparity in life-style expectations' that is drawing people away from rural places in large numbers. The available data doesn't permit a wholly satisfactory answer to this question, though it is clear that certain broad forces are at work – agricultural restructuring, rural deindustrialization and the presence of changing 'educational attainments and resulting aspirations' that cannot be satisfied in the rural areas. What has also come into play is the 'enhanced mobility' brought by rising rates of private car ownership and better roads and commuter rail networks. Haase differs from some other contributors to this volume in taking the view that stricter planning controls on one-off housing have impacted negatively on population growth possibilities in the countryside.

What sustainability might mean to young people in rural Ireland is the question Brian McGrath poses. To come to terms with the question two dimensions – 'a reasonable lifestyle and decent livelihood standards' – are paid particular attention. Putting the two together we hear how 'secure and meaningful employment provides the main ingredient of a sustainable *livelihood* while the possession of social capital is necessary for achieving a sustainable *lifestyle*'. McGrath emphasizes the importance not only of 'objective conditions' but of how these are subjectively perceived. Thus lifestyles and livelihoods may be broadly perceived as either constrained or enhanced under rural conditions. Based on survey data McGrath shows that while Irish rural youth may have 'a generally more positive view of their communities than their urban counterparts', they also have to endure 'limited recreation and opportunities for social engagement' and that such limitations can render lifestyle 'a heavily problematic feature of growing up in rural Ireland'.

Whether young people can avail of 'sustainable livelihoods' in rural areas tends to be contingent on factors such as 'proximity to employment opportunities, transport, access to childcare, educational credentials, housing opportunities, family and friendship networks'. It is not just proximity to employment but the nature of that employment that is at issue here. Outside agriculture the rural economy tends to be dominated by small firms providing employment that is relatively lower paid, less rewarding and less demanding of educational credentials. Something that can especially militate against 'sustainable livelihoods' is the 'restricted opportunities in secondary labour markets' often found in rural areas.

Gender differences feature very prominently in McGrath's account. We hear how boys in rural areas feel much safer and report stronger trust relations than is the case with their urban counterparts and with rural girls. The lack of youth leisure provisions is also markedly different for boys and girls. Religious and political leisure activities are available to youth in NI, but these again remain strongly gendered. Gender differences are further evident when the amount of evening and after school time spent with friends is considered; while this is lower for both girls and boys compared with their urban counterparts, it is significantly lower for rural girls. McGrath can also point to research indicating that rural restructuring is causing difficulties for young men obliged to grapple with a changing understanding of rural masculinity. Turning to livelihood chances there is a higher educational achievement rate for rural children, especially farm children. Once again girls gain higher educational qualifications than boys and McGrath contends that this is part of a conscious strategy to ensure good livelihood chances, most likely requiring exit from the countryside. Limited rural childcare facilities have implications for women's opportunities in rural labour markets. All in all McGrath sees public policy as playing a key role in determining whether youth can have better lifestyle and livelihood prospects in the countryside.

Eamon O'Shea's account of rural aging and public policy begins with a discussion of the 'cumulative cycles of decline' to which rural areas are prone. These cycles come into play as out-migration prompted by poor employment opportunities reduce the population, unbalance the age structure and depress local economic demand, thereby causing further decline in employment and social service provision. Using this model of cumulative decline as his starting point, O'Shea profiles rural older populations and considers their broad health needs. He then examines how things currently stand in rural Ireland in relation to how the elderly fare in relation to social care provision, transport, housing and the available technology.

O'Shea's final topic – policy and practice – makes a case for a rebalanced public policy that would give greater weight to social equity and less weight to economic efficiency. The problem to date had been that 'the visible hand of moral leadership has too often been absent as a counter-balance to the invisible hand of the market in public policy-making'. Here the potential for 'social entrepreneurship' among the old is seen as immense as 'older people are likely to have the skills, experience, wisdom and established social networks necessary to harness economic and social activity in local areas'. Policies based on 'rural proofing' and that seek to stimulate 'social entrepreneurship' among the old are therefore urgently needed if the position of the elderly is to improve and a new and dynamic dimension is to be added to 'the social economy sector'.

Sally Shortall and Anne Byrne's chapter examines how gendered divisions in rural society might impact on the politics of sustainability in rural Ireland. A review of anthropological and sociological studies shows that while gender roles may not often have been overtly discussed; there was some conception of how these could contribute to a viable rural society. Work by Viney and Messenger in the late 1960s was the first to overtly discuss how women's dissatisfaction with their lives could threaten rural viability, a theme returned to by O'Hara in the late 1990s in a study of women on farms. More recent research has come to focus on the difficulties for men of coming to terms with renegotiated masculine gender roles. In examining the role women play in rural development activities North and South, Shortall and Byrne note that favourable equality legislation has yet to translate into gender equality. They conclude with the observation that rural sustainability is more likely to occur when gender equality is taken seriously.

In the South there has always been some political and public support for the Irish language; indeed many Irish people see the Irish language as an integral element of their national identity. In contrast the language question in the North is more complicated in view of the historical and contemporary identification of the Irish language with Irish nationalism. Since the Belfast Agreement there has been more political support for the Irish language, but the political context is still very different. The Irish language planning process, according to Seosamh Mac Donnacha and Conchúr Ó Giollagáin, would need to achieve two things to deliver 'sustainable language planning outcomes'. There would need to be both increased Irish language usage among the population in general and 'intergenerational increases in the number of first language Irish-speakers'.

As Irish has failed to establish itself as a 'social and community' language outside the Gaeltachtaí (or Irish-speaking regions), its survival as 'a living community language' will critically depend on the maintenance of the remaining Gaeltachtaí, all of which (with the exception of the Shaw's Road Gaeltacht in Belfast) are found in rural areas. Located mainly in the western seaboard counties of Donegal, Galway, Kerry and Cork many of the rural Gaeltachtaí have historically been disadvantaged economically and have suffered serious population loss in the twentieth century. For long sociolinguists have sought to measure the extent and understand the dynamics of the language shift away from Irish. Thus Ó Riagáin's early work in the Kerry Gaeltacht points to the importance of the weakening of localized networks as 'people may only reside in rural areas, but work, attend school, and shop elsewhere'. In-migration and return migration, as well as the expansion of short- and long-term holiday homeownership, have posed other threats as non-Irish speakers, or less than fluent Irish speakers, settle in Gaeltacht communities. Where but one parent is a native Irish speaker 'the language of the household tends to be English'.

The latest research suggests that the advance of English is continuing strongly and that 'well over half of the current Gaeltacht population live in areas which are little different from the rest of the country in linguistic terms'. It further concludes that Gaeltacht school children are experiencing 'a school-based socialization process that is predisposed towards the use of English'. The linguistic imbalance between English and Irish, in other words, is continuing to grow. What is to be done at the planning level? If the state is serious about arresting the linguistic decline of the remaining Gaeltachtaí, then the creation or designation of one agency with overall responsibility for 'language planning and maintenance issues in the Gaeltacht' is considered crucial.

Aspects of the relationship between civil society and sustainable development are considered in three chapters. What can the environmental movement contribute to sustainable development in Ireland? Very little is the answer John Barry and Peter Doran give to this question unless it adopts an ambitious '"triple bottom line" conception of sustainable development' that assigns significantly less weight to the economic dimension of development and more weight to the ecological and social dimensions. Put bluntly, if Irish environmental campaigning is to make real headway then it has to confront 'the political economy of *unsustainable* development' that became more entrenched during the Celtic Tiger years of accelerated growth.

The question then becomes: is the Irish environmental movement up to such an ambitious challenge? Here Barry and Doran suggest that the 'localized campaigns that have typified the Irish environmental movement's myriad of mobilizations against specific state-backed industrial and infrastructural projects' are best seen as having their origins in an experience of imbalanced development that has stimulated an 'environmental justice' movement and an 'environmentalism of the poor'. Consequently, while post-materialist values are of some importance to environmental campaigning, they are not the leading element in view of Ireland's experience of colonialism, de-colonization and post-colonialism.

Given the formidable power of the orthodox political economy model, and the way it can rely on the backing of the British and Irish states, making headway is by no means assured. Ranged against environmental campaigners are not just 'major state and business/corporate interests', but to some extent farming interests (where GM crop growing is concerned) as well. There is a long history of those opposed to the demands of environmental campaigning groups dismissing them as NIMBYist and as 'as irrational, anti-progress, selfish and endangering the economic competitiveness of the national or local economy'.

It is in such a contentious context that Irish environmental campaigning has to address the 'denial of voice to local interests in resource use or infrastructural decision-making processes'. Ultimately contesting this 'denial of voice' is to be construed as a struggle for democratization. To make progress Irish environmental campaigning groups will need to undergo continued politicization and radicalization, as is seen to be happening in the current 'Shell to Sea' campaign and in anti-infrastructural projects and anti-incinerator protests. For Barry and Doran the green movement's demand for a radical alternative to the orthodox model of political economy is always likely to be attended by divisions and conflicts. In such a politically contentious context (and notwithstanding that the Green Party is now a party of Government in the south) the green movement has no choice but to become more political and to identify strong allies in pursuing a radical environmental politics.

Tony Varley's discussion asks whether community-based collective action might conceivably be a means of countering patterns of imbalanced development in the countryside. To frame the problem conceptually he introduces the optimistic communitarian populist suggestion that collective action on the part of relatively powerless organized community interests can be a potential form of countervailing power that can deliver real benefits to rural communities. Within this framework he compares the fortunes of two would-be alliances of community groups, Muintir na Tíre (People of the Land) in the RoI and the Rural Community Network (RCN) in NI, each of which has sought to defend rural communities and to improve their survival chances in a situation where these are perceived to be threatened by various forms of imbalanced development. Whether community-based collective action can deliver on its potential to become a counterbalancing form of power is seen to depend heavily on how effectively organized community interests (at local and supra-local levels) can mobilize internal resources and exploit external opportunities (in particular those arising in the state sphere). What emerges is that the RCN, for a number of reasons, has been more advantaged of late in the resources and opportunities available to it and more adept at mobilizing and exploiting these than Muintir na Tíre.

Rural dwellers, by virtue of their location, the centralization of paid employment, services and recreational outlets in towns and cities and an inadequate public transport system are typically obliged to drive (or be driven in) private cars. 'Community transport', the subject of Henrike Rau's and Colleen Hennessy's chapter, has been presented as one way of dealing with the 'access' problem while cutting dependence on private car usage. Yet the burden of provision of community transport for typically resource-short community groups can be onerous. A question particularly pursued by Rau and Hennessy is whether the responsibility of providing transport services has limited the ability of the community and voluntary sector to attend to wider community development issues and to exercise an advocacy role.

For community transport to work well, Rau and Hennessy argue, requires that it be seen as but one part of a comprehensive system of integrated provision that has to be orchestrated and adequately resourced by the state. This formulation of the problem throws into relief the 'slim state' and its tendency to withdraw (or reduce) public transport provision. From the evidence presented it is clear that the RoI fares appreciably worse in this regard than NI, where much change for the better has occurred since 1998. In arguing that a coherent national policy for community and non-community transport services is critical, it is contended that the problems of the mobile and the immobile socially excluded should not be treated in isolation in discussing the viability of rural community transport.

### References

Arensberg, C. M. and S. T. Kimball (2001) *Family and Community in Ireland.* 3rd edition. Ennis: CLASP Press.
Baker, S. (2006) *Sustainable Development.* London: Routledge.
Bryden, J. (2005) *Rural Development in the Enlarged EU.* Gibson Institute for Land, Food & Environment Research Paper Series, Vol. 1, Issue 1.

Commins, P. (2004) 'Poverty and Social Exclusion in Rural Areas: Characteristics, Processes and Research Issues', *Sociologia Ruralis*, 44, 1: 60–76.

Curtin C. and T. Varley (1997) 'Take Your Partners and Face the Music: the State, Community Groups and Area-based Partnerships in Rural Ireland', in Paul Brennan (ed.), *L'Irlande: identites et modernite*. Lille: Centre de Gestion des Revues, Université Charles-de-Gaulle, pp. 141–55.

Davis, J. (ed.) (1999) *Rural Change in Ireland*. Belfast: Institute of Irish Studies, Queen's University Belfast.

Dresner, S. (2002) *The Principles of Sustainability*. London: Earthscan.

Edwards, B., M. Goodwin, S. Pemberton and M. Woods (2000) *Partnership Working in Rural Regeneration: Governance and Empowerment*. London: Policy Press.

Kearney, B., G. E. Boyle and J. A. Walsh (1994) *EU LEADER 1 Initiative in Ireland: Evaluation and Recommendations*. Dublin: Department of Agriculture, Food and Forestry.

Kitching, G. (1989) *Development and Underdevelopment in Historical Perspective: Populism, Nationalism and Industiralization*. London: Routledge.

Lipton, M. (1977) *Why Poor People Stay Poor: A Study of Urban Bias in World Development*. London: Temple Smith.

Lowe, P. and N. Ward (2007) 'Sustainable Rural Economies: Some Lessons from the English Experience', *Sustainable Development*, 15: 307–17.

Luke, T. (2005) 'Neither Sustainable nor Development: Reconsidering Sustainability in Development', *Sustainable Development*, 13: 228–38.

Marsden, T. (1999) 'Rural Futures: The Consumption Countryside and its Regulation', *Sociologia Ruralis*, 39, 4: 501–19.

Marx, K. and F. Engels (1997) [1848] *The Communist Manifesto*. Oxford: Oxford University Press.

McAreavey, R. (2006) 'Getting Close to the Action: The Micro-politics of Rural Development', *Sociologia Ruralis*, 46, 2: 85–103.

McDonagh, J. (2007) 'Rural Development', in B. Bartley and R. Kitchin (eds), *Understanding Contemporary Ireland*. London: Pluto Press, pp. 88–99.

Redclift, M. (2005) 'Sustainable Development (1987–2005): An Oxymoron Comes of Age', *Sustainable Development*, 13: 212–27.

Roberts, B. R. (1995) *The Making of Citizens: Cities of Peasants Revisited*. London: Arnold.

Scott, M. (2004) 'Building Institutional Capacity in Rural Northern Ireland: The Role of Partnership Governance in the LEADER II programme', *Journal of Rural Studies*, 20, 1: 49–59.

Shortall, S. (2008) 'Are Rural Development Programmes Socially Inclusive? Social Inclusion, Civic Engagement, Participation and Social Capital: Exploring the Differences', *Journal of Rural Studies*, 24: 450–57.

Shucksmith, M. (2000) 'Endogenous Development, Social Capital and Social Inclusion: Perspectives from LEADER in the UK', *Sociologia Ruralis*, 40, 2: 208–19.

So, A. Y. (1990) *Social Change and Development.* London and New Delhi: Sage.
World Commission on Environment and Development (WCED) (1987) *Our Common Future.* Oxford: Oxford University Press.

# PART I
# Policy and Planning
# for Sustainability

Chapter 2

# A Legal Framework for Sustainable Development in Rural Areas of the Republic of Ireland and Northern Ireland

Yvonne Scannell and Sharon Turner

## Introduction

There is little doubt that since its endorsement at the United Nations Conference on Environment and Development in Rio in 1992, the concept of sustainable development has acquired a global political and policy currency. Seventeen years later, achieving sustainability is almost universally accepted as one of the central policy objectives of the international community. Within the European Union's (EU's) legal order the principle has acquired a constitutional status with promoting sustainable development now identified as one of the fundamental objectives of the Union. However, while sustainable development has undoubtedly had a profound political impact, its traction on specific policy choices and legal frameworks is much more uneven and still relatively diffuse. This chapter will examine the nature and scope of the legal commitment to achieving sustainable development on the island of Ireland, focusing on its application to the highly charged issue of rural development. It begins by tracing the evolution of the principle of sustainable development from its international origins and gradual integration into the EU's legal order, to its more recent embedding into the domestic legal frameworks on the island of Ireland. The chapter then examines the practical application of the principle in relation to the issue of rural housing which provides ones of the most potent litmus tests of Government commitment to achieving sustainable development on the island.

### Integrating sustainable development into legal frameworks

Much has been written concerning the development of the concept of sustainable development, tracing its origins from the 1972 Stockholm Conference on the Human Environment and the Bruntland Report published by the World Commission on Environment and Development in 1987, through to its political and legal crystallization at the 1992 Earth Summit at Rio (Laverty and Meadowcroft 2000; Stallworthy 2002,). Suffice it for present purposes to say that the major outputs of the Rio meeting, namely the Rio Declaration on Environment and Development and Agenda 21, placed the concept of sustainable development at the heart of international policy on the environment and provided a detailed blueprint for implementation at domestic and local level. Unlike the other 'Rio Treaties' on climate change, desertification and biodiversity, neither the Rio Declaration nor Agenda 21 are legally binding. In effect, while the concept of sustainable development has undoubtedly become the organizing concept around which international law and policy on the environment is now evolving, it has essentially remained a creature of 'soft law' – more akin to a policy or political commitment than an obligation or objective with legal force.

In contrast, the concept of sustainable development acquired a comparatively greater standing within the legal systems of the EU. Although the United Kingdom (UK) and the Republic of Ireland (RoI) are signatories of the Rio agreements as sovereign states, the embedding of sustainable development in the legal frameworks governing the Irish countryside has occurred principally in response to initiatives adopted at EU level. Despite the legally imprecise nature of the Bruntland formulation of sustainable development adopted by the EU, its approach to embedding and promoting sustainability in Europe has relied significantly on the rule of law and legal processes. In the same year as the European Commission (EC) signed the Rio Declaration and Agenda 21, the Community enshrined the concept of sustainable development within the EC Treaty. In 1992 the Maastricht Treaty amended Article 2 of the EC Treaty to include the promotion of 'a harmonious and balanced development of economic activities, sustainable and non-inflationary growth respecting the environment' amongst the fundamental objectives of the EC. It also amended the Environmental Title of the EC Treaty (then Article 130r(2)) to provide that environmental protection requirements must be integrated into the definition and implementation of Community policies, in particular with a view to promoting sustainable development. Five years later the Treaty of Amsterdam significantly strengthened the nature of the EU's legal commitment to promoting sustainability. Following considerable criticism of the European formulation of sustainability as linked to economic growth, the Treaty of Amsterdam amended Article 2 in 1998 to require the Community throughout its territories, to 'promote... the harmonious, balanced and sustainable development of economic activities'. It also moved the environmental integration obligation from the specific Environmental Title and embedded it centre stage within the opening sequences of the EC Treaty to Article 3(c) (now Article 6). Although the Treaty does not define the Community's conception of the term sustainable development, its elevation to constitutional objective undoubtedly provided a powerful legal symbol of the Community's commitment to promoting sustainability in Europe. This constitutionalization of the objective of sustainable development combined with the Treaty status of the allied environmental integration obligation have also enabled the European Court to interpret EU environmental Directives in an expansive manner thereby entrenching the principle of sustainable development within the EU's legal *acquis* (Bell and McGillivray 2005).

In addition to enshrining it within the Community's constitutional Treaty, the development of EU law and policy on the environment since the Rio meeting has been progressively aligned with the concept and principles of sustainable development. In so far as rural issues are concerned, the re-orientation of EU environmental law and policy towards the Rio agenda since 1993 is undoubtedly the major driver forcing legal frameworks governing the countryside to reflect the principle of sustainable development. However, it is worth adding that the increasing integration of environmental considerations into the design and delivery of the EU's Common Agricultural Policy (CAP) means that the reformed CAP is also likely to become a further important driver for change of this nature. In the immediate wake of the 'Earth Summit', the Community published its Fifth Action Programme on the Environment, entitled *Towards Sustainability*, designed to guide policy development in this sphere from 1993–2001. This clear focus is continued in the Sixth Environmental Action Programme, *Environment 2010: Our Future, Our Choice*, adopted in 2002. Three major themes have dominated the Fifth and Sixth programmes; namely: (a) a recognition that environmental protection is a 'shared responsibility' amongst all societal actors, and in particular the importance of supporting effective public participation in environmental governance; (b) the need to ensure Member State compliance with the core framework of existing EU legislation on the environment; and (c) the importance of ensuring policy coherence and, particularly, achieving the integration of environmental considerations into other EU policy sectors.

Consistent with its policy emphasis on building support for environmental citizenship, the EU has adopted a range of Directives creating procedural rights and obligations designed to strengthen the public's general right of access to environmental information and rights to participate in decision-making concerning the operation of key Directives such as the EIA, Waste Framework, IPPC and Nitrates Directives – all of which have strong rural applications in NI and the Republic.[1] More latterly, key environmental Directives have been further amended to create transboundary participatory rights for citizens of neighbouring Member States and require Member States to ensure a more sophisticated and active form of environmental citizenship reflected for example in the requirements of the Water Framework Directive (Macrory and Turner 2002). In contrast, the EU has made comparatively modest progress in building consensus around the equivalent need to harmonize and widen domestic rights of access to environmental justice. The intrinsic connection between the principle of public participation set down in Principle 10 of the Rio Declaration and rights of access to environmental justice is well established. Indeed the Aarhus Convention,[2] which is widely considered to be the most ambitious and legally binding elaboration of Principle 10 of the Rio Declaration developed thus far, requires

---

1   These rights were integrated into these Directives by Directive 2003/35/EC OJ 2003 L175 25.6.2003.

2   The Arhus Convention on access to information, public participation in decision-making and access to justice in environmental matters was approved by the EU by Regulation (EC) No 1367/2006, OJ L264, 25.9.2006. The text of the Convention is available at http://www.unece.org/env/pp.

signatories to make expansive provision for ensuring public access to justice. Despite signing and approving the Aarhus Convention,[3] the Commission's proposals for a general harmonizing Directive[4] have not been supported by Member States.

The only legislative action taken thus far by the EU has been the introduction of Directive 2003/35/EC[5] which seeks to integrate the Aarhus provision for wide access to environmental justice into the specific contexts of challenges to decisions made under the EIA and IPPC Directives. Member States are required to ensure that the public concerned has access to a means of review before a court or other impartial or independent body. More specifically this procedure must be 'consistent with the objective of giving the public concerned wide access to justice' and access must be 'fair, equitable, timely and not prohibitively expensive'. Although the obligation imposed on Member States in this regard is limited in its application, vague in important respects (Ryall 2007) and circumscribed by conditions pertaining in national legal systems, it nevertheless signals the EU's intention to begin the legal integration of the Aarhus requirements into the EU's environmental law framework.

The policy emphasis on ensuring regard for the rule of EU environmental law has undoubtedly resulted in the Commission adopting a more vigorous approach to monitoring Member State compliance with Community environmental law since the mid-1990s. This increased emphasis on compliance with EU environmental law has undoubtedly been felt by both Governments on the island of Ireland, particularly in relation to Directives governing the countryside such as the Habitats and Nitrates Directive. Although the EU's infraction process is notoriously slow, the true extent of the political and policy impact of litigation by the Commission is largely hidden. Although very few financial penalties have thus far been imposed on Member States under Article 228EC, in reality the threat of EU fines has a very real impact. Very few of the Articles 228EC proceedings opened by the Commission are referred to the European Court because compliance is usually induced prior to this step being required (Turner 2006a). As is discussed in the context of Chapter 3 concerning implementation of the Nitrates Directive on the island of Ireland, when faced with the unpalatable prospect of paying potentially large-scale fines due to failure to implement the Directive correctly, both Governments ultimately overcame their deep-seated political resistance to imposing EU controls on pollution by agricultural nutrients on their powerful agricultural industries. Infraction litigation has also forced improved compliance with the Habitats Directive on the island, notably driving the expansion of designations of Natura 2000 sites; halting damaging activities on sensitive terrestrial and marine sites; and, most recently, challenging the granting of permission for the installation of experimental tidal turbines in Strangford Lough.

---

3   Council Decision 2005/370/EC.
4   COM 2003 (0624) final.
5   OJ L156 25.6.2003.

*The development of sustainable development strategies in Northern Ireland and the Republic*

Consistent with its commitments as a signatory of the Rio Declaration and Agenda 21, the RoI in 1997 published a national strategy for implementing sustainable development entitled *Sustainable Development: A Strategy for Ireland*. In preparation for the World Summit on Sustainable Development in Johannesburg in 2002 RoI published *Making Ireland's Development Sustainable*, which reviewed national progress in implementing this objective and set out plans for future action. Guidelines on what the concept implies for local authorities in the Republic were issued by the Minister for the Environment in 1995 and all local authorities prepared Local Agenda 21s setting out their policies to promote sustainable developments throughout their jurisdictions, including rural areas. A National Sustainable Development Partnership, 'Comhar', was established in 1999 to promote the national agenda for sustainable development, evaluate progress in this regard, assist in devising suitable mechanisms and advise on their implementation and to contribute to the formation of a national consensus for sustainable development. The partnership agreement, between the Government and the social partners, *Towards 2016*, committed the Government to a review of RoI's national sustainable development strategy in 2007. More specifically, special provision was made to promote sustainable rural development. The State committed to 'rural proofing' all national policies to ensure the assessment of the likely impacts of policy proposals on the economic, social, cultural and environmental well being of rural communities.

On the other side of the border development of a dedicated sustainable development strategy for Northern Ireland (NI) proved to be a far more protracted process. The UK led signatories of the Rio Declaration with the development of a national sustainable development strategy in 1994,[6] and followed this in 1999 with a detailed White Paper, *A Better Quality of Life: A Strategy for Sustainable Development for the UK*, setting out how sustainability would be achieved. Although both documents adopted a UK-wide focus, their coverage of the challenges and priorities for NI was superficial to say the least. The Welsh Assembly led the way amongst the devolved administrations in publishing a separate strategy for Wales in 1999, followed by Scotland in 2002. The first devolved *Programme for Government* in NI affirmed the new Government's commitment to sustainable development in 2001. This was followed later that year by the adoption of the *Regional Development Strategy 2025* which was explicitly based on the concept of sustainable development. However, Wales and Scotland had moved on to publish second iterations of their sustainable development strategies (Wales in 2003, and Scotland in 2005) before NI finally published its first strategy in 2006. Indeed Jonathan Porritt's characterization of this as a 'constipated process' during his

---

6   *Sustainable Development: The UK Strategy.*

address at the launch of the Northern strategy vividly captured the tortured nature of policy development in this context.

However, it is interesting to note the distinctive emphasis within the Northern strategy on the importance of governance for sustainable development. This focus essentially arose from the impacts of significant under-investment in, and distortion of key elements of the arrangements for environmental governance due to thirty years of direct rule and serious civil disorder (Morrow and Turner 1998; Turner 2006a and b). In February 2006, months prior to the launch of the strategy, the Direct Rule Minister for the Environment (then Lord Rooker) launched an independent Review of Environmental Governance tasked to address all publicly funded elements of the governance regime. Their report, *Foundations for the Future, a Review of Environmental Governance in NI*,[7] published in June 2007, confirmed that without significant reform of its system of environmental governance, the transition towards sustainability in NI would be impossible.

Although development of the Northern strategy, like that in the Republic, was led by the Department of the Environment (DOE), responsibility for policy leadership and the production of the NI Implementation Plan was transferred to the Office of the First Minister and Deputy First Minister (OFMDFM) immediately after the strategy was launched. Ostensibly this move was justified on the grounds that sustainable development should lie at the heart of Government and therefore with the department tasked with central policy co-ordination. While the transfer to OFMDFM was welcomed by the UK Sustainable Development Commission, the malaise that has characterized the subsequent development of the Implementation Plan and stakeholder forum reveals the myth that OFMDFM has the policy capacity or influence to act as a proxy Cabinet Office. Despite its title, *A Positive Step*, which suggests a discernable degree of policy movement, the NI Implementation Plan does little more than collate existing departmental targets set out in their respective corporate plans. In terms of the institutional infrastructure surrounding the strategy or implementation plan, OFMDFM is said to be considering the merits of creating a Stakeholder Forum but as yet no announcement has been made. Similarly, while the remit of the UK Sustainable Development Commission as Government's 'critical friend' in this context extends to the region, there has been no agreement as yet to extend to NI the new watchdog function recently conferred on the Commission for Great Britain.

*Steps to incorporate the principle of sustainable development into Irish legal frameworks*

The legal integration of sustainable development on the island of Ireland has followed a broadly similar pattern in both NI and the RoI. In addition to their obligation to implement EU legislation designed to promote sustainability, the Governments on both sides of the border have taken additional 'home grown'

---

7   http://www.regni.info/.

steps to integrate sustainable development across their respective domestic legal frameworks. The most important of these has undoubtedly been in the context of town and country planning legislation, an area of law and policy making that remains largely outside the remit of the EU's competence but which has a fundamental impact on the development of the rural environment. However, it should also be noted that despite the very belated introduction of a regional strategy for sustainable development, NI has arguably advanced further than the Republic in terms of imposing a general legal obligation to contribute to sustainable development on all public bodies. The obligation in the Republic, while widespread, is somewhat more fragmented as is illustrated below.

The Oireachtas took its first step in the legal integration of sustainable development with the adoption of the Environmental Protection Agency Act 1992. Enacted only a year after Rio, section 52(2)(b) of the Act provides that in carrying out its functions the Environmental Protection Agency (EPA) shall have regard to the need for a high standard of environmental protection and the need to promote sustainable and environmentally sound development, processes or operations. Since then the principle has been incorporated into a wide range of other environmental legislation[8] but the key legislative instrument incorporating it is undoubtedly the Planning and Development Act 2000 (hereafter the Planning Act). The concept of sustainable development is central to the objectives of the Planning Act. The Preamble, which sets forth the motivation for the Act, states that it is 'An Act...to provide, in the interests of the common good for proper planning and sustainable development including the provision of housing'. Requirements relating to sustainable development permeate the Act and the concept is central to the core obligations of all planning authorities, in particular their obligation in Section 9 to make development plans providing for the proper planning and sustainable developments of their areas and their obligation in Section 34 in dealing with all applications for planning permissions, to have regard to the 'proper planning and sustainable development of their areas'. In addition, Section 69 of the Local Government Act 2001 obliges all local authorities to have regard to the need for a high standard of environmental and heritage protection and the need for sustainable development when carrying out their functions under that Act and any other legislation. This obligation, which is binding on both elected local politicians and executives in local authorities, means that the requirement to achieve sustainable development is a core function of all local authorities whatever the capacity in which they are acting.

Consistent with the environmental integration obligation inherent in the principle of sustainable development, great care was taken in the RoI Planning and

---

8    Other references to sustainable development appear in the Dublin Docklands Development Authority Act, 1997; the Urban Renewal Act, 1998; the Fisheries (Amendment) Act, 1999; the Town Renewal Act, 2000; the Local Government Act, 2001; the Planning and Development (Amendment) Act 2002; the Sustainable Energy Act, 2002; the Protection of the Environment Act 2003; the Fisheries (Amendment) Act, 2003; the Planning and Development (Strategic Infrastructure) Act 2006 and the Water Services Act, 2007.

Development Act 2000 to ensure that planning and other policies are integrated. There are specific and unambiguous obligations in the Act requiring planning authorities to ascertain and have regard to other sectoral policies when carrying out their functions and specifically when making development plans[9] and decisions on planning applications.[10] So, for example, section 11(3)(c) of the Act requires planning authorities to consult with the providers of energy, telecommunications, transport and any other relevant infrastructure, and of education, health, policing and other services, in order to ascertain any long-term plans for the provision of infrastructure and services in the area of the planning authority. The infrastructure providers are statutorily obliged to furnish the necessary information to the planning authority. Numerous statutory bodies with environmental responsibilities and An Taisce must be specifically informed of, and sent copies of applications received for permissions for developments of particular interest to them, and also given time to make submissions on the applications.[11] Special rights to appeal decisions on planning applications to An Bord Pleanála is given to other policy stakeholders if they are not properly informed about proposals for developments liable to affect their interests when planning applications are first lodged.[12]

Not surprisingly given its belated adoption of a regional strategy for sustainable development, NI has only recently begun the process of integrating this concept within its legal frameworks. In a rare moment of policy leadership, NI has introduced the first general sustainability duty for public bodies in the UK and RoI. Under the Northern Ireland (Miscellaneous Provisions) Act 2006 all public bodies in the region are obliged exercise their functions in a manner considered 'best calculated to contribute to the achievement of sustainable development... except to the extent that...any such action is not reasonably practicable in all the circumstances of the case'. Although subject to the caveat of what is reasonable in the circumstances, and despite concerns as to its confusing formulation, this general obligation undoubtedly represents an important legal commitment to infusing the principle of sustainable development in decision-making across all tiers of government and public-sector action.

In so far as planning legislation is concerned, the first phase in integrating the concept of sustainable development into the legislative framework began with the adoption of the Regional Development Strategy 2025 (RDS) in 2001. Although adopted five years before publication of the NI sustainable development Strategy, the RDS affirmed the devolved administration's commitment to promoting sustainable development and states that the development strategy is specifically designed to reflect UK-wide and international commitments to balanced and sustainable development. In particular, the RDS states that the application of

---

9    Planning and Development Act 2000, s. 11(3) (c).

10    Planning and Development Regulations 2001, art.28.

11    Planning and Development Regulations 2001, art.28.

12    Planning and Development Act 2000, s. 37(1) (6) (a), as amended by the Planning and Development (Amendment) Act 2002.

the principles of sustainable development must lie at the heart of future rural development.[13] The RDS was given a statutory basis by the Strategic Planning (NI) Order 1999, which placed NI Departments under a legal duty to have regard to the RDS when exercising any functions relating to development. In addition Article 28 of the Planning (Amendment) (NI) Order 2003 requires that the Department of Regional Development (DRD), the lead department responsible for the RDS, must explicitly affirm that development plans proposed by the Department of the Environment are in conformity with the RDS. Following a comprehensive review of the planning system, DOE launched a second phase of reform in 2003. This resulted in the adoption of the Planning Reform (NI) Order 2006 which recognizes the primacy of development plans within the planning system and requires both DOE and the Northern Ireland Planning Appeals Commission to exercise their functions in relation to the making of development plans with the objective of contributing to sustainable development.

While this change represents an important legal recognition of the centrality of sustainable development to the planning process, it does not identify sustainable development as the statutory purpose or central organizing principle underpinning the system of development control in NI. Nor has this requirement come into legal force as yet. One can only assume that this is because DOE must also amend Planning Policy Statement 1: General Principles, which merely identifies sustainable development as one of a number of 'key themes' influencing the planning process and makes clear that the system is currently underpinned by a presumption in favour of development. It should also be pointed out that despite NI's avowed commitment to integrating sustainable development within the plan-making system, the immediate impact of this change is likely to be very modest indeed. Although the Government is currently preparing new development plans for NI, this process has become chronically delayed due to a surge in planning applications as post-conflict recovery gathers momentum and landowners seek consents prior to the anticipated introduction of more strict controls on rural development (Turner 2006a). As a result, numerous planning decisions are being made in the absence of new development plans, much less plans that take account of sustainable development.

Consistent with the principle of sustainable development, both the RDS and Northern Ireland Sustainable Development Strategy emphasize the importance of co-ordinated policy making. However, it very much remains to be seen whether the NI administration can deliver the level of policy integration required to deliver sustainable development. Despite its small size, eleven central Government departments have been created in order to meet the political exigencies of devolution on the basis of power sharing. Although the Review of Public Administration is expected to rationalize the equally splintered nature of local government, it is unlikely that the architecture of central government will be rationalized until devolution has become more embedded. In the meantime,

---

13   Chapter 8.

policy responsibility for the environment is fragmented across nine departments and consequently the process of brokering policy integration for the purposes of sustainable development is slow and cumbersome.[14] It is possible that the legal obligation requiring development plans to be in conformity with the RDS will assist in overcoming the worst excesses of this structural fragmentation: namely, the confusing split of responsibility for planning policy development between DOE and DRD. Arguably the specific requirement imposed on DOE to submit proposed development plans to the Department of Regional Development for confirmation of conformity will help to avoid damaging differences of opinion evolving between these two departments. That said, a recent decision by the High Court in NI may force Government in NI to take early action to resolve the fragmentation of planning policy responsibility. In *Application by Omagh District Council for Judicial Review*,[15] the draft Planning Policy Statement 14 (PPS14) governing rural development (discussed further below) was quashed on the grounds that DRD lacked the power to issue planning policy statements under NI planning legislation. Justice Gillen ruled that DRD's powers were confined to development of the Regional Development Strategy and associated guidance, whereas sole power to issue planning policy statements, such as PPS 14, rested with DOE. While this ruling may not force the Executive to merge planning policy responsibility within DOE as recommended by the Review of Environmental Governance, it certainly highlights – and in a very contentious context – the confusion within and outside Government concerning the demarcation of planning policy responsibility between these two departments.

In so far as policy making concerning rural NI is concerned the RDS emphasizes the need for a co-ordinated and integrated approach to policy development at all levels and to this end emphasizes rural proofing of policy proposals. A non-statutory system of rural proofing was introduced in NI in 2002; however, five years later there are widespread concerns that this process has not evolved beyond a formulaic 'tick box' exercise. The non-statutory and closed nature of the consultation relationship between the NI Planning Service (as the plan-making authority) and NI's environmental regulator, represents a further important barrier to credible decision-making in this context. Because both the Planning Service and Environment and Heritage Service (EHS) are non-executive agencies of DOE, EHS, unlike the Environment Agency and An Taisce in the Republic, does not have the status of statutory consultee in relation to planning decisions. Although there is consultation between the two departmental agencies as a matter of practice, it is associated with a serious lack of transparency because it effectively occurs entirely within central Government and therefore behind closed doors. Furthermore, as pressure on the Planning Service has escalated as economic recovery gathers

---

14    The impact of environmental policy fragmentation was also discussed by the report of the Review of Environmental Governance, *Foundations for the Future* (2007, pp. 43–5), supra note 8.

15    Unreported, Gillen J, 7/9/07, GILC5915.

pace, there are increasing concerns that this consultation relationship has become seriously strained.[16]

Consistent with the emphasis on citizen participation and good governance as fundamental tenets of the principle of sustainable development, planning and associated freedom of information legislation on both sides of the border has made extensive provision for participatory rights. Although this process has been driven to a considerable extent by the need to implement successive waves of EU directives conferring participatory rights, Governments on both sides of the border have taken significant 'home-grown' action to underpin this ethic in their respective planning systems.

The Irish Planning and Development Act 2000 (and the subordinate legislation made under the Act) and, indeed, most framework environmental legislation make extensive and detailed provision for public notification of proposed developments, freedom of access to information held by decision-makers, and confer public and non-governmental organization (NGO) rights to participate in decision-making on various land-use plans and in decisions on applications for planning permissions and other planning approvals. Although integration of these participatory rights are now required under EU law, many of these provisions predated the advent of EC rules in this regard. Indeed legislation in the RoI creating public participation rights in environmental decision making and in enforcing environmental laws were at one stage probably the most progressive in the EU and are still more progressive than in NI. In RoI, any person who makes a valid written submission or observation in relation to a planning application may appeal any planning decision to an independent An Bord Pleanála. Although participation rights are somewhat more extensive where private sector (as distinct from local authority or state) development is concerned, they are very well known and used, and public participation in environmental decision making in the RoI is very extensive.

In sharp contrast, decades of democratic deficit under Direct Rule in NI and its highly-centralized system of development control have significantly inhibited the development of environmental citizenship in the North (Turner 2006a).[17] During the first phase of devolution the NI Assembly indicated strong support for the introduction of third party rights of appeal during its consideration of the Planning (Amendment) Bill in 2001. This proposal was rejected by the Direct Rule administration following the suspension of devolution and consequently the right to appeal planning decisions is still confined to the applicant for permission. In 2007 the Review of Environmental Governance in Northern Ireland[18] identified the legacy of public marginalization from and disengagement with development

---

16  This matter was discussed by the Report of the Review of Environmental Governance; supra note 15, at p. 71.

17  The scale of this disengagement is vividly captured by the report of the National Trust Planning Commission in 2004, The National Trust (2004), *A Sense of Place: Planning for the Future in NI.*

18  http://www.regni.info/.

control processes as a key governance challenge facing the newly restored devolved administration. While the review team emphasized their sensitivity to the escalating regional pressure for post-conflict recovery and regeneration, they nevertheless urged the new Government to take specific steps to remedy the distinctive legacy of the region's constitutional history. The report acknowledged that the return of devolution and political accountability, and the planned repatriation of development control to local government would each make significant contributions in this respect; it did, however, recommend that Government should revisit the arguments in favour of third party rights of appeal in planning. As a viable first step towards restoring public confidence in the planning system, the report recommended that the proposed Environmental Protection Agency for Northern Ireland should be conferred with powers to challenge planning decisions in the public interest as a means of channelling that third party challenge process.

The recent introduction of a statutory obligation to provide statements of community involvement in planning processes under the Planning Reform (NI) Order 2006, and the planned introduction of a Community Planning obligation as part of the Review of Public Administration in NI[19] indicated the Direct Rule administration's intention to follow then UK policy concerning public participation in planning. However, it is also clear from the planning White Paper, *Planning For a Sustainable Future*,[20] published in May 2007 that UK central Government has embraced the ethic underpinning the recent Barker and Eddington Reports; namely, that reform of the planning system was required to ensure faster and more efficient decision making. There is little doubt that the planning system in NI is straining to cope with the pressures generated by the process of post-conflict economic recovery. However, it very much remains to be seen whether local political representatives will follow the flow of UK policy development which is expected to constrain rights of public participation or follow the RoI example and adopt third party rights of appeal. Either way, decision-making in this regard will send important signals to society in NI concerning the regional administration's commitment to environmental citizenship and governance for sustainable development.

Last but not least, there is the vexed question of access to environmental justice. As already stated, with the exception of stipulations requiring the provision of Aarhus levels of access to justice to challenge decision-making concerning the operation of the EIA and IPPC Directives, the EU Commission has failed to gain support for a general directive harmonizing procedural and financial rules governing access to environmental justice across the EU. At the time of writing neither Government on the island of Ireland has taken action to implement the access to justice amendments made to the EIA and IPPC Directives. However, both the UK and RoI Governments take the view that the availability of judicial review (combined with the additional availability of third party rights of appeal in

---

19   http://www.rpani.gov.uk/.
20   CM 7120.

RoI) satisfies their Aarhus Convention obligations to provide wide access to justice to challenge environmental and planning decisions on substantive and procedural grounds before a court of law or other independent and impartial tribunal. It is clear however, that this is a highly-problematic stance. Government's reliance on judicial review for Aarhus purposes has been the subject of considerable controversy in the UK. Although courts in the UK have progressively relaxed the rules on standing to enable a wide range of interested parties, including representational groups, to take judicial review challenges in the public interest, it is argued that the uncertainty inherent in judicial decision making is inconsistent with the Aarhus requirement for 'wide' access to justice. Similarly, the costs associated with taking a judicial review are regarded as being incompatible with the Aarhus requirement that access to environmental justice should not be 'prohibitively expensive'.[21]

Although RoI currently stands as the only EU Member State yet to ratify the Aarhus Convention, its planning system is unique within these isles in conferring third party rights of appeal. Under Part III of the Planning and Development Act 2000, both the applicant for consent and third party 'objectors' have a right to appeal a decision concerning an application for development consent taken by a local planning authority to the Planning Appeals Board (An Bord Pleanála). Appeals to An Bord Pleanála are undoubtedly easier to take, faster and less expensive than a judicial review action, and to that extent indicate a likely procedural compatibility with Aarhus and the requirements of EU Directive 2003/35/EC. However, because the scope of the board's power to determine questions of law is limited in certain respects, it is questionable whether this appeals procedure can be regarded as representing a full response to Aarhus and the EU Directive in that they stipulate a public entitlement to challenge the 'substantive or procedural legality of the contested decision' (Ryall 2007, p. 193).[22] In certain circumstances those seeking to contest planning decisions will therefore be forced to take a judicial review. Quite apart from well-rehearsed concerns about the general shortcomings of judicial review as a response to Aarhus and the Directive, there are specific aspects of the Irish arrangements governing the operation of this remedy in the planning context that, if anything, raise further doubts as to its compatibility with the growing emphasis on ensuring wide public access to environmental justice.

Access to judicial review in the planning context in RoI is governed by special rules set down in Section 50 of the Planning and Development Act 2000 as amended by the Planning and Development (Strategic Infrastructure) Act 2006.

---

21  For a discussion of this issue see for example, Coalition for Access to Justice for the Environment, 'Briefing: Access to Environmental Justice' (July 2004); Castle et al. (2004); Carnwarth (1999).

22  A key example of the limited nature of the Board review powers concerns the interpretation of EU law. In *O'Brien v South Tipperary County Council and An Bord Pleanála* (unreported, High Court, 22 October 2002), it was made clear that only the High Court has the jurisdiction to determine whether Ireland has implemented the EIA Directive.

Prior to 2000, applicants were required to show a 'sufficient interest' in the subject matter of the planning decision – much the same as the UK approach to *locus standi*. Under Section 50 the bar has been raised in that applicants in RoI are now required to show a 'substantial interest' and 'substantial grounds'. Although Section 50(4)(d) provides that a substantial interest is not confined to 'an interest in land or other financial interest', it is clear that this change has stimulated a shift in judicial opinion concerning the threshold for establishing *locus standi* to challenge planning decisions. Whereas the decisions in *ESB v Gormley*,[23] *Chambers v An Bord Pleanála*,[24] *Fallon v An Bord Pleanála*,[25] *Mc Bride v Galway Corporation*[26] and *Lancefort v An Bord Pleanála*[27] reflected the liberal approach to standing for local residents and environmental NGOs adopted by the UK judiciary, the post-s.50 decision in *O'Shea v Kerry County Council*[28] involved a much closer assessment of the impact of the planning decision on the applicant as a local resident than hitherto had been practised, before denying the applicant standing.

Two years later in *Harrington v An Bord Pleanála*[29] the High Court emphasized that Section 50 reflected a clear legislative intention to restrict the criteria governing challenges to planning decisions. However, while the court stated that it would adopt a 'rigorous approach' to assessing whether a substantial interest existed, it also noted that the requirement must not be applied 'in such a restrictive manner that no serious legal issue legitimately raised by an applicant could be ventilated or which would have as its effect the inability of the courts to check a clear and serious abuse of process by the relevant authorities'.[30] Although the decision in *Harrington* is cited as an example of a more restrictive approach to standing (Ryall 2007), Macken J.'s ruling is also regarded as potentially easing the effect of this position by suggesting that where an applicant fails to satisfy the 'substantial interest' threshold, access to judicial review could still be established where 'substantial grounds' for challenge are demonstrated.

In 2007, Section 50 was considered once again by the High Court in *Peter Sweetman v An Bord Pleanála, the Attorney General and Clare County Council*.[31] Clarke J. held that it was 'certainly open to argument that it will be necessary to construe the term 'substantial interest' in a manner which does not infringe the Directive', and that 'it follows, therefore, that the term 'substantial interest' needs to be construed having regard to the requirement that there be wide access to justice'. Although the precise scope of, and relationship between, the concepts

---

23   [1985] IR 129.
24   [1992] 1 IR 134.
25   [1992] 2 IR 380.
26   [1998] IR 485.
27   [1999] 2 I.R. 270.
28   [2003] 4 IR 143.
29   [2005] IEHC 344.
30   pp. 312–13.
31   High Court, 26 April 2007.

of substantial interest and substantial grounds have yet to be resolved, it is clear that the Irish legislature has recently sought to restrict the wide public access to environmental justice afforded by the courts in the context of establishing *locus standi* for judicial review challenges to planning decisions. Pending clarification from the Supreme Court, it appears likely that two different interpretations of the requirements for standing to sue in environmental cases may emerge depending on whether or not the issues involved in a case involve an aspect of EU law. Although this judicial uncertainty is undesirable,[32] a degree of rebalancing in favour of the Aarhus agenda is reflected in the amendment to Section 50 recently introduced by the Planning and Development (Strategic Infrastructure) Act 2006. Section 13 of the 2006 Act makes special provision for environmental NGOs challenging planning decisions and approvals involving the EIA Directive 85/337/EEC when they satisfy the conditions in Section 50A(3) of the Planning and Development Act 2000. NGOs which have pursued aims or objectives relating to the promotion of environmental protection for 12 months, who qualify to make planning appeals and who satisfy conditions (if any) prescribed by Ministerial regulations now have standing to sue without having to prove that they have a substantial interest in the matter. Although this access is limited to specific forms of appeals and planning approvals, and it remains to be seen what conditions may be prescribed by the Minister, this new provision appears to be an attempt by the RoI Government to implement the requirements of Directive 2003/35/EC.

Moving on from concerns as to the threshold for establishing *locus standi*, judicial review is also problematic in terms of the restrictions it imposes on challenges to the merits of decision – in other words, challenges on substantive grounds. Both the Aarhus Convention and the EU Directive require that members of the public with standing should be entitled to challenge the 'substantive or procedural legality' of a decision. However, courts in RoI and the UK are extremely deferential to decisions by administrative bodies. The Supreme Court decision in *O'Keeffe v An Bord Pleanála*[33] essentially ruled that the courts should not override decisions taken by planning authorities unless they are manifestly unreasonable. This position is so restrictive that it effectively frustrates the right to challenge environmental decisions on substantive grounds. This means that those campaigning to ensure that all decisions address sustainability issues will rarely succeed if they challenge regulatory decisions. It also means that decisions on the merits of planning applications by planning authorities or An Bord Pleanála are virtually unassailable in RoI unless there are procedural irregularities in the manner in which they were made. Thus far only about four planning decisions have ever

---

32  It should be noted however, that the High Court in *Harding v Cork County Council (No.2)* [2006] IEHC 295 gave leave to appeal to the Supreme Court on the grounds that the ruling on *locus standi* involved a point of law of exceptional public importance; the outcome of this appeal was not available at the time of writing. See next paragraph where standing is also different depending on whether or not EU law is involved.

33  [1993] 1 I.R. 39.

been overturned in RoI because the courts found that they were unreasonable. However, it is worth noting that the High Court decision in *Peter Sweetman v An Bord Pleanála, the Attorney General and Clare County Council* reflects a potential scope for softening this position due to EU requirements for widened access to justice. Clarke J. ruled that there are substantial grounds for arguing that a higher level of scrutiny should be applied in examining the merits of cases covered by the EIA and IPPC Directives (both of which have been amended to take account of Aarhus requirements), but noted that this can be accommodated within the ambit of existing Irish judicial review law.[34]

Nonetheless, to date no court has overturned the findings of local authorities or An Bord Pleanála on what constitutes sustainable development. A very similar position pertains in the UK where the courts, acting under a modified *Wednesbury* doctrine, also defer significantly to the administrative expertise of those who make planning decisions. So, for example, in *Fairlie v The Secretary of State for the Environment*,[35] the English Court of Appeal held that it was unlikely that the Secretary of State had misunderstood the concept of sustainable development when he had refused planning permission to a group of subsistence farmers who wanted to erect tents on their lands. The farmers argued that their proposals were sustainable because they would not impact on the ability of future generations to meet their needs while the Secretary of State considered their proposals unsustainable because the proposed development would not support higher living standards for current and future generations.

Last, but by no means least there is the thorny issue of costs as a barrier to wide access to environmental justice. At administrative level An Bord Pleanála has recently been given statutory power to attach conditions to approvals for what is termed 'strategic infrastructure development' (usually waste, energy or environmental infrastructure developments), requiring the developer to pay the reasonable costs of third parties who have participated in the approval process; but, curiously, it has not been given this power where the applicant for the approval is a local authority.[36] It has also got power to direct the payment of a contribution towards the costs to persons who have appeared in oral hearings held in connection with certain compulsory purchase orders. The legal costs aspect of the Aarhus requirements and of Directive 2003/35/EC, in so far as they apply to judicial proceedings, have been the subject of litigation in RoI in *Friends of the Curragh Environment v An Bord Pleanála.*[37] Unfortunately the argument concerning the costs of litigation was deemed premature because the Directive had not been transposed into Irish law at

---

34   High Court, Clarke J. 26 April 2007.

35   [1997] EWCA Civ.1677.

36   Planning and Development Act 2000, s.37h (2) (c) inserted by Planning and Development (Strategic Infrastructure) Act 2006, s. 3. Compare s. 37(2) (c) with section 175(5) substituted by Planning and Development (Strategic Infrastructure) Act 2006, s.34. and note that local authorities may not be required to pay reasonable costs.

37   High Court, 14 July 2004.

the time the case was taken and the court appeared to have considered (somewhat unusually) that the term 'costs' meant mere transaction costs such as document filing charges, not the costs of hiring lawyers.

More recently in *Peter Sweetman v An Bord Pleanála, Ireland, the Attorney General and Clare County Council*,[38] Clarke J. considered that the costs referred to meant 'costs' as conventionally understood, but held that the requirements as to costs in the Aarhus Convention and in Article 9(3) of Directive 2003/35/EC do not require immunity from exposure to the sort of costs that arise in Irish judicial review proceedings. He considered, somewhat unrealistically in the authors' view, that the court's discretion not to award costs against unsuccessful public interest litigants, or to award costs to unsuccessful public interest applicants for judicial review, meant that applicants for judicial review in RoI would not be exposed to excessive costs. A similar stance has been taken by courts in the UK (Bell and McGillivray 2005). It is clear, however, that while courts in both jurisdictions have been willing to make no award of costs against public interest litigants, representational groups and individual litigants still take a significant financial risk in taking judicial review proceedings. Furthermore, in some cases, unless litigants are able to make the financial undertakings in damages necessary to obtain an injunction suspending any further action prior to the review hearing, victory at the hearing may ultimately be pyrrhic.

*The absence of a legal definition of sustainable development*

The sustainable development strategies on both sides of the border adopt the original Bruntland definition of sustainable development; namely, that sustainable development means 'development that meets the needs of the present, without compromising the ability of future generations to meet their own needs'. However, while this formulation has also been adopted by the other UK countries and the EU, thus far none have sought to provide a statutory definition of the principle of sustainable development. The principle of sustainable development is undoubtedly a fluid and evolving concept leading many to argue that it is unsuited to precise legal definition. Quite apart from the significant ambiguity inherent in even the widely-adopted Bruntland formulation, the potential pitfalls of distilling the principle down to provide a sufficiently precise definition for legislative drafting purposes were explicitly acknowledged by the then Minister for the Environment, Mr. Noel Dempsey T.D., during the steering of the Planning and Development Bill, 1999 through the Oireachtas. He justified the absence of a statutory definition of the concept in the following terms:

> The question arose in the Seanad of giving a concrete definition of sustainable development in the Bill. I gave a good deal of thought to this but felt in the end that it was such a dynamic and all embracing concept, and one which will evolve

---

38   High Court, 26 April 2007.

over time, that any legal definition would tend to restrict and stifle it. Infusing the concept through the Bill, as we have done, gives effect to it in a holistic and comprehensive way.

The courts in NI and the Republic have been as reluctant as politicians to define sustainable development, and thus far neither has attempted to provide a judicial definition. However, in the first attempt to give a legal EU definition of the term in *R v Secretary of State for the Environment ex parte Corporate Shipping Ltd*[39] Advocate General Leger makes it clear that sustainable development does:

> ...not mean that the interests of the environment must necessarily and systematically prevail over the interests defended in the context of the other policies pursued by the Community ...On the contrary, it emphasizes the necessary balance between various interests which sometimes clash, but which must be reconciled.

In this regard it is interesting that the Royal Society for the Protection of Birds (RSPB)[40] in NI urged Government to adopt a statutory definition of sustainable development that clearly reflected the centrality of environmental protection within that mediation process as was recommended by the Royal Commission on Environmental Pollution in its 2002 report, *Environmental Planning*.[41] The Commission stated that at the heart of the definition of sustainability must be a fundamental recognition that the environment can impose constraints on human actions; that this will sometimes lead to hard choices; but that the goal of protecting and enhancing the environment must be fundamental.[42] Thus far, the RoI and the UK have adopted the approach of requiring decision makers to follow the non-statutory guidance published by Government concerning the meaning of the principle of sustainable development. Guidance has not yet been produced in NI concerning the practical implications of taking this principle into account in decision making concerning planning. Similarly, although the Planning and Development Act 2000 is eight years old, the Department of the Environment in the Republic has also failed to provide comprehensive guidance on this matter. In the absence of formal guidance or a statutory definition to the contrary it is fairly safe to say that legislative duties to 'promote' or 'have regard' to the principle of sustainable development do not necessarily mean that environmental interests must always triumph over other interests. However, it is clear that they must always be taken into account and balanced with other interests. Any decision that fails to do this when the law requires that it should be done is potentially

---

39    [2000] ECR-1 9235; [2001] 1 CMLR 19.

40    RSPB(NI) Response to the DOE Consultation concerning the Draft Planning Reform (NI) Order, December 2005.

41    Cm 5459.

42    Ibid. at p. 38.

invalid. Nonetheless, although concern for the environment must now be included in economic cost-benefit calculations, once this has been done, even in a token manner, it is highly unlikely that the courts in the Republic and NI will question whether the correct balance has been achieved.

## Sustainable development in practice: The question of rural housing

Our discussion so far has provided an overview of the extent to which the principle of sustainable development has been integrated into key legal and policy frameworks, and in particular those governing rural NI and the RoI. The task now is to consider the extent to which this evolving policy and legal rhetoric has been put into practice. There is little doubt that while both Governments have made important if uneven advances in terms of embedding the concept of sustainable development into their legal and policy frameworks, they have baulked at the prospect of applying this principle in practice in the highly sensitive context of the rural environment. Although several significant examples of this pattern exist – spanning nature conservation, water pollution and access to the countryside – the most graphic instances of this resistance arise in relation to the implementation of the Nitrates Directive and policy development concerning rural housing. The Irish experience of implementing the Nitrates Directive, which was only achieved at the eleventh hour forced by advanced Article 228EC litigation against both Governments, is considered separately in Chapter 3. The question of rural housing will therefore be addressed in the present chapter.

Policy making concerning rural housing is without doubt one of the most sensitive political issues facing the future of rural NI and the RoI. In so far as RoI is concerned, many of the concerns about single dwellings in the countryside were stated succinctly in 1997 with the publication of *Sustainable Development: A Strategy for Ireland*. Almost uniquely in Europe, many people with no ostensible connection to the countryside live in isolated rural dwellings frequently with private sanitation and water supplies. This phenomenon has excited passions on both sides of a heated debate. One-off houses in rural areas accounted for 43 per cent of the 68,819 new homes built in 2003 – 36 per cent more than in 2000. Apart from the fact that many of these houses are not connected to sanitary facilities or public water supplies, by many criteria the construction, design and siting of many rural houses is defective and does not meet modern standards for sustainable development.

Long before the phrase sustainable development entered the legal lexicon, Irish social policy was to encourage rural settlements. Article 45.2(v) of the Constitution which expresses some of the Directive Principles of Social Policy for Ireland states: 'The State shall, in particular, direct its policy towards securing: ... (v) That there may be established on the land in economic security as many families as in the circumstances shall be practicable'. This policy has been implemented by many fiscal and other benefits conferred on rural dwellers, especially farmers[43] and by a lack of any real disincentives to those who wish to locate in rural areas. Environmentalists argue that once-off rural housing is contrary to the principles of sustainable development mostly because of the impact of individual septic tanks on groundwaters and of dispersed housing on the landscape. Others point to more indirect environmental and social effects – low density settlement patterns undermine public transport systems and are extremely car-dependent, leading to more traffic, air pollution and energy consumption.[44] Most rural dwellers do not work in the countryside and therefore often travel long distances by car to work in urban centres, the exhaust emissions from which contribute to climate change. Infrastructural costs are greater because providing utilities and social services to occupants of dispersed housing is more expensive than in urban areas. Occupants of one-off housing, particularly the elderly and incapacitated and children are isolated from formal and informal social contacts and supports. The sale of sites for rural housing has also led to the commoditization of the countryside and the fragmentation of agricultural units.

However, in determining what is or is not sustainable development in the context of planning legislation, the environment and the cost of public and social services are not the only concerns. The maintenance of a permanent population in the countryside and of rural economic activity independent of the tourist sector is arguably a component of sustainable rural development. Facilitating housing in rural areas, including one-off housing, is one way of ensuring that there are future generations in rural areas. Conway (2003, p. 145) comments: 'A variety of studies have highlighted the key role that housing can play in the regeneration of rural areas'. Arguments about the higher cost of services and utilities are answered by what may be termed 'the house at the end of the valley' argument.[45] It is argued that rural housing is very affordable, it enables the younger generation to live near relatives thus ensuring intergenerational social supports, it allows farmers to live on their farms so that they can tend to livestock and crops more easily, and it satisfies the aspirations of emigrants returning to their roots. It also enables landowners (mainly farmers) to sell land to supplement declining incomes, provides rural employment, sustains declining rural communities and supports their distinctive cultures. Without rural housing, large areas of our countryside might soon be deserted.

---

43   See e.g. Capital Consolidation Act, 1997, s.603A (exempting transfers of sites valued at less than €254,000 by farmers to their children from capital gains taxes) and subsidies to farmers under the Rural Environmental Protection Schemes.

44   See McDonald, *The Irish Times*, 6 August 2001: 'Planners warn on dangers of "one-off housing" in countryside'; '...rural housing is predominantly and increasingly car-dependent, with consequential increases in greenhouse gas emissions, as well as generating more pressure on rural roads and more demand for parking in towns'.

45   Nix (2003, p. 82) describes this as arguing that where utility lines, pipelines and post are already delivered to a house at the end of a valley, there can be no argument against ribbon development on the road leading to that house. However he goes on to contend that this argument overlooks the fact that the 'house at the end of the valley' is usually served at shoe-string capacity.

Analyzed in terms of the language of sustainable development, there is little doubt that the question of rural housing raises an acute clash of economic, social, cultural and environmental interests. While rural housing promotes a certain form of economic growth and social cohesion, and ensures that there will be future generations in an area to enjoy their environmental inheritance, there are also serious questions concerning despoliation of rural landscapes, pollution to groundwater which is almost impossible to remediate, the overstretching of social and other public services and a significant section of the population isolated from adequate social supports.[46]

In an attempt to resolve political pressures generated by the one-off housing debacle in the RoI, the Minister for the Environment, Heritage and Local Government attempted to give policy guidance to planning authorities on the problem in 'Sustainable Rural Housing – Guidelines for Planning Authorities' (hereinafter the 'Ministerial Guidelines').[47] Drawing on the National Spatial Strategy (NSS) the Ministerial Guidelines published in April 2005 recognize four different types of rural areas, and their differing needs. Appendix 3 to the Guidelines sets recommended development-plan objectives for each area. For example, in what are termed 'Structurally Weak Rural Areas' the key development plan objective should be the need to accommodate any demand for permanent residential development, subject to good practice in design, location and the protection of landscape and environment. However, in what are termed 'Areas under Strong Urban Influence' the development plan should direct urban-generated housing to areas zoned for new housing development in urban centres in the planning authority's area, subject to meeting 'the housing requirements of the rural community as identified by the planning authority in the light of local conditions'.[48] McDonald and Nix (2005, p. 85) have argued this means that: 'Essentially, councillors opt to ban one-off housing but exempt their own electorate from that ban'. Truly an Irish solution to an Irish problem.

The NSS indicates that in order to secure co-ordinated and sustainable development, new housing in rural areas that are under development pressure should generally be confined to persons with roots in or links to those areas.

---

46   The Strategic Planning Guidelines for the Greater Dublin Area recognize the value of rural areas of counties adjoining the Dublin area in providing an amenity resource, and a strategic resource base for food production, water supply, and other supplies of natural resources. The Guidelines designate large areas of County Meath as Strategic Green Belts wherein sporadic and dispersed development is described as unsustainable and recommend that it should be subject to strict control. The Guidelines envisage that land uses in such Green Belts should be primarily rural, including agricultural, forestry and recreational uses. They recommend that other forms of development, including housing and employment activities, should be to serve local needs only.

47   Minister for the Environment, Heritage and Local Government Sustainable Rural Housing – Guidelines for Planning Authorities (http://www.irishspatialstrategy.ie/Rural%2 0Planning%20Guidelines%2013505.pdf).

48   Ibid. 53.

These include: persons working full time in rural areas, sons and daughters of families living in rural areas who want to live near their parents and returning or retiring emigrants. In order to combat vendors of sites fraudulently claiming an intention to settle in a rural area, the Ministerial Guidelines state that planning permissions granted to applicants with roots or links to an area should normally be conditioned to require that the dwelling should be occupied by the applicant (or members of his/her immediate family) for a specified period. This unusually restrictive type of permission, which impinges on the marketability of affected houses, is permitted by Section 39(2) of the Planning and Development Act 2000. This enables a planning authority to attach a condition (known as an occupancy condition) to planning permissions for a dwelling house specifying that only persons of a particular class or description may live in it. Occupancy conditions may be embodied in an agreement under Section 47 of the Act. Planners claim that these conditions are difficult to enforce and there is some anecdotal evidence that they are often disregarded.[49]

An important criticism of the Ministerial Guidelines has been by way of legal analysis questioning the compatibility of the recommended type of occupancy condition with the Constitution, EU law and the European Convention on Human Rights. So, while occupancy conditions are aimed at promoting sustainable development, the method chosen to achieve this is legally questionable and it may be that giving effect to higher constitutional values will make it impossible to implement the best ways of achieving sustainable development. Doyle and Keating, the authors of a report on occupancy conditions,[50] consider that conditions restricting occupancy of rural houses to persons who work in rural areas are probably legal, as are conditions restricting occupancy in Gaeltacht areas to persons who speak Irish, if not applied uniformly over a large area. However, they consider that bloodline conditions privileging occupants who are sons, daughters or relatives of rural dwellers are probably contrary to the Constitution as well as to EU law and the European Convention on Human Rights, and that conditions privileging returning emigrants may contravene EU law and the European Convention of Human Rights. Notwithstanding these views, the solution proposed in the Ministerial Guidelines is one which appeals more to those who place greater value on the economic, social and cultural pillars of sustainable development rather than the environmental one, and their values are expressed in the manner in which the Ministerial Guidelines are applied. Very recently, the EU Commission has questioned this privileging of persons with local connections and litigation on the matter is contemplated.

The elected members of local authorities are required by law to 'have regard' to the guidelines on rural settlement policies when they are making their development

---

49    'Planners have been "lied to, deceived and hoodwinked"' *The Western People*, 7 September, 2005.

50    Law Reform Committee of the Law Society, 'Discriminatory Planning Conditions: The case for reform' (Law Society of Ireland, February 2005.

plans. This means that they have to take them into account but it does not mean that they must adhere to them.[51] Planning officers are required to 'have regard' to the development plans (which in turn must have had regard to the Ministerial Guidelines) when deciding on applications for planning permissions for single houses in the countryside.[52] In practice, however, both elected members of local authorities and many planning officers have for various reasons (good and bad) failed to halt the progressive despoliation of the countryside by one-off houses.

Not surprisingly given its similar pattern of land ownership and largely rural culture, the issue of single dwellings in the countryside is also a highly contentious issue North of the border, and indeed threatened reform of its traditionally permissive approach to development of this nature was used as one of a number of political levers to force local parties to resume the reins of devolved power. In contrast to the United Kingdom, the Department of the Environment in NI has presided over a rural planning policy which today permits almost three times more single dwellings to be built in the NI countryside each year than occurs in total throughout the rest of the UK.[53] Single dwellings in the countryside now account for half of all dwellings constructed in the region[54] which is rapidly leading to the suburbanization of the region's countryside. Despite the fact that serious concerns have been expressed as to the negative impact of this policy for over two decades in terms of damage to the environment and landscape, implications for regional transport policies, the cost of providing of services to dispersed dwellings and impact on rural communities, it remained unchanged.[55]

The first indication of a willingness to embrace a more environmentally sustainable policy came with the publication of the *Regional Development Strategy 2025* in 2001, which emphasized the need to place sustainable development principles at the heart of future rural development in NI. Hopes of a more environmentally-sustainable approach to rural development appeared to evaporate four years later with the publication of an initial *Issues Paper* in 2004 by the same Government

---

51    *Keane and Naughton v An Bord Pleanála* [1995] I.C.L.Y. 411. Murphy J. in the High Court held that the duty to 'have regard' to Ministerial policies means to 'take account of these matters, not necessarily to regard them as crucial'. In *McEvoy v Meath County Council* [2003] I.R. 208. Quirk J. found that the requirement 'to have regard to' particular concerns (in that case, the Strategic Planning Guidelines) meant 'informing oneself fully of and giving reasonable consideration to such concerns'.

52    Planning and Development Act 2000, s. 28.

53    PPS 14, Sustainable Development in the Countryside: Issues Paper, (Department of Regional Development) at p. 4. http://www.planningni.gov.uk/AreaPlans_Policy/PPS/pps14/issues_paper.pdf.

54    National Trust Planning Commission, *A Sense of Place: Planning for the Future in Northern Ireland* (2004), at p. 24 para 3.5.2.http://www.nationaltrust.org.uk/main/w-ni-sense_of_place-summary.pdf.

55    House of Commons Select Committee on the Environment (1990), supra note 5; and the House of Commons Northern Ireland Affairs Committee, *The Planning System in Northern Ireland*, HC 53 Session 1995–96; and Ibid at para 3.5.6 et seq.

department as a precursor to a full consultation document setting out formal proposals for reform. Despite the progressive nature of the Regional Development Strategy in terms of its endorsement of integrated spatial planning, the Issues Paper indicated a willingness to deviate significantly from current policy on rural development. Then in an apparent *volte face* essentially the opposite approach was taken in the draft PPS 14 published in March 2006. To considerable political outcry, Government proposed the introduction of a presumption against rural development, subject to very limited exceptions, effectively embracing many of the concerns expressed by environmentalists during the earlier consultation process. However, it is important to emphasize that the decision to publish proposals for strict controls on rural development must be understood as part of a wider political strategy to break the political stalemate surrounding constitutional negotiations in NI. Whereas previous direct rule administrations had effectively suspended policy making in NI pending the anticipated return of devolution, the then Secretary of State, Peter Hain adopted an explicit strategy of pushing ahead with major policy decisions, most controversially, to abolish the '11-plus' school transfer test, introduce water charges and impose restrictions on rural development.

In effect, the Hain administration took the view that a policy limbo simply facilitated political procrastination and deadlock in constitutional negotiations. Hain's determination to push ahead with policy changes on highly contentious issues is widely accredited with forcing the Northern electorate to put pressure on local political representatives to assume the reins of devolved power and halt the proposed changes. Devolution was restored in March 2007. As was discussed earlier, the High Court of NI has quashed draft PPS14 on the ground that DRD lacked the power to issue planning policy statements. The proposed PPS14 had in any event become the subject of a review by the new Government. Although policy responsibility for bringing forward the revised PPS14 now rests with DOE, it remains to be seen whether the new Government will adopt the solution employed in the Republic, or develop an alternative solution that allows better alignment with UK and regional commitments to sustainable development.

It goes without saying that there is no easy solution to the problem of rural housing in either NI or the Republic. The future control of rural development on the island will undoubtedly be a key litmus test of both Governments' willingness to embrace integrated policy development. It is arguable that the pragmatic Ministerial Guidelines adopted in the Republic strike the right balance but require elected members of local authorities to adhere to their legal mandates and to comply conscientiously with the Ministerial Guidelines when deciding on local policies for rural housing in development plans. Some argue that while not ideal, this solution is preferable to a rigidly enforced bureaucratic ban on rural housing. Furthermore, other environmental mitigation measures should be adopted to ensure that every effort is made when permission is given for rural houses that they satisfy tests for low impact developments so that they either enhance or do not significantly diminish rural environmental quality. Conditions should require that they are aesthetically pleasing, appropriately sited having regard to the local landscape and settlement patterns, designed by qualified architects, incorporate waste minimization and energy conservation systems and that their construction does not involve the unnecessary destruction of hedgerows and general biodiversity. Others argue that a more sophisticated 'criteria-based' approach should be adopted to assess the compatibility of proposed development with Government's sustainability objectives[56] – for example, a more formal adoption of the concept of natural capital (Owens 1995; Oleweiler 2006) in strategies to provide for affordable rural housing and employment. It is furthermore suggested that it is important to co-ordinate the objectives of rural planning policy with policies for rural employment, transport, affordable housing, and rural service provision, and in particular to integrate specific targets and objectives into national or regional sustainable development strategies on the island.

Other indirect inducements to deter rural housing could also be promoted. If living in urban areas were more attractive, only those who genuinely need to live in the countryside would be willing to forgo access to the recreational, educational, social and cultural facilities available in urban areas. Attempts in the Republic to increase the attractiveness of small towns as places to live in the Town Renewal Rural Scheme have had a limited success to date and more efforts are needed in this regard. If the two Governments were really serious in their intent to minimize the unsustainable amount of rural housing, they would also integrate planning policy with their fiscal and other policies. Increased stamp duties and higher local charges for waste, water or other public services, reflecting the true costs of providing these services in rural areas, could also act as more appropriate disincentives to inappropriate rural housing than the condemnations of the aesthetic professions.

**Conclusion**

There is little doubt that the national, EU and global emphasis on sustainable development is likely to intensify in the years to come with far-reaching implications for the rural environment on the island of Ireland. While the embedding of sustainable development within legal frameworks is as yet in its early stages, as environmental pressures become more acute, it is likely that legal frameworks and processes will play an increasingly important role in the process of articulating and ensuring implementation of the principle of sustainable development. While the two Governments on the island of Ireland have started to respond to their international commitments as signatories of the Rio Declaration, the integration of sustainable development into their domestic legal frameworks – and particularly in relation to the rural environment – has been largely driven by the EU. Quite apart from the Treaty amendments, EU action has led to the creation of increasingly ambitious procedural rights for individuals, strengthened the scope and impact

---

56   Friends of the Earth briefing Paper on PPS 14 (June 2006) at para 4.1.2.

of EU environmental law, stimulated significant policy integration between agriculture and environment and intensified infraction action in key areas. There is little doubt but that these trends will continue.

However, policy development on rural housing on the island of Ireland is a key litmus test of the two Governments' willingness to implement the principle of sustainable development in the rural environment. Although both Governments have taken the independent initiative to integrate regard for sustainable development into their respective legislative frameworks governing planning controls, it remains to be seen whether the political will exists to counter the trenchant opposition that exists to the imposition of more environmentally-sustainable controls on rural housing on the island. At present planning policy falls outside the EU's sphere of competence. However, the environmental impacts of continuing with a highly permissive policy on rural housing will, almost certainly, render compliance with established EU environmental Directives more difficult and more expensive for both Governments (particularly in the context of the demanding standards required under the Water Framework Directive, Urban Waste Water Treatment Directive and the Habitats Directive), and, potentially, as cumulative development impacts increasingly trigger the EIA Directive. Consequently, in the absence of political will from within the island, it is possible that the need to contain the risk of EU infraction in relation to these key Directives may ultimately force both Governments to strengthen legislative controls and adopt formal guidance ensuring an environmentally sustainable approach to decision making concerning rural housing.

### References

Bell, S. and D. McGillivray (2005) *Environmental Law.* 6th edn. Oxford: Oxford University Press.

Bruntland, G. (1987) World *Commission on Environment and Development, Our Common Future.* Oxford: Oxford University Press.

Carnwarth, R. (1999) 'Environmental Litigation – A Way Through the Maze?', *Journal of Environmental Law*, 11: 3–14.

Castle, P., M. Day, C. Hatton and P. Stookes (2004) *A Report by the Environmental Justice Project.* London: Environmental Justice Project.

CEMAT (2003) *Ljubljana Declaration on the Territorial Dimension of Sustainable Development*, 17 September 2003.

COM (2001) *A Sustainable Europe for a Better World: A European Union Strategy for Sustainable Development 2002.* Final.

Convention on Access to Information, Public Participation in Decision-making and Access to Justice in Environmental Matters (1998). Aarhus.

Conway, M. (2003) 'Building at the Crossroads: Rural Housing Policy in Northern Ireland', in J. Greer and M. Murray (eds), *Rural Planning and Development in Northern Ireland. Dublin: Institute of Public Administration*, pp. 145–68.

Cordonier Segger, M. and A. Khalfan. (2004) *Sustainable Development Law: Principles, Practices and Prospects*. Oxford: Oxford University Press.

Department of the Environment and Local Government (1997) *Sustainable Development- a Strategy for Ireland in 1997*. Dublin: Stationery Office.

European Landscape Convention 2000 (2000) Strasbourg: Council of Europe.

Greer, S. and M. Murray (eds). (2003) *Rural Planning and Development in Northern Ireland*. Dublin: Institute of Public Administration.

Government of Ireland (2000) *National Development Plan 2000–2006*. Dublin: Stationery Office.

Government of Ireland (2002) *Making Ireland's Development Sustainable*. Dublin: Stationery Office.

Government of Ireland (2005) *Sustainable Rural Housing – Guidelines for Planning Authorities*. Dublin: Stationery Office.

Lafferty, W. and J. Meadowcroft (eds) (2000) *Implementing Sustainable Development*. Oxford: Oxford University Press.

Law Society of Ireland (2005) *Discriminatory Planning Conditions: The Case for Reform*.

Macrory, R. and S. Turner (2002) 'Participatory Rights, Transboundary Environmental Governance & EC Law', *Common Market Law Review*, 39: 489–522.

McDonald, F. and J. Nix (2005) *Chaos at the Crossroads*. Kinsale: Gandon Books.

Morrow, K. and S. Turner (1998) 'The More Things Change, The More They Stay The Same: Environmental Law, Policy & Funding In Northern Ireland', *Journal of Environmental Law*, 10: 41–59.

Nix, J. (2003) 'Urban Sprawl, One-Off Housing and Planning Policy: More to Do, But How?', *Irish Student Law Review*, 10: 78–93.

Olweiler, N. (2006), 'Environmental Sustainability for Urban Areas: The Role of Natural Capital Indicators', *Cities*, 23, 3: 184–95.

Owens, S. (1995) *Planning Settlements Naturally*. London: Nathaniel Lichfield and Partners.

Ryall, A. (2007) 'Environmental Justice and the EIA Directive: The Implications of the Aarhus Convention', in J. Holder and D. McGillivray (eds), *Taking Stock of Environmental Assessment*. Abingdon: Rutledge Cavendish, pp. 191–219.

Scannell, Y. (2005) *Environmental and Land Use Law*. Dublin: Thomson, Round Hall.

Stallworthy, M. (2002) *Sustainability, Law Use and Environment: A Legal Analysis*. London: Cavendish.

Turner, S. (2006a) 'Transforming Environmental Governance in Northern Ireland, Part One: The Process of Policy Renewal', *Journal of Environmental Law*, 18, 1: 55–87.

Turner, S. (2006b) 'Transforming Environmental Governance in Northern Ireland, Part Two: The Case of Environmental Regulation', *Journal of Environmental Law*, 18, 2: 245–75.

Chapter 3

# Environmental Lessons for Rural Ireland from the European Union: How Great Expectations in Brussels get Dashed in Bangor and Belmullet

Brendan Flynn

## Introduction

In their classic study of policy implementation, Pressman and Wildavsky (1984) entitled their book, *Implementation: How Great Expectations in Washington are Dashed in Oakland*. This chapter draws something of its inspiration from that tradition of policy analysis. The focus is placed squarely upon how the European Union (EU) has interacted with the environmental policy regimes of both the Republic of Ireland (RoI) and Northern Ireland (NI). An assessment is offered of the impact of the EU regarding environmental issues, especially those connected with rural areas. Has the EU managed to modernize official thinking on environmental questions, especially in the sense of disseminating best practice? Or has it been largely an ineffectual force for change, and perhaps even counter-productive?

The argument of this chapter is straightforward: the EU's environmental impact has been patchy. There has unquestionably been a dissemination of best practice on environmental policies within both the RoI, and also NI. Yet that process appears to be quite superficial and weak. The EU's good intentions have been met by Irish administrative mismanagement and politicking, especially in the RoI. Notwithstanding environmental absurdities which the EU has promoted (such as paying farmers to over-stock and then de-stock sheep), for the most part the EU influence has been very positive. Without the influence of EU environmental policy, the push for higher environmental standards would likely have been weaker over the last three decades.

What is also evident is that the implementation of environmental policies on both sides of the border has been inadequate, and at times woefully so. With regard to the rural environment, the implementation of the Birds and Habitats Directives has proven difficult, especially so in the Republic. The Nitrates Directive remains a political hot-topic in both the Republic and Northern Ireland. Its implementation has been successively delayed, due to lobbying by organized farming groups.

Incredibly, it was only by the summer of 2006, that the RoI had implemented this Directive. It was originally supposed to be in force since 1995!

The EU on paper then has set out great ambitions for environmental policy. If these laws were followed and properly implemented there is little question but that ecological problems in Ireland would be significantly reduced. In reality, local politics and weak local institutions have heavily diluted that promise through poor implementation.

Structural developments need also to be considered. The 1990s saw major transformations of political and economic fortunes on both sides of the border. In the case of the Republic, the 'Celtic Tiger' became a phenomenon and was accompanied by a new era of political stability and apparent consensus. Partnership arrangements were forged between various coalition Governments, employers, and unions, giving the RoI a neo-corporatist policy framework. In NI, the social and economic benefits of a major reduction of violence from the mid 1990s were also significant. By the late 1990s, Belfast was experiencing its own building boom and much heightened economic activity.

Against the backdrop of economic growth, rural communities have been unquestionably left behind. This trend is perhaps clearer in the Republic, where much of the 'Celtic Tiger' effect has been concentrated in distinctively urban and suburban locations, especially clustered around Dublin, with demographic growth in the counties around Dublin being especially strong (+15 per cent since 2002) (Central Statistics Office [CSO] 2006, p. 11). Parts of rural ROI have come under intense development pressure from urban workers who seek to live in and commute from countryside locations anything up to 50–60 km away from their place of work. The result is ROI's lopsided spatial development: a bloated greater Dublin region of suburban sprawl, allied with several provincial cities and small towns each ringed by their mini-versions of the ubiquitous sprawl phenomenon.

One important difference between NI and the Republic has been that in NI, policy measures have been much influenced by distinctive United Kingdom (UK) thinking on sustainable development. In particular the key idea of a multi-use 'countryside' ideal[1] has formed a tangible centrepiece of their regulatory efforts (see Page 1999; Marsden et al. 1993; Bishop and Philips 2004). This ideal welds landscape, farming, recreation, and biodiversity interests together into a common conception of 'countryside'. In the Republic such an enveloping focus has been less obvious. Instead, the interests of commercial farming have politically predominated to a much greater extent. In the RoI, the term 'rural' is still simply equated with farming interests. Within the wider United Kingdom debate on rural sustainability that mindset has at least been challenged.

The EU has unquestionably provided a modern template of environmental laws that are basically sound. These have been let down by poor implementation in both

---

1 See, for example, Page (1999), Bishop and Philips (2004) and Marsden et al. (1993).

NI and the Republic. Policy failures have been largely sourced within Ireland and not in Brussels. Moreover, it would be unfair, to expect too much of the EU.

As a political system, the EU lacks the comprehensive authority and legitimacy to decisively engage with substantive social and economic policies. The EU has only very minimal scope to engage with national budgetary allocations or local institutional practices. As a result there has been an overemphasis on legal instruments and comparatively little use made of fiscal tools within environmental policy. In light of this it can be argued that the focus on environmental laws has been overly restrictive; what is needed are broader social and economic measures.

Although much law making is done by the EU, only limited fiscal powers are held by the EU authorities. Moreover, law enforcement is mostly a national if not local responsibility (Keleman 2000, pp. 139–42). The EU's scope to fund (or 'bribe') better implementation, so common in the US experience of environmental policy, is therefore much reduced.

The upshot of all this is that the EU can at best coax good practice from national governments. It can also occasionally choose to litigate to get member states to refrain from their worst excesses; the European Court of Justice (ECJ) has found against member states in several cases of appalling non-implementation. Yet, considering the sheer volume and complexity of EU environmental law, these are only a few cases for each state. And the fact that it can take easily up to five years before a complaint ends up before the ECJ tends to limit its scope to force countries to speed up implementation.

As long as the constitutional nature of the EU remains basically confederal, this phenomenon of poor implementation will likely remain a systematic and structural weakness. As the effective remedies for poor implementation are mostly local, they are therefore ultimately an internal responsibility.

To elaborate on these more general arguments, this chapter will present a general discussion on the reception of EU environmental laws, and then offer two more detailed case studies: one on the Nitrates Directives and a second on the related Birds and Habitats Directives.

### The reception of EU environmental law in Northern Ireland and the Republic

The great weakness of EU environmental policy has always been its implementation by national authorities (Glachant 2001). Almost every EU member state has cases where there has been slowness to enforce some environmental law, and in a few instances political sensitivities have led to naked refusal and evasion of legal responsibilities. For example, this has been the case with France over the Birds and Habitats Directives (Szarka 2002), but also with RoI over the Nitrates Directive. RoI's implementation record can be described as basically worsening throughout the 1990s, so that several court cases were taken by the Commission against it, something that before this was rare (Flynn 2006, pp. 138–50).

What is further noticeable about the RoI experience of implementation is the fact that litigation has mostly failed to stir the Irish authorities. Dublin Governments are not very politically embarrassed when the ECJ finds a failure with Irish implementation, in the same way say that any Danish Government would be. Such failures simply do not make front-page news, and their political fall-out is relatively limited. Apparently, ruling politicians fear much more the political pressure from their farmers' lobby groups over certain key environmental laws.

Only in the last few years has the cumulative number of negative verdicts begun to stack up and take effect. These have included findings that the RoI has failed to correctly implement the Directives on Environmental Impact Assessment (EIA) in 1999, on nature protection sites in 2001, and on wild birds, water pollution, and industrial hazards (all in 2002); there have also been failures in relation to shellfish water protection (2003), nitrates (2004) and waste management (2005).[2] The Commission has also threatened to seek fines against RoI for non-adherence to previous Court rulings on nitrates (in 2004). This is a very serious breach of trust and law. So far, last minute deals have managed to fend off that particular scenario of heavy fines. One wonders for how long.

The situation with regard to implementing EU environmental laws is not much better in NI. Indeed Macrory has argued that 'Northern Ireland has also gained a reputation for late transposition of European Community Directives concerning the environment' (Macrory 2004, p. 3). Within a number of cases concerning UK failures to implement EU environmental laws properly, the situation in NI has been referred to as one element of the proceedings. NI's distinctive legal machinery can be slower, as part of the UK legal order, to respond to EU directives. Indeed with devolution of powers to Scotland, Wales and NI since the late 1990s there is now something of a co-ordination problem in transposing EU directives properly and on time for the entire UK.

---

2   The exact references for these cases are: Case 392/96, (EIA), Commission of the European Communities v Ireland, Judgment of 21st September 1999. *European Court Reports*, (1999), p. I-05901; Case 67/99, (natura sites), Commission of the European Communities v Ireland, Judgment of 11th September, 2001. *European Court Reports*, (2001), p. I-05757; Case 117/00, (wildbirds), Commission of the European Communities v Ireland, Judgment of 13th June, 2002. *European Court Reports*, (2002), p. I-05335; Case 316/00, (water pollution*)*, Commission of the European Communities v Ireland, Judgment of 14th March, 2002. *European Court Reports*, (2002*)*, p. I-10527; Case 394/00 (Seveso/industrial hazards), Commission of the European Communities v Ireland, Judgment of 17th January 2002. *European Court Reports*, (2002), p. I-00581; Case 67/02 (Shellfish water pollution), Commission of the European Communities v Ireland, Judgment of 11th September, 2003. *European Court Reports*, (2003), p. I-09019; Case 396/01 (nitrates), Commission of the European Communities v Ireland, Judgment of 11th March, 2004. *European Court Reports*, (2004), p. I-02315; and Case 494/01 (waste management), Commission of the European Communities v Ireland, Judgment of 26th April, 2005. *European Court Reports*, (2005), p. I-03331.

For example, the UK was collectively found to have improperly implemented aspects of the Habitats Directive in 2005,[3] on the grounds that their regulations were not legally precise enough. This verdict partially applied to the Northern Ireland 1995 habitats regulations. In another case,[4] while the EU's Integrated Pollution Prevention and Control (IPPC) Directive had been transposed properly and on time in England, Wales and Scotland through legislation, the same had not been done for NI before the deadline of 2000.

### Implementing the Nitrates Directive in the face of farmers' power

The saga of the Nitrates Directive is worth recounting in a little detail as it reveals how implementation, both in NI and the RoI, was effectively stymied out of sensitivity to organized farmers' interests and their lobbying. This directive was originally supposed to come into effect in 1995. As a law it was initially quite limited in its scope and a product of Danish and German worries during the late 1980s. EU member states are required to monitor ground and surface waters, as well as estuaries, all to ensure that the standard World Health Organization (WHO) safety limit of 50mg/l is not exceeded. Where levels are above this measure, or likely to become so, Nitrate Vulnerable Zones (NVZ) must be designated and an action plan must be submitted to the Commission. After a vital case involving France in 2003,[5] the ECJ has much widened the scope of the Nitrates Directive. Effectively this means that it applies now to any freshwaters or coastal waters where eutrophication from agricultural nutrients is likely to occur.

Yet by the end of the 1990s no substantive work was done on nitrates in Dublin, other than a voluntary code of good agricultural practice being published in 1996 (a similar code was published in NI in 2003 and posted to 32,500 NI farmers). A *de facto* strategy of dragging out the implementation process emerged. In addition to this, a curiously stubborn view took hold in Dublin that there was actually *no* real nitrates pollution problem in Irish waters. Therefore the official mindset was that the Directive didn't simply apply at all to RoI! At the time, the focus was upon fish kills and phosphate pollution in Irish freshwaters.

By 2001 the Commission had lost its patience, and began legally challenging the RoI to respond to the Nitrates Directive. In subsequent litigation the Irish position was completely undermined when the Commission could simply point to Irish Environmental Protection Agency (EPA) data which revealed relatively high nitrates levels in at least a few locations. RoI was then found to be plainly in

---

3  See Case C-6/04, *Commission v UK*.

4  See Case C-39/01, *Commission v UK*.

5  Case 258/00, *Commission v France*, has had the effect of considerably broadening the scope of the directive and narrowing the ability of member states to avoid their obligations.

breach of the Nitrates Directive in 2004 mainly because she had failed to designate any NVZs.[6]

The same situation also emerged with regard to NI as part of the UK. Again the Commission pushed all the way to litigate the issue before the ECJ. In their judgment the ECJ revealingly noted that by the late 1990s not one NVZ had been properly designated within NI, 'despite the fact that at least one area had been (previously) identified… (and that) while three zones were designated for Northern Ireland as at 11 January 1999, such designation, like that relating to the whole of the United Kingdom, is based on an incorrect definition of waters'.[7]

What can explain this bureaucratic tardiness and sloth? That question is especially pertinent given that these verdicts provoked rapid if belated action in both Northern Ireland and the Republic. The official Governmental machinery suddenly lurched into activity and in the RoI the *entire* national territory was designated a NVZ in 2004. Northern Ireland has adopted much the same approach, the entire territory of the six counties being designated a NVZ in 2005. Previously seven discrete 'candidate' NVZs had been identified in NI, even though by then only about half of the entire NI fresh water body had been designated as eutrophic, chiefly Loughs Neagh and Lough Erne.

In both cases these moves seems odd given that the RoI argument had always been the nitrate pollution was a rare event confined to a few isolated locations. It seems this approach was taken as a 'belt and braces' precaution after the scope of the original Nitrates Directive was widened by a 2003 ruling of the ECJ. In other words, after years of inaction, go-slow, and denial, panic mode had set in.

Yet if one looks for a deeper political explanation of this sorry saga it is clear that the patchy detail of implementation has been littered with sensitivities to farmers' lobby groups, foremost among them the formidable Irish Farmers Association (IFA) and the Ulster Farmers Union (UFU). The reason why the nitrates issue was left on the back burner was due to political sensitivity to the farmers' lobby.

For example, at least three different Nitrate Action Plans (NAPs) were prepared by the RoI's Government between summer 2002 and December 2004. This was all part of a process designed to placate the Commission but even more so, Irish farmers. After heavy lobbying, Denis Brosnan, a captain of agribusiness, was even asked to chair one separate action plan. This was done in order to placate and reassure farmers that the Directive would not threaten their businesses.

The critical issue in these plans became the limits of manure and slurries which could be applied per land unit per annum. In general, the original Directive allowed for a level of 170kgs of N per hectare. Through derogations, it was possibly to seek

---

6   See Case C-396/00, *Commission of the European Communities v Ireland*, available at http://www.europa.eu.int/eurolex/.

7   See for example Case C-69/99, *Commission v UK*, at para. 18. This was the UK Nitrates case. The UK had failed to comply with aspects of the Directive in 1998, by limiting its focus upon a narrow definition of surface waters and ground waters – those intended for drinking supplies.

a limit of up to 250kgs, but this had to be justified on scientific grounds and cleared with the Commission as well as a special committee of national experts, and only after the coming into force of the Directive's terms.

In most of the Republic's NAPs submitted, aggressive and foolish attempts were made to ensure that the upper limit of 250kgs (or 230kgs) would be available to some categories of Irish farmer from the very outset, even though it is clear such derogation cannot be pre-granted by a member state itself. For that reason alone, the Commission rejected these plans. It was only in autumn 2005 that an action plan was tortuously agreed with the Commission. By February 2006, some 11 years after the implementation deadline, a set of regulations finally came into force which partially implemented the Nitrates Directive in the RoI. The NI Nitrate Action Plan (NAP) was only finalized in spring 2005, a level of delay that is similar to that of the Republic.

However, bitter farmer opposition has continued. In the Republic the original Nitrates Regulations of late 2005 were simply not accepted by farmers' groups. One result was a boycott of Teagasc, the state farm advisory service, in February 2006. Farm leaders alleged that this agency had failed to ensure more generous allowances than the 170kg level through proper 'scientific' advice. In the face of such pressure some parts of the 2006 Irish Nitrate Regulations were suspended and further scientific investigations were promised. This was all because of a heated and orchestrated campaign against the Nitrates Directive. The IFA portrayed it as a 'draconian' law.[8] Farmers had once again proven their ability to slow down and suspend an important EU directive.

Equivalent regulations were enacted in NI a little earlier, in 2003, although these met with less resistance there.[9] In part this may be simply a function of the lesser influence that the UFU enjoys compared with the IFA, and also the complexities of UK/NI legislative co-ordination.

In truth the negative impact of the Nitrates Directive is likely to fall heavily upon only a quite specific set of farmers: expanding dairy herds and intensive pig and poultry operators. Given the relatively fragmented and small-scale land ownership pattern in RoI, dairy farmers who wish to expand their operations cannot feasibly buy much more land to make larger holdings and thus deliver economies of scale. Instead they must use much more intensively the grassland available to them, which means chemical fertilizers and animal wastes become structurally an integral part of such a farming model. The last thing these small cohorts of expanding dairy farmers

---

8   This suspension only applied to measures relating to phosphorous, although subsequent Teagasc scientific advice in March of 2006 advised the RoI Government to also moderate some of its regulations on nitrates. These findings were communicated to the Commission and in August 2006 yet another set of modified regulations were produced on nitrates, which apparently met many farmers objections and relaxed conditions generally.

9   These were the *Protection of Water Against Agricultural Nitrate Pollution Regulations (Northern Ireland)*, 2003 and also the *Control of Pollution (Silage, Slurry and Agricultural Fuel Oil) Regulations (Northern Ireland)*, 2003.

want is a limit on their field nutrient balances, which in fact is what the Nitrates Regulations (2005) represent. In NI the impact of the directives is similar in that it raises problems largely for intensive operators only. Their Department of Agriculture and Rural Development (DARD) has estimated that about 90 per cent of NI farmers were already under the 170kg/ha limit for nitrogen manures (DARD 2006).

By mid 2006, as it became clear that the Nitrates Directive must be complied with, the tactical position of Irish farmers' groups switched to seeking the most favorable set of rules applying its terms, and of course subsidies to help offset any costs of compliance. A farm waste management grant scheme was put in place as a substantial 'carrot' to overcome resistance.

Another strategy has been to shift the blame for instances of eutrophication away from agriculture. The IFA have repeatedly alleged that much water pollution is caused by local authorities' sewerage works. As if in echo, the Ulster Farmers' Union has begun to challenge the water pollution record of NI's local authorities as well. By August/September 2006, the Irish regulations on nitrates and phosphates had been revised yet again, in general to liberalize their impact on farmers. Moreover, the Irish Government has continued to promise farmers they will get their derogation from the full rigors of the directive (MacConnell 2006).

Politically, the point of interest here is that implementation in RoI has been unquestionably held up simply due to political sensitivities. It is of course true that there are many complexities associated with the exact implementation of the Directive. Farmers do have some valid points about rigidities and ambiguities in the Nitrates Regulations. Yet the real story is that the political power of the farmers' lobby has been enough to force over a decade of delay. Under such pressure, the UK attempted to 'try on' a narrow definition of the Directive's scope, whereas in the RoI a high-risk route was taken, of gambling that litigation might reveal the Directive did not really apply to the Irish situation. The very fact that Irish authorities would even contemplate such a risky approach reveals that as regards rural environmental issues, the political power of the farmers' lobby has an unquestioned ability to delay EU legislation and weaken it.

### Implementing directives on wildbirds and habitats: Legal formalism versus land ownership

Serious political implementation problems have also been a feature of the saga of the Birds and Habitats Directives in RoI. Here the problem was a little more subtle: instead of just refusing to implement the directives in question the Irish authorities did so in a half-hearted way. They simply did not designate enough lands for ecological protection, adopting a minimalist and gradualist approach.

The reason for this, at one level, was due to legal complexities and to cultural sensitivities over land rights, a strong feature of rural Ireland. However, at another level, it is also clear that land designations were simply resisted because landowners refused to accept restrictions on certain types of land use. They also feared losses on resale values, and hoped for compensation payments. In other words, much of the conflict was about money. Indeed it is possible to portray it as another form of lobby group 'rent seeking'. One detailed account of the Irish implementation process concluded: 'Buying off the farmers was the key to unlocking the transposition process' (Laffan and O'Mahony 2004, p. 12).

The result is that the RoI has been unquestionably one of the poorer performances on habitats and bird protection compared with other EU states. One official assessment in 2004, by the European Environment Agency (EEA), indicated that RoI came *fifth* from the bottom among 15 states for the degree to which nature sites had been designated that are considered sufficient to protect habitats and species under the Habitats Directive.[10] About 15 per cent of the necessary sites remained to be protected by this assessment (EEA 2004).

Notwithstanding improvements, the Irish performance is still mediocre. For example, even when protected lands are measured in terms of per capita statistics (hectares per 1,000 head), RoI still comes second worst, after Belgium (ibid.). The Commission's own summer 2006 assessment of performance in implementing the Birds and Habitats Directives placed RoI in 9th position among the EU15 states, for the number of sites designated representing a given percentage of national territory.[11] In RoI's case it was around 10 per cent for Heritage sites but notably below 5 per cent for bird protection sites, which are called Special Protection Areas (SPAs) in the jargon of the Birds Directive.

Indeed RoI had the lowest percentage of its national territory designated for bird protection of the EU15 states. As of 2006 Ireland had about 135 SPAs designated for birds protection (of which seven were sites awaiting full legal designation), and 424 Special Areas of Conservation (SACs). Some 500 of these sites were also co-designated with the Irish category of 'National Heritage Area' (NHA) of which there was a total of 1247 (EPA 2006, p. 49).

The institutional story here concerning the implementation of the Habitats and Birds Directives is also revealing. In the Republic, responsibility was quickly assigned in the mid 1990s to Dúchas, the National Heritage Service, reporting to the new Department of Arts, Culture, and the Gaeltacht. The trouble with this setup was that Dúchas was a relatively weak 'Cinderalla' agency, working with an equally weak and new Department which itself had to fight its institutional corner with the powerful and established entities of the Department of Agriculture and the Department of Environment and Local Government.

In the end, Dúchas became institutionally isolated as political conflict grew over the two EU directives on habitats and birds during the mid 1990s. Environmentalists were complaining to the Commission about the slowness of efforts to designate lands

---

10   See EEA/European Environment Agency (2004) Indicator Fact Sheet: (BDIV10e) EU Habitats Directive: sufficiency of Member States proposals of protected sites.

11   See the table reproduced from Commission data in: Environmental Protection Agency (2006) *Environment in Focus 2006* (Wexford: EPA), p. 48. Available at http://www.epa.ie.

for protection or whether enough land had been so protected. Political opposition was also coming from landowners and farmers. This was even more politically intense and sharp than Commission complaints. By the late 1990s Dúchas was on the receiving end of criticism from every quarter. A change of Government in 1997, meant that the political state of play shifted firmly to meeting farmers' demands for cash compensation, although this took some time to be finalized. An amendment of the Wildlife Act of 1976 did not emerge until 2000 (the Wildlife (Amendment) Act 2000). This was vital to providing a statutory basis for conservation activity and a solid legal footing for the details of land designation (Comerford 2001).

Considering that the Habitats Directive was supposed to be implemented by 1994, and the Birds Directive by the early 1980s, this absence of a solid legal footing for the activities of Dúchas is both remarkable but also revealing of political priorities.

Looking back now after more than a decade of trying to protect habitats and birds through the device of legal designation, one is tempted to draw the conclusion that there is something intrinsically weak about that approach. In particular it demands huge administrative efforts for very limited gains. It is not clear if getting parcels of lands 'designated' means that much. The deeper structural economic forces which are driving rural development in NI and the Republic remain the core threat to natural habitats.

Such threats include: the demand for one-off rural housing; riparian property development and marinas; for road building; quarrying and aggregates; the growing popularity of golf; the intensification of dairying; and increased forestry and other novel economic users of land and waters. These drivers of change are not in any way removed or tamed by the mere fact of legal designation. It is *these* economic activities that threaten natural zones. To control the economic forces at work it would seem logical that a more appropriate response lies with economic (and especially fiscal) instruments rather than just laws.

Of course, for constitutional reasons the EU has only minimal competence in such matters, and thus is doomed to have to rely on the somewhat naive idea that legal designation alone will really protect ecologically sensitive habitats. In this way, the EU laws in question should be seen as amounting to a very limited baseline effort at biodiversity protection. A much more proactive engagement is called for than merely enforcing EU laws, which anyhow has not even been done properly.

Unfortunately, it is clear that the whole habitats and birds issue simply did not (and does not) matter enough for key Irish decision-makers. The opposition by landowners and farmers was (and remains) too intense in any case to merit much by way of very determined action, without major compensatory side-payments.

By the end of the 1990s, within Dúchas, morale had plummeted, and the agency retreated to an entirely defensive posture.[12] It was therefore little surprise

---

12   For example, they became very reluctant to engage in wide public consultation, having witnessed how public meetings became shouting matches against their harried staff. Also there were not unfounded fears that consultation could alert landowners to impending

when Dúchas was effectively abolished in 2003 and its staff and personnel incorporated into the newly styled Department of Environment, Heritage, and Local Government (DEHLG).

In political terms, they had effectively been made a scapegoat for implementation failures which were not really their fault. Their euphemistically labelled 'incorporation' into the DEHLG was certainly interpreted as an obvious political victory for farmers: an agency they distrusted, if not despised, was *de facto* abolished.[13] The stark lesson this must have provided for anyone working within the Irish civil service on environmental policy matters was also surely chilling. Tough implementation of EU nature laws brought no bureaucratic rewards within the Irish system of governance. Indeed quite the opposite lesson could plausibly be learned.

It is of course not obvious that the abolition of Dúchas made sense. Should personnel tasked with the sensitive and highly technical job of designating lands which are ecologically sensitive under EU laws, be under such direct ministerial supervision, or should they be in a separate and independent state agency? Equally the Department of Environment and Local Government has at times its own agenda in furthering particular developments that might be in conflict with the need for nature protection. Certainly some scope for conflicts of interest exists between defending bird and natural habitats and its other departmental duties such as drainage, flood defence, coastal foreshore works, and water supply projects.

In NI the implementation of the Habitats and Birds Directive has been a little better than the experience of the RoI. By the end of 2006 roughly 10 per cent of the total land area of NI had been designated in compliance with these two EU directives.[14] This is broadly similar with the most recent level of designation in the RoI, although it does not include candidate or potential sites nor protected lands under UK domestic laws or international conventions.

Within NI there has been less obvious organized political controversy from farming interests and rural landowners. This is not to say, however, that there have not been failings there as well. The relevant legislation on NI's habitats was initially agreed in 1995, but these were subsequently found by the ECJ to be imprecise. It is only comparatively recently that improved legislation was agreed in the form of the Environmental Impact Assessment (Uncultivated Land and Semi Natural Areas) Regulations (Northern Ireland) Order 2001 and the Environment

---

designations. In some cases this could trigger destruction of vital habitats as owners sought to take pre-emptive action, a trend very common in the cases of quarries or drainage projects.

13   For a sample of the mostly negative media commentary, see McDonald (2003), Viney (2003), Editor/Irish Times (2003) and Battersby (2003).

14   This has been estimated based on data taken from the UK's Joint Nature Conservation Committee (JNCC) website. The JNCC is a statutory advisory body on nature conservation and biodiversity. About 61,250 hectares have been designated as SACs whereas about 81,114 hectares have been designated as SPAs. Taken together these two figures suggest a total designated land-mass of 1423.64 Km2 under the two directives, out of a total NI territory of 14,144 Km2. For more details see: http://www.jncc.gov.uk/ProtectedSites/SACselection/SAC_list.asp?Country=NI.

(Northern Ireland) Order 2002. These new legislative measures were required precisely because the initial measures were inadequate. Whereas the institutional machinery to implement EU environmental directives has been modernized in the RoI, in some ways improvement is less evident within NI.

The most significant institutional issue within NI remains the absence of a truly independent environmental inspectorate or protection agency, which the Republic has (although environmentalists in the RoI bitterly dispute its independence). The Environmental and Heritage Service is not independent of the Department of Environment (DoE).

As with the RoI, it was not clear just how effective the mechanism of legal designation was proving in protecting biodiversity in a more substantive way. In terms of concrete outcomes, the Royal Society for the Protection of Birds (NI) has estimated that more than 50 per cent of the Yellowhammer, Lapwing, and Curlew populations have been lost over the last 25 years. Legal designations have not protected them (Royal Society for the Protection of Birds [RSPB] 2005).

As of late summer 2006, the Commission was actively investigating RoI Government activities regarding the still lingering implementation of the Habitats Directive, and litigation was ongoing. In particular there was controversy over whether the proposed Shell gas pipeline in Mayo, had been allowed to proceed without due regard for the considerations required under EU nature laws (Siggins 2006). On 23 September 2006, the Advocate General of the European Court of Justice issued a preliminary ruling that was strongly critical of RoI Government's efforts to implement the Habitats Directive (Mahony 2006). This judgment pointed to a lack of expert knowledge of species populations among the authorities, a failure to conduct proper impact studies of developments on habitats, and a failure to develop species protection plans for those at risk. In other words, the Irish experience of poor implementation is still ongoing and systematic. It is not just a problem of bureaucratic de-prioritization and learning.

**Conclusions: Dashing good intentions in the future?**

In the longer term, it is an open question as to what extent rural communities on either side of the border can manage a genuinely sustainable form of rural development. It is hardly sustainable development to become *de facto* exurban dormitories, although this is what census data appears to be suggesting is in fact happening. Nor does an increasingly globalized food production system necessarily offer much promise of sustainability either. In fact, food production might well get more intensive in future. Certainly the responses of the intensive dairy and pig sector to the Nitrates Directive show just how hostile such interests could be to ambitious environmental laws. Unless addressed intelligently, such interests could easily slip into advancing an anti-environmental agenda plain and simple. The fury over the Nitrates, Habitats and Birds Directives shows how good environmental intentions in Brussels end up dashed on the ground in Ireland.

Yet it is also clear that there are divisions between the industrialized agribusiness sector and the larger (but more fragmented) group of increasingly part time marginal farmers. The latter are more likely to participate in environmental schemes as income supplements. They therefore could be coaxed to become environmental advocates.

We should in consequence never lose sight of the fact that perhaps one of the greatest areas of potential for promoting rural sustainability, surely lies with continually reforming the Common Agricultural Policy (CAP). This is not least because that is where the bulk of funding possibilities remain. The innovation of a 'single farm payment' will not last forever and it is a moot question what will replace it. European farmers remain powerful enough to require subsidies as part of any future reform.

It seems plausible that in order to justify future public financial support, the environmental rationale has still much mileage left in it. It is just that the various schemes have lacked a more genuine environmental ambition and logic to date. The Rural Environmental Protection Schemes (REPS) (in RoI) have generally lacked focus: relatively small amounts of money have been parceled out to a large cohort of farmers who are typically low scale marginal producers of limited ecological threat. By 2002 as much as 27 per cent of agricultural land in the Republic was under REPS (Fields 2002).

In NI the main agri-environment scheme has been the Environmentally Sensitive Areas Scheme (ESAS), which was introduced in 1988, well before the REPS scheme and borrowing from extensive English experience of the 1980s. Another noteworthy difference from the RoI, is that this scheme was more targeted at vulnerable and important areas, compared with the REPS. Five areas were designated as zones where ESAS payments could be made, representing roughly 20 per cent of the land area of NI.[15]

The latter type of essentially socio-economic instruments, so routine within the CAP, stand in contrast to the largely legalistic approach which the EU has followed as regards environmental policy. Perhaps, one lesson to be learned would be to integrate EU-level funding with EU-level legal norms more closely and in a systematic manner. Only such a savvy approach has the promise of unblocking the scope for powerful domestic interests to politically dilute or even derail the implementation process.

In other words, one reason why the EU's implementation of environmental laws has been so patchy lies in its own institutional weakness. It remains a brittle confederation enjoying only limited powers over member states. Yet, when the resources of the still-significant CAP are placed alongside the more formal legal responsibilities now in place, perhaps the EU may well have more clout at its disposal than is realized. It is just a question of co-coordinating its efforts more intelligently – in effect to become a stronger type of confederation.

---

15   See DARDNI, 'Agri-environment schemes', available at http://www.ruralni.gov. uk/index/environment/countryside_management_main/scheme.htm.

As if in some kind of institutional echo of the EU's institutional woes, one could also note here the relative failure to explore a joint NI-RoI dimension to rural environmental issues. Both jurisdictions have arguably much scope to learn from each other. Indeed, under the original Mitchell draft of an agreement for power-sharing in NI (1998), it was envisaged that Ni-RoI co-operation would include under the heading of implementation bodies, a joint Environmental Protection body (Macrory 2004, p. 8). However, the final version of the Northern Ireland Peace Agreement (1998) did not produce such a body. Instead the environment was listed merely as an area of co-operation. To date then links remain at the level of joint ministerial dialogue and some tangible co-operation on water quality under the demanding EU Water Framework Directive of 2000. Once again, institutional limitations raise their head here.

Finally, one can speculate that even if EU environmental policies for the rural environment were to be much more integrated with CAP fiscal resources, or better enforced, the ability of the domestic institutional system and policy actors to skew, manipulate, or more baldly resist EU environmental laws will very likely still remain strong. At a more profound level in both NI and the Republic the political and social salience of environmental questions remains relatively low. Perhaps future generations will have different preferences, and the low priority accorded to environmental issues will change. However, implementation remains the core rural environmental policy problem to date, both in NI and the Republic. Pressman and Wildavsky would have been quite unsurprised.

## References

Battersby, E. (2003) 'Knocking down Dúchas', *The Irish Times*, 19 April.

Bishop, K. and A. Philips (eds) (2004) *Countryside Planning: New Approaches to Management and Conservation.* London: Earthscan.

Central Statistics Office (2006) *Census 2006: Preliminary Report.* Dublin: CSO. Available at http://www.cso.ie/census/documents/2006PreliminaryReport.pdf.

Comerford, H. (2001) *Wildlife Legislation 1976–2000.* Dublin: Round Hall/Sweet and Maxwell.

Department of Agriculture and Rural Development (Northern Ireland) (2006) *Impact of N and P controls on land requirements.* Available at: http://www.dard.org.

Editor/*Irish Times* (2003) 'End of Dúchas', *The Irish Times*, 28 April.

Environmental Protection Agency (2006) *Environment in Focus 2006.* Wexford: EPA.

European Environment Agency (2004) *Indicator Fact Sheet: (BDIV10e): EU Habitats Directive: Sufficiency of Member States Proposals of Protected Sites.*

Fields, S./An Taisce (2002) *Monitoring and Evaluation of the Rural Environmental Protection Scheme.* Dublin: An Taisce.

Flynn, B. (2006) *The Blame Game: Rethinking Ireland's Sustainable Development and Environmental Performance.* Dublin: Irish Academic Press.

Glachant, M. (2001) *Implementing European Environmental Policy: The Impacts of Directives in the Member States.* London: Edward Elgar.

Kelemen, D. R. (2000) 'Regulatory Federalism; EU Environmental Regulation in Comparative Perspective', *Journal of Public Policy*, 20, 3: 133–67.

Laffan, B. and J. O'Mahony (2004) 'Mis-fit, Politicisation and Europeanisation: The Implementation of the Habitats Directive in Ireland', *OEUE Occasional Paper*, 1.3, September. Available at: www.oeue.net/papers/ireland-implement ationofthehab.pdf.

MacConnell, S. (2006) 'Nitrates Dispute 'to be Resolved', *The Irish Times*, 21 September.

Macrory, R. (2004) *Transparency and Trust: Reshaping Environmental Governance in Northern Ireland. Executive Summary.* London: Centre for Law and Environment, UCL. Available at: http://www.rspb.org.uk/Images/macrory_tcm5-112128.pdf.

Mahony, H. (2006) 'Government Breached EU Law on Natural Habitats', *The Irish Times*, 23 September.

Marsden, T., J. Murdoch, P. Lowe, R. J. C., Munton and A. Flynn (1993) *Constructing the Countryside.* London: UCL Press.

McDonald, F. (2003) 'Break-up of Dúchas may be Unlawful', *The Irish Times*, 23 April.

Page, R. (1999) 'Restoring the Countryside', in Barnett, A. and R. Scruton (eds), *Town and Country.* London: Verso, pp. 99–108.

Pressman, J. and A. Wildavsky (eds) (1984) *Implementation: How Great Expectations in Washington are Dashed in Oakland.* 3rd edition. Berkeley: University of California Press.

Royal Society for the Protection of Birds (Northern Ireland) (2005) 'Farmers and Wildlife Lose Out', Press Release, 7th November 2005. Available at: http://www.rspb.org.uk/nireland/farming/dardcuts.asp.

Siggins, L. (2006) 'Commission Queries Pipeline', *The Irish Times*, 27 July.

Szarka, J. (2002) *The Shaping of Environmental Policy in France.* New York: Bergham.

Viney, M. (2003) 'Brooding on the Politics of Conservation', *The Irish Times*, 10 May.

Chapter 4

# Governance for Regional Sustainable Development: Building Institutional Capacity in the Republic of Ireland and Northern Ireland

Gerard Mullally and Brian Motherway[1]

## Introduction

Sustainable development, due to its huge ambition and its diverse interpretations, always risks becoming something between a marketing slogan and an evangelical, utopian doctrine. Everyone is in favour of it, but the devil is in the detail. Yet, rooted firmly in the Brundtland Commission's report *Our Common Future* (World Commission on Environment and Development [WCED] 1987), sustainable development defined as 'development that meets the needs of the present without compromising the ability of future generations to meet their own needs' (WCED 1987) has mobilized collective actors in different sectors and at different levels of society and has shifted the ground of environmental debate considerably over just two decades. Contemporary debate, as Lightfoot and Burchell (2005, pp. 77–8) observe:

> has focused less on the existence of an environmental crisis, and more on the nature of environmental responsibility, the predominant focus for that responsibility and the best methods of undertaking it.

Taking responsibility, in the sense employed here by Lightfoot and Burchell, implies not just the functional governance *of* sustainable development, but also refers normatively to 'governance *for* sustainable development' (Lafferty 2004; Meadowcroft, Farrell and Spangenberg 2005). Meadowcroft et al. define 'governance *for* sustainable development' as the deliberate adjustment of practices of governance and of the structures that regulate societal interactions in order to

1 The authors would like to acknowledge the contribution of: the research team: Aveen Henry, Jillian Murphy and Gillian Weyman at the Cleaner Production Promotion Unit, UCC.

ensure that social development proceeds along a sustainable trajectory through a process of adaptation (2005, p. 5).

Institutions for sustainable development now stretch across multiple scales of governance from the international (United Nations [UN], Organization for Economic Cooperation and Development [OECD]) to the supra-regional (European Union [EU]) and to the national, local and regional levels in Ireland (Mullally 2004). Our focus here is on the development of institutional capacities for sustainable development on the regional and local scales as an access point to the politics of sustainable development in rural Ireland. This chapter takes an all-island perspective for a number of reasons:

1. Despite a history of conflict and division on the island of Ireland, the jurisdictions of both the Republic of Ireland (RoI) and Northern Ireland (NI), share common environmental, social and economic challenges (Ellis et al. 2004).
2. Strategies for sustainable development in both jurisdictions were conceived against a background of profound social change: rapid economic growth through the 1990s in the Republic and moves to create a 'post-conflict' Northern Ireland in the aftermath of the 'Northern Ireland Peace Agreement' (Department of the Environment and Local Government [DoELG] 2002; Department of the Environment [NI] [DoENI] 2006).
3. Both jurisdictions are linked through membership of the EU which has provided an important external lever on the governance of sustainable development on the island of Ireland (Turner 2006; O'Mahony 2007).
4. Processes of transformation in local and regional governance, and the modernization of environmental governance are underway in both jurisdictions, albeit at a different pace (Mullally 2004; Turner 2006).
5. There is political recognition and agreement on the importance of a cross-border or 'all-island' dimension to sustainable development (DoELG 2002; DoENI 2006).

Our discussion begins by considering the question of governance with a specific emphasis on the importance of institutional design and the fostering of institutional capacity for sustainable development. This is followed by an examination of some of the existing institutions engaged in governance *of* sustainable development and the changes that have taken place over the last decade or so. The focus then turns to strategies for sustainable development on the island of Ireland as a way of tracing the intended approach to the creation of institutions *for* sustainable development. In order to consider the evolution of institutional capacities for the governance of sustainable development the focus then switches to the experience of Local Agenda 21 (LA21) – one of the main vehicles for pursuing local and regional sustainable development on the island of Ireland – in the decade following the Earth Summit in Rio de Janeiro in 1992.

Our argument will be that despite the recent creation of institutions claiming to represent the evolution of governance *for* sustainable development on the island of Ireland, the evidence suggests that institutional capacities for sustainable development are heavily conditioned and shaped by existing patterns of governance and subject to many of the same constraints to innovation experienced in other policy domains. Therefore, we contend that experimentation with new institutional forms for realizing sustainable development remain at the level of an emergent regime nested within a dominant system that is contingent on the ongoing shift from government to governance. Despite the creation of significant strategic, integrative and participative capacities on both sides of the Irish border, we are still at the stage of developing capacities for the governance *of* sustainable development that in effect means it will remain secondary to the established priorities of socio-economic development.

## Governance and institutions for sustainable development

Governance for sustainable development is concerned not *only* with the design and implementation of government policy, but also with collective processes of monitoring, reflection, debate and decision that establish the orientation for policy (Meadowcroft et al. 2005, p. 5). The specific model of governance for sustainable development considered here is the 'Rio model of governance' that emerged from the Earth Summit in 1992 (Jänicke 2006, p. 1).

Recent contributions to the debate on deliberative democracy caution that the emergent emphasis on horizontal forms of coordination should be regarded as a complement to, rather than a replacement for vertical forms of coordination (Lafferty 2004). Lafferty (2004) conceives of governance for sustainable development as referring *both* to 'vertical environmental policy integration' (across levels of governance) and 'horizontal environmental policy integration' (across sectors). The general shift observed is increasingly reflected upon in studies of Irish governance in general (Larkin 2004; Adshead 2006) and environmental governance in particular (Murray 2006; O'Mahony 2007).

In the context of sustainable development the shift towards governance has meant the embrace of softer 'steering mechanisms' than just 'command-and-control' regulation (Flynn 2007), while on the other hand there has been a growing emphasis on decentralization and the mobilization of civil society (Lafferty 2004). A broad-ranging programme for social change, like sustainable development, needs intentional institutional transformation, which in turn requires institutional design: 'at all levels of social deliberation and action, including policymaking, planning and programme design and implementation' (Alexander 2006, p. 2).

Alexander's (2006) characterization of institutional design here sits well with the idea of governance for sustainable development since it encompasses both the democratic imperatives highlighted by Meadowcroft et al. (2005) as well as the responsibility of governments for the realization of the substantive goals of sustainable development established by the World Commission for Environment and Development and programmed by Agenda 21 (Lafferty 2004).

The Agenda 21 model of multi-level, multi-sectoral and multi-stakeholder governance takes account of the extreme complexity of the environmental field (Jänicke 2006, p. 1). Yet, 'ambitious strategies need adequate capacities', where capacity can be defined by the limits of possible action within a given political, economic and informational opportunity structure (ibid., p. 7). The key issue here is really about developing institutional capacity to steer societal development within the parameters of ecological sustainability (Meadowcroft 2004, pp. 163–4). A key question is how participation relates to policy integration for sustainable development (Steurer 2005). Steurer (2005) argues that participation is about integration in indirect policy fields, such as public governance in general and administrative policy in particular. However, participation is only one condition of governance for sustainable development, since the state also needs not just to develop its participative capacity but also its integrative and strategic action capacities as well (Meadowcroft 2005). We are therefore faced with one of the characteristic challenges of the sustainable development problematic: reconciling the substantive goals of a global programme with its procedural aspirations in national, regional and local contexts.

## Institutions and the governance *of* sustainable development

When compared with the EU experience in general, local government in the Republic and Northern Ireland has a high level of central government control, weak financial independence, a narrow range of powers and few locally elected representatives (Harris 2005). This section provides us with a vantage point from which to understand the path dependencies of the integrative, strategic and participative capacities for sustainable development in both jurisdictions. Thus in each case, we examine the recent development of environmental governance, the nature of central-local relations, the emergence of local development partnerships and the relative openness to public participation.

*Republic of Ireland (RoI)*

The Department of Local Government was transformed into the Department of the Environment in 1978 (now the Department of Environment, Heritage and Local Government) and was assigned a leading policy role in promoting the protection and improvement of the physical environment. The responsibility for the implementation of environmental legislation was, however, placed on local authorities. Local government in the RoI principally consists of 34 major local authorities, the City and County Councils, which typically tend to serve a larger population than many of their European counterparts. Local authorities in the RoI derive their power and function from central government and are regarded as executive agencies of Government departments charged with implementing central policy.

In the 1990s, central government embarked on a range of reforms to redress the perceived weaknesses of local government in the areas of environmental governance and the governance of local development. The modernization of environmental governance began with the publication of the *Environmental Action Programme* in 1990. The policy programme committed the Irish Government to the integration of environmental considerations into all policy areas and significantly acknowledged the principle of sustainable development. The Environmental Protection Agency Act 1992 provided the legal basis for the establishment of an independent statutory authority for the protection of the environment. At the same time as some of the environmental functions of local government were ceded to the Environmental Protection Agency (EPA) a number of other substantial changes were taking place in sub-national governance in the RoI.

The Republic of Ireland in the 1990s was characterized by experimentation with a new localism in an otherwise centralist system of public policy (Adshead and Quinn cited in Mullally 2004). The introduction of Community Initiatives designed to complement Structural Funds, and to ensure that local and regional government would have direct access to funding created a new impetus for local development. Reviews of the impact of LEADER partnerships from the perspective of promoting sustainable rural development have been mixed (Moseley et al. 2001; Meldon et al. 2004). However, there was a growing perception in the 1990s that the local systems of government and development were being progressively divorced, and that local development agencies were gaining considerable autonomy. In 1998, the Task Force on the Integration of Local Government and Local Development Systems highlighted the existing overlaps in the activities of local government, the state agencies and local development agencies and identified the need for integration. It proposed the creation of local development boards known as City and County Development Boards (CDBs) that would be linked to (but separate from) local government – though under the auspices of the Director of Community and Enterprise within local government (Mullally 2004). The purpose of these bodies was to increase the coordination, cooperation and integration of existing bodies through the creation of long-term strategies (Adshead and McInerney 2006). Since their creation in 2000, the emphasis of the CDBs has been on their role in improving participative democracy at the local (county) level (Meldon et al. 2004).

The EU was instrumental in the creation of eight NUTS III regional authorities in 1994, and subsequently two NUTS II regional assemblies in 1999. In the case of the former, their role lies primarily in the coordination of their constituent local authorities. In the case of the latter, their designation as assemblies is not comparable with the implication that the title confers on regional assemblies in the UK. Representatives are nominated rather than being directly elected, though the assemblies do have some discretion over the disbursement of funding under the regional dimensions of the National Development Plan. However, regions have become involved in several pan-European projects and networks promoting LA21 and regional sustainable development (Mullally 2004).

Partnership, according to Sommers and Bradfield (2006, p. 69), is particularly attractive as a mode of governance because it 'spreads risk in times of policy shift, changing priorities and the uncertainties of aims, purposes and practices'. The success of the social partnership model at the national level has, therefore, resulted in a 'coordination reflex' in the RoI's governance (O'Mahony 2007, p. 281), which in turn has been replicated at the local level (Larkin 2004). Many government departments now engage in public consultation on policy matters, but participation in environmental decision-making tends to remain largely adversarial at the implementation level (O'Mahony 2007).

If we extend the notion of institutional capacity beyond the formal structures of government to the domain of civil society we have to acknowledge the comparative weakness of the RoI's environmental movement in relation to many of its European counterparts (Garavan 2006; Murray 2006). In fact, environmentalism has recently been identified among the weakest sub-sectors in the voluntary sector in the Republic (Hughes et al. 2007). Since the turn of the century the creation of specific institutions for sustainable development appears to have opened the way for stakeholder participation at a number of levels of governance (see below).

*Northern Ireland (NI)*

The policy context for sustainable development in Northern Ireland is partially determined by the larger United Kingdom (UK) context in which it is located. Some 26 years of direct rule from Westminster have shaped the direction of environmental policy in the region. The impact of EU litigation and the brief restoration of devolution, however, have spurred the modernization of legislation and environmental policy making following decades of neglect (Turner 2006). Since the instigation of direct rule the region has enjoyed 'a unique level of structural integration in terms of functional responsibilities for planning and environmental policy, with both falling within DoENI's remit until the realignment of functions under devolution' (Turner 2006, p. 77). A number of different Government departments have direct responsibility for environmental issues; in terms of the horizontal integration of environmental governance, a central role is played by the Department of the Environment together with its key body for implementing environmental policy and law, the Environment and Heritage Service (EHS) (Macrory 2004). In terms of the vertical integration of environmental governance, the Department of the Environment, which is responsible for planning the whole of Northern Ireland, must consult district councils on the preparation of overall development plans as well as on individual planning applications (Callanan 2004). Turner (2006) argues that the potential of this level of structural integration was never fully realized as a result of decades of marginalization of the environment as a policy priority in Northern Ireland. Under the realignment of functions, responsibility for strategic planning was transferred to the Department of Regional

Development, thus leaving the Department of the Environment with responsibility for Area Plans, many (but not all) planning policy statements and operational control of the planning process.

Despite the suspension of the devolved assembly in 2002, Government departments continued to follow through the terms of the Northern Ireland Peace Agreement (NIPA) under direct rule (Graham and Nash 2006). In 2006, the Department of the Environment published the sustainable development strategy for Northern Ireland which provides a regional framework for guiding the governance of sustainable development. The document explicitly acknowledges the challenge of advancing sustainable development in a post-conflict society (DoENI 2006).

The Local Government (Northern Ireland) Act, 1972, divided Northern Ireland into 26 district council areas and the functions of former local authorities that were regional in character were transferred to central Government departments (Knox 2003). Today, there is no intermediate tier between the district councils and the Northern Ireland Assembly and Executive, although the latter could be considered a purely regional form of government (Callanan 2004). Local government structures and functions are currently being examined under the Review of Public Administration, set up by the Northern Ireland Executive in 2002.

Throughout the 1980s and 1990s, just like the situation in the UK and in the Republic, a series of local partnerships began to emerge (Moseley et al. 2001). Although Northern Ireland was just above the threshold for Structural Funding, its particular problems were taken into account and the region was deemed eligible for Structural Funds and other Community Initiatives e.g. LEADER and INTERREG (Dubnick and Meehan 2004). In terms of 'rural governance' LEADER partnerships in Northern Ireland differed somewhat from their counterparts in the Republic because of the leadership role adopted by local authorities (Moseley et al. 2001). In their survey of LEADER based projects in Northern Ireland they found evidence of community involvement, economic regeneration and a commitment to 'integrated' sustainable development (ibid., p. 189). Dubnick and Meehan (2004) point out that in the case of INTERREG, the potential for integrative governance was only realized when the management of projects were undertaken by cross-border partnerships. In looking at specific cross-border cooperation for sustainable development through the Foyle Basin Council (Derry, Donegal) and the Sliabh Beagh Partnership (Fermanagh, Monaghan, Tyrone), Ellis et al. (2004) suggest that it may be easier for partnerships than it is for local government (or indeed regional government) to overcome the political and administrative constraints that limit greater levels of cooperation and sharing of experience.

District partnerships were established in each council area under the EU Special Support Programme for Peace and Reconciliation (usually called PEACE I). Local Strategic Partnerships (LSPs) have been set up as successors to district partnerships to administer a second round of funding under PEACE II (Callanan 2004). The 26 LSPs are responsible for the delivery of PEACE II funding in each district council area; and for the development of Integrated Area Plans based on public participation and encompassing the economic, social and environmental needs of the area (Ellis et al. 2004). While there are parallels with the CDBs in the Republic in terms of composition, the Local Strategic Partnerships are not integrated within the 'normal regulative and administrative functions of local government' (ibid.). Ellis et al. (2004) further point out that a change in the emphasis under PEACE II towards reconciliation activities foreclosed a source of funding which had been crucial for supporting local sustainable development initiatives in Northern Ireland.

Morison (2001, p. 296) suggests that 'characterized as it was by the absence of a nexus between the local political process and mechanisms of government, direct rule in some ways allowed the [voluntary] sector to act as alternative site of politics and as an unofficial opposition'. Dubnick and Meehan (2004) argue that, in the absence of regular interaction with elected representatives, civil servants administering the region have enjoyed a unique level of consultation with citizens. But as Morison (2001, p. 299) notes: 'in the post-agreement situation the voluntary sector is in a different position ... the exact nature of the role that the sector will play in the future remains unclear'.

## Strategies, institutions and governance *for* sustainable development

A key outcome of the Earth Summit in Rio de Janeiro in 1992 was the obligation placed on governments to devise national strategies for sustainable development. The OECD (2006) points out that the integration of the three dimensions of sustainable development is one of the most difficult balances to achieve in formulating a strategy. Therefore, a good way of gaining a perspective on the development of institutional capacities for steering sustainable development is through the window of sustainable development strategies that specify the strategic, integrative and participatory intentions of governments.

*The sustainable development strategy in the Republic of Ireland*

*Sustainable Development: A Strategy for Ireland* was published in April 1997. There is no doubt that the prime motivation for developing the strategy was to respond to the UNCED process and obligations under Agenda 21. A recent assessment of progress on Agenda 21 points out that the focus on integrating environment into various policy sectors (agriculture, forestry, marine resources, energy, industry, transport, tourism and trade) provided a re-balancing of the previous situation where environment was generally not well integrated into national policy (Mullally 2004).

Niestroy (2005) notes that the lead role of the Department of the Environment and Local Government (now Environment, Heritage and Local Government) in relation to sustainable development policy is as of yet uncontested in the Irish context. One of the central components of the horizontal integration of sustainable development was the creation of an environmental network of government departments, in which the environmental units of the relevant ministries participate (ibid.). However, one of the most significant innovations in this regard was the creation of the National Sustainable Development Partnership.

The National Sustainable Development Partnership – Comhar – was established 'to advance the national agenda for sustainable development, to evaluate progress in this regard, to assist in devising suitable mechanisms and advising on their implementation, and to contribute to the formation of a national consensus in these regards' (www.comhar-ndsp.ie). Although Comhar is a specific adaptation of the Irish model of social partnership, it is one step removed from the bargaining contexts of more mainstream social partnership institutions such as the National Economic and Social Council and the National Economic and Social Forum (Flynn 2007). According to Flynn (2007, p. 178), 'institutionally, Comhar is a marginal entity even if its contribution has been laudable'.

The Department of the Environment, Heritage and Local Government undertook a review of the implementation of sustainable development in RoI prior to the Johannesburg World Summit on Sustainable Development (DoELG 2002). Niestroy (2005, p. 184) sees this document more as an attempt to review the strategy relative to the experience of the Celtic Tiger economic boom than an attempt to revise it since the '1997 strategy remains the pre-eminent statement of sustainable development policies in Ireland'. The review also identified key policy and cross-sectoral priorities for the next decade. Among the cross-sectoral priorities identified the National Spatial Strategy and LA21 are of particular interest. The National Spatial Strategy establishes the basis for regional sustainable development in RoI; and the reaffirmation of LA21 – after it had begun to wane in other countries – points to the specific approach to the governance of local sustainable development in the RoI (Jonas et al. 2004; Mullally 2004).

The National Spatial Development Strategy (DOELG 2002) provides a twenty-year planning framework designed to deliver a more balanced social, economic and physical development between the regions, the aim of which is to realize economic and social progress in a manner consistent with environmental sustainability. Regional and local authorities are required to implement the National Spatial Strategy through regional planning guidelines and local development plans and strategies that have to be consistent with the overall framework. A key indication of the lack of integration with the National Sustainable Development Strategy lies in the areas of rural housing or 'one-off housing' in the countryside where a perception of lax controls is central to the debate (Niestroy 2005).

In terms of developing the strategic, integrative and participative capacities of governance for sustainability at the sub-national level the key institutional design that has emerged in recent times is the City/County Development Boards (CDBs). The CDBs are not only the key vehicle for the implementation of LA21 in RoI; they also represent the localization of partnership approach that has dominated Irish governance since the 1990s (Larkin 2004). The centrality of the partnership approach is outlined in the Irish Government's report to the Johannesburg Summit (DoELG 2002, p. 105):

> [S]ustainable development is not solely about government and what it will do; rather it is about all parties involved – government, social partners, NGOs, individual citizens – in their different roles and capacities, making the right decisions and taking the right actions in partnership with each other.

## *The sustainable development strategy in Northern Ireland*

The Northern Ireland Sustainable Development Strategy, First Steps towards Sustainable Development, attributes its origins to 'the first UK strategy for sustainable development A Better Quality of Life' introduced in 1999 and the creation of devolved administration with related responsibilities for sustainable development (DOENI 2006, p. 6). The UK framework for sustainable development, One Future, Different Paths (2005), 'recognized the need for a consistent approach across the UK and provides the framework under which each of the Devolved Administrations will translate its aims and objectives into actions based on their different responsibilities, needs and views' (DoENI 2006, p. 6). While Northern Ireland is broadly in line with the principles of the UK framework it specifically added 'Governance for sustainable development' as a key priority for the region with a commitment to:

> ensure that sustainability is properly recognized as the overarching policy framework for building a post-conflict society in Northern Ireland and that social and environmental objectives are incorporated into the decision making process alongside economic objectives (ibid., p. 126).

Given the structural location of Northern Ireland within the UK, *First Steps towards Sustainable Development* is very much a strategy for regional sustainable development. It was presaged by the Regional Development Strategy adopted in 2001 which, according to Turner (2006), gave rise to the hope that a new era of integrated environmental planning was dawning. However, the strategy ultimately failed to tackle the glaring policy weaknesses represented by the region's permissive rural development policy. Again it seems that concerns surrounding rural settlement patterns and practices in the North, just as in the Republic of Ireland, remain divorced from the Sustainable Development Strategy.

The Sustainable Development Commission in the UK was established in 2000 to advise and provide critical feedback on Sustainable Development to the UK Government as well as to the First Ministers of the devolved administrations and the Secretary of State in Northern Ireland. The Northern Ireland Commission played an important role in the development and delivery of the Northern Ireland Strategy (www.sd-commission.org.uk). One of the notable differences of the strategy from the experience in the Republic is the central responsibility borne by the Office of the First Minister.

The Sustainable Development Strategy speaks explicitly of sustainable communities: building community capacity and effective participation in decision-making, and of the need to '*consider* use of consultative and stakeholder forums to allow citizens to be involved in decision making on sustainable development issues at local level' (ibid., p. 77 [our italics]). Significantly, there is no mention at all of LA21 in relation to local sustainable development in the Northern Ireland Strategy for Sustainable Development. However, in 2007 the Northern Ireland Environment Forum was inaugurated with the support of the Environment and Heritage Service and was dedicated to a strategic overview of environmental issues facing Northern Ireland's policy makers and decision-takers.

*The cross-border dimension*

While the cross-border institutional dimension of governance for sustainable development has significant roots in the NIPA, the European dimension through LEADER, INTERREG and more recently through PEACE I and PEACE II is just as important. As these initiatives matured so too did the possibilities of 'integrative governance' through the development of capacities for bottom-up steering (Dubnick and Meehan 2004). In terms of sustainable development there is recognition of the cross border dimension in the Sustainable Development strategies in both jurisdictions (DOELG 2002; DOENI 2006). However, neither strategy even entertains the scenario of a joint Environmental Protection Body outlined by Macrory (2004, p. 6) or the more modest proposal to convene and institutionalize an all-island local sustainable development roundtable (Ellis et al. 2004).

**Learning from Local Agenda 21?**

Throughout the 1990s LA21 was the main vehicle for the development of the ideals of Rio into practical models of local governance. While National Strategies for Sustainable Development exist in most countries in the world, a total of 113 countries had initiated at least 6,400 LA 21 processes by 2002 (Jänicke 2006).

In the Republic of Ireland, the first official local-level institutional response to the sustainable development project was, as in most states, inspired by LA21. The most recent evaluation of progress on LA21 on the island of Ireland was funded by the Centre for Cross Border Studies and published in 2004. This study found that on the island of Ireland 54 per cent of local authorities have 'begun a process of LA21' – about 58 per cent in the North and 50 per cent in the Republic. It is notable here that even among the local authorities stating they have a LA21 process in place only 32 per cent engaged in participation with the community and only 14 per cent claimed that they went on to implement an action plan (Ellis et al. 2004).

There are some differences in the progress of LA21 in NI and the RoI, but in fact the extent of similarity is probably the most striking feature. Much of the language and issues are the same, and quantitative progress is also similar.

In Northern Ireland LA21 is more likely to be the specific responsibility of a dedicated officer, leading to more vision statements and action plans, whereas in the Republic responsibility tends to be shared, the process more integrated and hence harder to isolate. The cross-border research 'uncovered a fairly widespread view that much of what LA21 initially set out to do has now been mainstreamed as part of a broader approach to modernizing local government incorporating participation, integration of different policy areas (economic, social, etc) and the promotion of citizenship' (Ellis et al. 2004, p. 71)

The 'Rio model of governance' on which LA21 is based is essentially a voluntary process of policy innovation, lesson drawing and policy diffusion which often lacks the institutional strength to guarantee successful implementation (Jänicke 2006). Cooperative modes of steering often need the final responsibility and capacity of governments (ibid.). If we transpose this onto a comparatively weak institutional capacity of sub-national governance (environmental or otherwise), it is remarkable that LA21 persisted for over a decade on the island of Ireland and was imprinted on the modernization of local government to the extent that it was (Mullally 2004). Despite the mainstreaming of LA21 through the City and County Development Boards in the Republic (ibid.) and significant reforms in local government, we have not witnessed a devolution of powers or any great increase in local democracy (Harris 2005).

*Integration and disintegration: Lessons from LA21*

What is particularly evident in both the Republic and in Northern Ireland is what a central role the European Union has played in promoting sustainable development. Whether we are talking about specific rural development initiatives (LEADER) or ones that have implications for cross-border rural development (INTERREG), a number of things become clear. As they unfold they open the way for proactive integrative governance, and they create the opportunity to engage with sustainable development. Yet O'Mahony (2007, p. 281) points out that 'implementation is a *living* process of negotiation and bargaining even after decision making at the supranational level is completed'. Moreover, implementation is often a process that is not confined to a single level, but unfolds at multiple levels of governance (ibid.).

LA21, as an external initiative, has stimulated substantial change beyond what might have occurred in its absence. As it became incorporated into local government, LA21 was simultaneously detached from certain key priorities in local and regional governance: land use planning, waste management, water and energy and economic development (Jonas et al. 2004). This has been replaced by the growth of sectoral partnerships and co-ordinating mechanisms in these policy areas. Meanwhile, in the case of contested issues, such as land use policy and waste management, the partnership approach fails to address one of the key issues of governance: 'namely the ability to manage conflict and the lack of institutional decision-making capacity between partners' (Murray 2006, p. 448). As Adshead and McInerney (2006, p. 16) point out, 'the focus of civil society participation has been heavily on the creation of participation opportunities but only marginally concerned with participation outcomes'.

Horizontal structures like Comhar, and their vertical integration into structures like the European Environment and Sustainable Development Councils, allow for the convergence of the participative, strategic and integrative capacities of governance for sustainable development (Steurer 2005). In Northern Ireland the UK framework for Sustainable Development and the Commission for Sustainable Development have helped to improve the vertical integration of policy. However, the partnership approach to sustainable development is less institutionalized than in the RoI in spite of the fact that consultation processes related to sustainable regional development have been more comprehensive and inclusive in Northern Ireland. At the local level, the CDBs in the Republic and, to a lesser degree, the LSPs in NI should provide for the development of institutional capacities for sustainable development. However, we appear to be moving away from LA21 with its specific links to the Rio model of governance to more diffuse appeals to quality of life and community.

One of the indicators for the development of institutional capacities for sustainable development is the degree of integration between national- and sub-national-level partnership arrangements (Adshead 2006). Yet, the vertical integration of sustainable development from national to sub-national governance has hitherto remained underdeveloped (Niestroy 2005). There is a sense in the Republic of Ireland that the CDBs, despite representing an innovative form of public participation, are more about developing social trust than sustainable development, less about integration than institutional accommodation at a remove from real influence (Hughes et al. 2007). This is not confined to sustainable development policy and is evident in other policy domains such as 'social inclusion' (Asdhead and McInerney 2006). In Northern Ireland, the role of the voluntary sector in governance for sustainable development appears to have diminished somewhat from the situation in the 1990s where organizations promoting LA21 were at the forefront of the debate (Ellis et al. 2004).

**Conclusions**

The last decade has seen substantial policy and institutional innovation with regard to sustainable development in both Northern Ireland and the Republic of Ireland. Experimentation with horizontal forms of integration through innovative partnership arrangements in the name of sustainable development has proliferated at multiple levels of governance. Meanwhile, strategies for sustainable development have provided direction for the vertical integration of governance mechanisms with varying levels of success. On the regional to national, and national to European levels, we have witnessed the development of institutional capacities for sustainable development. Yet partnerships for sustainable development have remained the poor relation of social partnership bodies at the national level. Further down the vertical dimension of governance at the local level we find less of a sense of the capacity to influence sustainable development outcomes or indeed to advance the implementation of policies.

The paths taken towards sustainable development in both jurisdictions have to a large extent been shaped both by historical patterns of governance and the available opportunity structures for innovation conditioned by EU membership or, more often, EU funding. Although official discourse is increasingly couched in the language of governance *for* sustainable development, the institutional designs we have outlined here are path-dependent rather than path-creating. In other words, what we are witnessing is the continuing development of the governance *of* sustainable development. In this respect, we must concur with other studies (Adshead and McInerney 2006; Murray 2006; Flynn 2007) that there is a lack of fit between the 'software' of new governance relationships and the hardware of existing institutional structures. These design flaws are compounded on an all-island basis by the fact we are dealing with two distinct operating systems. The revised Strategy for Sustainable Development in the Republic (due for publication) will tell whether these systems can become more compatible. Meanwhile, we will have to look to specific instances of cross-border sectoral cooperation e.g. renewable electricity and waste management, for indications of how the strategic and integrative capacities for sustainable development can develop on the island of Ireland.

Governance *for* sustainable development does not simply mean the development of institutional capacity in terms of structures and strategies for multi-level governance; it also requires corresponding capacity-building in civil society. With the demise of LA21, the impetus for developing the participative capacities of governance is more likely to emerge from internal processes like the Review of Governance in the North and the Task Force for Active Citizenship in the South.

## References

Adshead, M. (2006) 'New Modes of Governance and the Irish Case: Finding Evidence for Explanations of Social Partnership', *Economic and Social Review*, 37, 3: 319–42.

Adshead, M. and C. McInerney (2006) 'Mind the Gap – An Examination of Policy Rhetoric and Performance in Irish Governance Efforts to Combat Social Exclusion' in *Governments and Communities in Partnership: From Theory to Practice*, University of Melbourne available at http://www.public-policy.unimelb.edu.au/conference06/presentations.html.

Alexander, E. R. (2006) 'Institutional Design for Sustainable Development', *Town Planning Review*, 77, 1: 1–27.

Callanan, M. (2004) 'Local and Regional Government in Transition', in N. Collins and T. Cradden (eds), *Political Issues in Ireland Today*, Manchester University Press: Manchester, pp. 56–78.

Department of the Environment (Northern Ireland) (2006) *First Steps Towards Sustainable Development: A Sustainable Development Strategy for Northern Ireland*, Belfast: DoENI.

Department of the Environment and Local Government (2002) *Making Ireland's Development Sustainable: Review, Assessment and Future Action*, Dublin: DOELG.

Dubnick, M. J. and E. M. Meehan (2004) *Integrative Governance in Northern Ireland*, Working Paper QU/GOV/16/2004, Belfast: Institute of Governance, Public Policy and Social Research, Queens University Belfast.

Ellis, G., B. Motherway, W. J. V. Neill and U. Hand (2004) *Towards a Green Isle? Local Sustainable Development on the Island of Ireland*, Armagh: Centre for Cross Border Studies.

European Commission (2004) *National Sustainable Development Strategies in the European Union: A First Analysis by the European Commission*, Brussels: European Commission, Commission Staff Working Document.

Flynn, B. (2007) *The Blame Game: Rethinking Ireland's Sustainable Development and Environmental Performance*, Dublin, Portland: Irish Academic Press.

Garavan, M. (2006) 'Seeking a Real Argument', in M.P. Corcoran and M. Peillon (eds),*Uncertain Ireland: A Sociological Chronicle, 2003–2004*, Dublin: Institute of Public Administration, pp. 73–90.

Graham, B. and C. Nash (2006) 'A Shared Future: Territoriality, Pluralism and Public Policy in Northern Ireland', *Political Geography*, 25: 253–78.

Harris, C. (ed.) (2005) *Engaging Citizens: The Case for Democratic Renewal in Ireland*, The Report of the Democracy Commission, Dublin: *tasc* at New Island.

Hughes, I., P. Clancy, C. Harris and D. Beetham (2007) *Power to the People? Assessing Democracy in Ireland*, Dublin: *tasc* at New Island.

Jänicke, M. (2006) *The "Rio Model" of Environmental Governance – A General Evaluation*, Berlin: Forschungsstelle Für Umweltpolitik (Environmental Policy Research Centre), Freie Universitat Berlin, FFU-Report 03-2006.

Jonas, A. E. G., A. While, and D. C. Gibbs (2004) 'State Modernisation and Local Strategic Selectivity after Local Agenda 21: Evidence from Three Northern English Localities', *Policy and Politics*, 32, 2: 151–68.

Knox, C. (2003) 'Northern Ireland Local Government', in M. Callanan and J. F Keogan (eds), *Local Government in Ireland: Inside Out*, Dublin: Institute of Public Administration, pp. 460–71.

Lafferty, W. (ed.) (2004), *Governance for Sustainable Development: The Challenge of Adapting Form to Function*, Cheltenham, Northampton: Edward Elgar.

Larkin, T. (2004) 'Participative Democracy: Some Implications for the Irish Polity', *Administration*, 52, 3: 43–56.

Lightfoot, S. and J. Burchell (2005) 'The European Union and the World Summit on Sustainable Development: Normative Power Europe in Action', *Journal of Common Market Studies*, 33, 1: 75–95.

Macrory, R. (2004) *Transparency and Trust: Reshaping Environmental Governance in Northern Ireland*, London: University College London, Centre for Law and the Environment.

Meadowcroft, J. (2004) 'Participation and Sustainable Development: Modes of Citizen, Community and Organizational Involvement', in W. Lafferty (ed.), *Governance for Sustainable Development: The Challenge of Adapting Form to Function*, Cheltenham, Northampton: Edward Elgar, pp. 162–90.

Meadowcroft, J., K. N. Farrell, and J. Spangenberg (2005) 'Developing a Framework for Sustainability Governance in the European Union', *International Journal for Sustainable Development*, 8, 1 & 2: 3–11.

Meldon, J., M. Kenny and J. Walsh (2004) 'Local Government, Local Development and Citizen Participation: Lessons from Ireland', in W. R. Lovan, M. Murray and R. Schaffer (eds), *Participatory Governance: Planning, Conflict Mediation and Public Decision Making in Civil Society*, Aldershot: Ashgate, pp. 147–64.

Morison, J. (2001) 'Democracy, Governance and Governmentality: Civic Public Space and Constitutional Renewal in Northern Ireland', *Oxford Journal of Legal Studies*, 21, 2: 287–310.

Moseley, M. M., T. Cherret and M. Cawley (2001) 'Local Partnerships for Rural Development: Ireland's Experience in Context', *Irish Geography*, 34, 2: 176–93.

Mullally, G. (2004) 'Shakespeare, the Structural Funds and Sustainable Development', *Innovation: the European Journal of Social Science Research*, 17, 1: 25–42.

Murray, M. (2006) 'Multi-level "Partnership" and Irish Waste Management: The Politics of Municipal Incineration', *Economic and Social Review*, 37, 3: 447–65.

Niestroy, I. (2005) *Sustaining Sustainability: A Benchmark Study on National Strategies for Sustainable Development and the Impact of Councils in Nine EU Member States*, Den Hagg: European Environment and Sustainable Development Advisory Councils (EEAC), EEAC Series, Background Study, No. 2.

OECD (2006) *Good Practices in the National Sustainable Development Strategies of OECD Countries*, Paris: Organisation for Economic Cooperation and Development.

O'Mahony, J. (2007) 'Europeanisation as Implementation: The Impact of the European Union on Environmental Policy Making in Ireland', *Irish Political Studies*, 22, 3: 265–85.

Turner, S. (2006) 'Transforming Environmental Governance in Northern Ireland. Part 1: The Process of Policy Renewal', *Journal of Environmental Law*, 18, 1: 55–87.

WCED (1987) *Our Common Future*, World Commission on Environment and Development, Oxford: Oxford University Press.

# Chapter 5

# Regional Planning and Sustainability

## Mark Scott

### Introduction

This chapter is concerned with the regional and spatial dimensions of rural sustainable development, drawing on the recent experiences of both the Republic of Ireland and Northern Ireland. Recent years have witnessed unprecedented interest in Europe in the formulation of spatial strategies for territorial development emphasizing the regional scale of policy delivery (Healey, Khakee, Motte and Needham 1997; Shaw, Roberts and Walsh 2000; Faludi 2001; McEldowney and Sterrett 2001). As Albrechts, Healey and Kunzmann note (2003), the motivations for these new efforts are varied, but the objectives have typically been to articulate a more coherent spatial logic for land-use management, resource protection, and investments in regeneration and infrastructure. Typically, therefore, spatial planning frameworks embrace a wider agenda than traditional regulatory approaches to land-use management in an attempt to secure integrated policy delivery and more effective linkages between national and local planning.

This chapter aims to examine two major initiatives in strategic spatial planning in Ireland, namely the publication of Northern Ireland's Regional Development Strategy (RDS) in 2001, and the Republic of Ireland's National Spatial Strategy (NSS) in 2002. The current wave of interest in Ireland in the formulation of spatial strategies for regional development provides a new point of reference for thinking about and shaping rural space, particularly as non-agricultural interests increasingly shape rural areas. In an increasingly 'post-agricultural' era (McDonagh 1998), rural sustainable development has become an increasingly contested arena. For example, housing in the countryside, environmental directives for landscape protection, potential wind-farm development, and access to farmland for recreation, have all been marked by high profile and polarized debates in the popular media. In this context, spatial strategies have the potential to offer a holistic approach to balancing the economic, social and environmental processes which shape Ireland's rural space. The first part of this chapter will examine key issues surrounding spatial planning and regional development followed by a review of regional planning policy in Ireland. The chapter then considers the role of regional planning within rural sustainable development, emphasizing three key aspects: spatially differentiated rural policies; the urban-rural relationship; and accommodating housing in the countryside, and concludes with insights relevant to regional planning and contested ruralities.

## Spatial planning, regionalism and sustainable development

The systems of land-use planning in the Republic of Ireland and Northern Ireland consist of a framework of development plans, prepared for county or sub county level, that form the basis for evaluating applications for development. Until recently, the regional dimension of planning practice was a missing tier within the policy framework. As Haughton and Counsell (2004a) observe, 'regions' tend to go in and out of fashion, both academically and in terms of policy practice, and from the late 1990s regional planning in Ireland began to emerge as an active arena. Indeed, regionalism has started to ascend the political agenda in many European countries, leading to growing experiments with policy devolution and political devolution, such as in France, Italy, Spain and the UK (Haughton and Counsell 2004b). In parallel, recent years have also seen a growing academic interest in regional debates, and as Murdoch et al. (2003) comment, although the term 'region' is often difficult to define, this spatial scale appears to be gaining new significance for economists, sociologists, political scientists and geographers. In particular, a growing body of literature has emerged relating to two key and related themes: regional economic development (see for example, Porter 2003; Cooke 2004; Kitson et al. 2004; Turok 2004; Ward and Jonas 2004); and regional governance (see for example, Giodano and Roller 2004; Gualini 2004; Goodwin et al. 2005; Jessop 2005).

The current enthusiasm for regionalism within planning debates undoubtedly owes much to European policy developments. Particularly important in the context of this chapter has been the European Union's (EU) growing interest in spatial planning to secure balanced and sustainable territorial development, culminating in the publication of the *European Spatial Development Perspective* (ESDP) in 1999. The ESDP (CSD 1999) provides a non-statutory framework for spatial development in the EU, providing a definition of new policy discourses, new knowledge forms and new policy options (Richardson 2000), which are being increasingly translated and applied into individual member states' national and regional policies. The key elements of the ESDP have been well documented elsewhere (see for example: Faludi 2000; Tewdwr-Jones and Williams 2001; Healey 2004) and can be distilled as (CSD 1999, p. 11):

- Development of a balanced and polycentric urban system and a new urban-rural partnership;
- Securing parity of access to infrastructure and knowledge;
- Sustainable development, prudent management and protection of nature and cultural heritage.

In contrast to the emphasis on *local* development in the 1990s, the current policy proposal of the EU is to tie rural areas much more into their urban and regional contexts. In this regard, the ESDP calls for the strengthening of the partnership between urban and rural areas to overcome 'the outdated dualism between city and countryside' (CSD 1999, p. 19) and to provide an integrated approach to regional problems. As Tewdwr-Jones and Williams (2001) argue, this focus on core-periphery (or urban-rural) relations necessitates an analysis of territory, rather than periphery, urban or rural alone. A regional approach often represents a more meaningful scale of action in terms of labour and housing markets, and of daily leisure activities (Healey 2002), and can encompass home-work relationships; central place relationships; relationships between metropolitan and urban centres in rural and intermediate areas; relationships between rural and urban enterprises; and rural areas as consumption areas for urban dwellers (Bengs and Zonneveld 2002).

## Regional planning in Ireland

Since the 1970s, regional planning can be described as a missing tier in Irish spatial policy (Bannon 2004). Previously, in the 1960s, there was a brief flirtation with regional planning in both parts of the island. In Northern Ireland, regional development and planning during the 1960s was based on the Matthew Plan (1963). The central aim of this plan was to demagnetize Belfast in terms of the dispersal of population and investment to selected growth centres, such as Craigavon and Antrim, while introducing urban containment and greenbelt policies for Belfast. Although targeted at the Belfast sub-region, the Matthew Plan set the physical planning context for the region as a whole (Gaffikin et al. 2001), leading to a marginalized position for the west of the Province and for rural communities (termed in the plan as the 'rural remainder').

In relation to rural Northern Ireland, the countryside was perceived largely in terms of landscape and amenity resulting in a regulatory protection ethos. As Gaffikin et al. record, the 1970s saw the pursuit of this agenda and in planning terms this was reflected in the comprehensive development of Belfast and the creation of new towns and growth centres. As the 1970s progressed, this policy of demagnetizing Belfast was endorsed with a more diffuse settlement strategy of multiple district towns, with the intention of spreading development more evenly between east and west of the Province (Neill and Gordon 2001). However, as Greer and Murray argue:

> Although the number of county towns and larger villages selected for growth in
> subsequent local plans did increase, this expansion could only be achieved by
> considerable population movement from smaller settlements and families living
> in the open countryside, a settlement pattern which personifies rural Northern
> Ireland (2003, p. 10).

Within this context, development control policies operated a presumption against development in the countryside, resulting in rural housing emerging as one of the most politically contentious features of planning policy.

A number of these themes can also be identified in relation to the Republic of Ireland. During the 1960s, there was a growing interest in regional planning as a tool for economic growth and development, culminating with the Buchanan Report in 1968. This report presented the argument for promoting growth centres at both national and regional levels. However, its recommendations became diluted as industrial policy increasingly favoured diffusion rather than concentration. In the early 1970s, the Industrial Development Authority implemented a policy of dispersing new industrial employment to small towns and rural areas in the early 1970s (Johnson 1994; Murray et al. 2003). This was followed by a period in the 1970s and 1980s when inter-regional policy was of diminishing importance (CEC 1999) and national economic rather than regional goals were the imperative. However, with the Republic of Ireland's well-documented impressive economic growth in the 1990s (see for example, Breathnach 1998; Walsh 2000; Clinch et al. 2002), the issue of regional balance within the State again emerged. Although Ireland can meaningfully be regarded as a region of the larger EU economy, the interest in the regional distribution of economic activity within the country remains high. Although it is clear that Dublin is the only city in Ireland that is of sufficient size to compete at a European level, Clinch et al. contend that: 'policy makers are continually faced with the question, explicitly or implicitly, how much national economic growth should be traded off for a better regional balance?' (2002, p. 96).

Recent planning initiatives in Ireland have been clearly influenced by European notions of spatial planning, which is wider in scope than traditional UK and Irish approaches to land-use regulation. Within this discourse (drawing on Jessop 2005), the region emerges as a crucial nodal scale for planning policy – though the drivers for regional policy formulation differ north and south. In Northern Ireland, interest in the regional dimension has been interlinked with the search for good governance and identifying the most appropriate scale for policy intervention. The regional aspect was further emphasized by political developments, primarily related to the peace process and the establishment of the Northern Ireland Assembly in 1998, and can be further contextualized in the wider UK debate concerning political devolution introduced by the New Labour Government (see for example, Jones et al. 2005). In contrast, the growing interest in regional planning in the Republic of Ireland resulted from economic realities and the functional role of territories in production and accumulation processes, in particular the accelerating dominance of the Greater Dublin Area. Emerging debates on regional disparities focused on achieving balanced economic, social and physical regional development, but were largely undertaken in the absence of a corresponding debate concerning regional governance or political devolution.

*The Northern Ireland Regional Development Strategy 2025*

The current regional planning framework in Northern Ireland is provided by *Shaping Our Future: The Regional Development Strategy* (RDS) 2025 (DRD

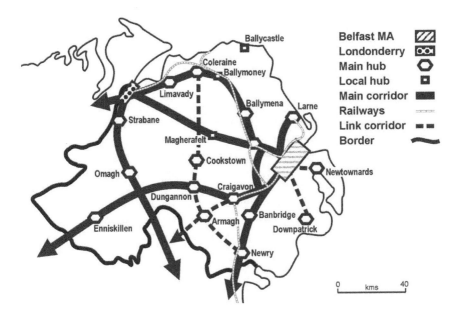

**Figure 5.1    The Spatial Development Strategy for Northern Ireland**
*Source*: DRD, 2001

2001), a statutory plan prepared by the Department of Regional Development (NI) and endorsed by the Northern Ireland Assembly in 2001. This political endorsement brought to an end a plan preparation process which had commenced in 1997 and was marked by an extensive participatory planning process involving over 500 community or interest groups in the plan's formulation (McEldowney and Sterrett 2001). This inclusive approach stands in contrast to the previous expert dependant and technocratic prescriptions of past regional planning (Albrechts et al. 2003; Murray and Greer 2003). The broad aim of the spatial strategy is to guide future development in order to 'promote a balanced and equitable pattern of sustainable development across the Region' (p. 41) and adopts a framework of interconnected hubs, corridors and gateways. Two regional *gateways* are identified – Belfast and Londonderry/Derry – in addition to a polycentric network of *hubs*, based on the main regional towns serving a strategic role as centres of employment and services for urban and rural communities. The key and link transport *corridors* provide the skeletal framework for future physical development (see Figure 5.1).

Perhaps the key challenge outlined by the RDS will be the accommodation of the projected housing growth for Northern Ireland. Out of a regional need of 160,000 dwellings for this period, the Strategy has allocated 51,000 to the Districts covered by the Belfast Metropolitan Area, of which 42,000 should be located within the existing built-up area. The RDS, therefore, at least in rhetoric, supports the concept of the 'compact city', establishing a regional target of 60 per cent of new housing to be located within existing urban areas (which contrasts with the recent level of achievement of less than 30 per cent).

*The Republic of Ireland's National Spatial Strategy 2020*

The need for a national spatial framework was identified in the Irish Government's National Development Plan (NDP) in 1999, establishing as a priority the goal of delivering more balanced regional development given the accelerating dominance of the Greater Dublin Area (GDA). Preparatory work on the NSS commenced in January 2000 and, while its publication was anticipated in late 2001, a general election during mid 2002 delayed its release until the end of the year. Planning is very much a political activity and thus the sensitivities attached to the possible designation (and non-designation) of growth centres would undoubtedly have placed the spatial strategy at the centre of political controversy in the run-up to voting day (Murray 2003).

The NSS sets out a twenty-year planning framework designed to achieve a better balance of social, economic, physical development and population growth on an inter-regional basis and comprises three key elements. Firstly, the NSS aims to promote a more efficient Greater Dublin Area which continues to build on its competitiveness and national role, while recognizing that it is not desirable for the city to continue to spread physically into the surrounding counties. Therefore, the NSS proposes the physical consolidation of Dublin supported by effective land-use policies for the urban area, such as increased brownfield development, and a more effective public transport system.

Secondly, the NSS designates strong 'gateways' in other regions. Balanced national growth and development is to be secured with the support of a small number of nationally significant urban centres which have the location, scale and critical mass to sustain strong levels of job growth in the regions. The National Development Plan 2000–2006 had previously designated Cork, Limerick/Shannon, Galway and Waterford as gateways, and the NSS further identified four new national level gateways: Dundalk, Sligo, and two 'linked' gateways of Letterkenny (linked to Derry in Northern Ireland) and Athlone/Tullamore/Mullingar (see Figure 5.2). Undoubtedly the designation of gateways was underpinned by political pragmatism. The gateways originally designated in the National Development Plan, with the exception of Galway, are located in the south and east of the State, which are the most prosperous regions in the Republic of Ireland. The designation of the four new gateways in the NSS allows for a more geographically inclusive process.

Thirdly, the Strategy also identifies nine medium sized 'hubs', which are to support and be supported by the gateways and will link out to wider rural areas. The hubs identified include Cavan, Ennis, Kilkenny, Mallow, Monaghan, Tuam and Wexford and two linked hubs comprising Ballina/Castlebar and Tralee/Kilarney. Along with these three elements the Strategy mentions the need to support the county and other town structure and to promote vibrant and diversified rural areas. The settlement hierarchy is further developed in its relationship to the

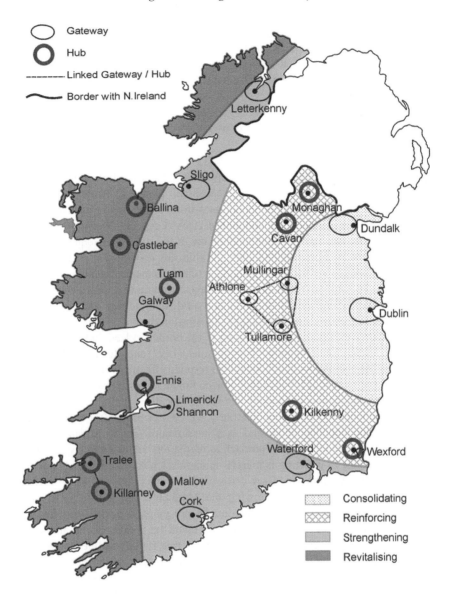

**Figure 5.2    The National Spatial Strategy**
*Source*: DOELG, 2002

proposed national transport framework based on radial corridors, linking corridors and international access points.

As Murray (2003) observes, the NSS is very much skeletal in design and thus in terms of implementation further work is acknowledged as being necessary. In this regard, provisions were made in the recent Planning and Development Act 2000 (Government of Ireland 2000) for the State's eight Regional Authorities to prepare statutory Regional Planning Guidelines (RPGs) to give full effect to the principles outlined in the NSS. The RPGs have now been completed for all Regional Authorities.

## Spatial planning and rural change

Historically, the fate of smaller settlements and rural areas in Ireland has received limited attention from economic and physical planners. Rural areas have often been 'perceived largely as scenic backdrops to the drama of urban based investment in infrastructure, industry and services' (Greer and Murray 1993, p. 3). This perspective was reinforced with the view of the rural arena equating solely with agriculture as a productivist space. Within this context, the principal rural planning challenge over the last few decades relates to the continuing controversy surrounding housing development in rural areas, leading to a vexed relationship between local planning authorities and many rural communities. The proliferation of dispersed single dwellings (or one-off housing) in the countryside has been an issue for many years both north and south of the border. Indeed, in the case of the Republic of Ireland, commentators such as Aalen (1997) and McGrath (1998) have argued that the planning system is unable to respond effectively to rural settlement growth. In a critique of rural planning, both commentators suggest policy is driven by the priorities of a few individuals, an intense localism, and the predominance of incremental decision-making. Similarly, Gallent et al. (2003) classify rural planning in the Republic of Ireland as a laissez-faire regime, suggesting that: 'the tradition of a more relaxed approach to regulation, and what many see as the underperformance in planning is merely an expression of Irish attitudes towards Government intervention' (p. 90).

Within Northern Ireland, following the Matthew Plan, policy prescription for rural housing during the 1970s favoured a presumption against new housing outside of selected settlements, unless need could be proven (for example, on employment or health grounds). However, as Sterrett (2003) outlines, opposition to the operation of what was widely regarded as a restrictive policy, particularly by district councils in the south and west, led to the Government appointing a Review Body (the Cockcroft Committee) in 1977, resulting in a short term relaxation of housing policy. This was followed by a period in the 1980s where policy was focused on aesthetic control of rural housing through location, siting and design guidelines, and in 1993, the Department of Environment's *A Planning Strategy for Rural Northern Ireland* again emphasized concerns with the visual impact of new housing development.

The renewed interest in regional planning has thus provided a timely opportunity to reformulate rural planning policies in line with the changing realities of rural living in contemporary Ireland. In relation to rural Northern Ireland, the

Regional Development Strategy establishes as a key aim to 'develop an attractive and prosperous rural area, based on a balanced and integrated approach to the development of town, village and countryside, in order to sustain a strong and vibrant rural community' (p. 93).

Similar themes can also be identified in the Republic of Ireland's NSS, which sets out in broad terms how rural areas will contribute to achieving balanced regional development. Three areas of policy are identified (p. 51). Firstly, the NSS highlights 'Strengthening the Rural Economy' as a key policy goal. The NSS recognizes that the role of traditional rural based sectors (agriculture, forestry and fishing) will continue to provide a base for the rural economy, but also outlines the importance of tailored responses in differing local contexts in relation to tourism, enterprise, local services and natural resource sectors. Secondly, the NSS identifies 'Strengthening Communities' as a policy area, in particular calling for new approaches to underpin the future vitality of rural communities. The NSS proposes two main types of responses: (1) settlement policies are needed that take account of varying rural development contexts (this is further discussed below); and (2) enhanced accessibility must be linked with an integrated settlement policy. Thirdly, the Strategy identifies the importance of 'Strengthening Environmental Qualities' of rural areas, and highlights the linkages of sensitive development and conservation of natural resources with the rural economy, in particular tourism development.

Therefore, on paper at least, both spatial strategies attempt to apply principles of sustainable development to rural planning by emphasizing the importance of environment, quality of life for rural communities and the rural economy. However, although a commitment to these three policy areas – economy, communities and environment – seems unquestionable at a national level, the incorporation of these broad goals into detailed planning policies at the local level is likely to be a contested arena. Research from Northern Ireland (see Murray and Greer 2000) and England (see Owen 1996) suggests that planners often favour restraint policies for rural settlement planning as a selective interpretation of what constitutes sustainable planning practice. In these cases, restrictive rural planning policies with goals such as reducing car dependency and landscape protection are often promoted rather than policies which are aimed at diversifying the economic base of rural areas or sustaining rural communities.

How these broad policy goals (and the mediation of policy objectives that are potentially conflicting) are translated into local authority development plans will therefore have profound effects on planning policies for rural areas, suggesting the need for enhanced understanding of the inter-relationships between economic, social and environmental processes within rural localities. The remainder of this chapter aims to review recent policy developments by 'unpacking' the RDS and the NSS and assessing the implications for the formulation of rural planning policies on three aspects of rural planning: (1) a spatially differentiated rural policy; (2) the conceptualization of the urban-rural relationship; and (3) accommodating housing in the countryside.

*Towards a spatially differentiated rural policy?*

A significant development in both spatial strategies is the recognition that rural areas are not homogenous spaces, but are increasingly characterized by diverse development and community contexts, suggesting the need for spatially differentiated rural policies. For example, the RDS for Northern Ireland outlines contrasting development pressures between the Belfast travel-to-work area and the rest of the region, resulting in a suite of policy measures that support the revitalization of declining settlements, while also adopting growth management policies for rapidly expanding small towns and villages.

A more sophisticated approach to assessing rurality can be found in the Republic of Ireland's spatial framework. The NSS provides a typology of rural areas to identify different types of rural areas and to reinforce the need for differing policy responses appropriate to local contexts (see Figure 5.3). The typology is based on a commissioned background report prepared by NUI Maynooth and Brady Shipman Martin (2000) who based their analysis on demographic structure, labour force characteristics, education and social class, sectoral employment profiles, performance of the farming sector and 'change' variables (e.g. population change, changes in numbers at work, etc.). The different types of rural areas identified in the NSS are as follows:

1.  Strong areas mainly located in the South and East where agriculture will remain strong, but where pressure for development is high and some rural settlements are under stress;
2.  Changing areas including parts of the Midlands, the Border, the South and West where population and agriculture employment have started to decline and where replacement employment is required;
3.  Weak areas including the more western parts of the Midlands, certain parts of the Border and mainly inland areas in the West, where population decline has been significant;
4.  Areas that are remote including parts of the west coast and the islands;
5.  Areas that are culturally distinct including parts of the west coast and the Gaeltacht which have a distinctive cultural heritage.

This typology is significant in that it appears to represent a first step towards developing a spatially defined rural policy rather than a sectoral (essentially agricultural) based approach which has predominated in the past. The typology provides the basis for a differentiated policy process which reflects the diversity of rural Ireland, enabling planning policies to be tailored to specific regions or localities. This is a belated recognition that new patterns of diversity and differentiation are emerging within the contemporary countryside (as outlined by Marsden 1999) and that the key to understanding rural areas is the avoidance of easy assumptions of homogeneity (McDonagh 2001). As asserted by McDonagh, rural areas in Ireland are dynamic and they have become arenas for conflict and

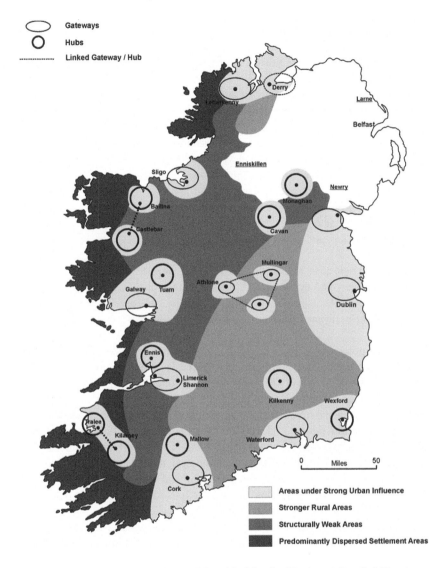

**Figure 5.3    Rural Policy Areas Identified in the National Spatial Strategy**
*Source*: DEHLG, 2004

tension, sites for consumption as well as production activities – however, not all rural areas have the same capacities or undergo change at the same time or pace. Planners at a local authority level must respond to this 'recasting' of rurality in the national spatial framework, by avoiding the 'one size fits all' approach which has been prevalent in rural settlement planning and recognize that planning policies for rural areas should reflect the diversity of the challenges facing rural communities.

*The urban-rural relationship*

A clear example of the adoption of the ESDP's spatial planning vocabulary can be seen in relation to the urban-rural partnership for territorial development. This challenge to the separation of urban and rural in spatial planning discourse has clearly been translated into both Irish spatial frameworks:

> National and international evidence also demonstrates that rural areas have a vital contribution to make to the achievement of balanced regional development. This involves utilizing and developing economic resources of rural areas … while at the same time capitalizing on and drawing strengths from vibrant neighbouring urban areas. In this way rural and urban areas are seen as working in partnership, rather than competing with each other (National Spatial Strategy, DOELG 2002, p. 36).

> Urban and rural areas have distinct roles, but it is important that these roles are complementary and that town and country maintain their distinctiveness and respective social and physical integrity in the sustainable development of each (Regional Development Strategy, DRD 2001, p. 86).

Through adopting a regional approach, both strategies recognize the growing complexity of urban-rural relations. For example, while major urban centres reinforce their role as drivers of the economy through agglomeration processes, many households in Ireland have expressed a consumer choice to live outside urban centres in accessible rural areas, resulting in urban decentralization and counterurbanization patterns of residential development (Gkartzios and Scott 2005). The primary mechanisms for developing the 'new' urban-rural partnership are, in the case of the Republic of Ireland, the designated gateways and hubs that 'have the capacity to support the stronger urban-rural structure needed to drive the development of these other regions' (p. 49), and in Northern Ireland, 'a polycentric network of hubs and clusters based on the main towns' (p. 43).

In some senses, the terms 'gateways' and 'hubs' have replaced an earlier lexicon of regional development in Ireland in designating 'growth centres' in the 1960s, acting as a public-friendly metaphor for a two-way interdependent relationship characterized by a complex 'space of flows' (drawing on Hadjimichalis 2003). This is an important recognition that the spatial dimensions of economic change and development cannot be reduced to a single urban-rural dichotomy (Commins et al. 2005). However, it also represents a key spatial planning challenge on two levels: firstly, will the gateways and hubs act as an effective counterbalance to increased development in the Greater Dublin Area and Belfast Metropolitan Area? And secondly, will the gateways and hubs act as effective development nodes capable of dispersing economic growth? Indeed, Healey (2002) suggests that the idea that towns and cities are the key development nodes in a region and that they disperse development around a territory needs serious questioning: 'each region is likely to have its own relational and distributive specificities. Failure to recognize these leads to the disjunction between policy imagery and lived realities' (p. 337). Although designated gateways and hubs may have the potential for growth, as a recent Rural Foresights Report (NUIM/UCD/Teagasc 2005) contends, the rural 'spillover' may be limited given the human resources and infrastructural deficits at a rural level. This suggests that for those rural areas outside the main urban catchment areas, specifically focused programmes for local economic development will be necessary. Furthermore, for designated gateways and hubs to develop the necessary critical mass needed to contribute to 'balanced regional development', it is probable that restrictive rural settlement policies will be required to facilitate the growth of larger settlements in the hierarchy, suggesting the submergence of rural interests within a city-region geography and demonstrating little affinity with rural communities.

*Accommodating housing in the countryside*

Rural housing emerged as one of the most contested features during the formulation of both the RDS and NSS, particularly as building projects involve often highly visible indicators of rural structural change. Analysis undertaken during the preparation of the National Spatial Strategy suggests that between 1996–1999 over one in three houses built in the Republic of Ireland have been one-off housing in the open countryside, and highlights that the issue of single applications for housing in rural areas has become a major concern for most local planning authorities (Spatial Planning Unit 2001). Similarly, with a comparative rural culture, but with a contrasting centralized planning regime with less political control, Northern Ireland has experienced similar rural housing trends with approximately 27 per cent of private house-building completions comprised of single houses in the open countryside each year (Sterrett 2003), with the number of single new dwellings being approved increasingly significantly from 1,790 in 1991/92 to 5,628 by 2002/03 (DRD 2004).

Both the RDS and NSS provide positive statements in relation to rural housing, recognizing the strong sense of belonging and sense of place in rural areas. The Northern Ireland strategy outlines a vibrant, living and working countryside as a key policy goal, outlining the need to accommodate new housing development to meet local housing need and to encourage the development of balanced rural communities by promoting housing choice and affordable housing in rural areas. However, as Greer and Murray (2003) note, while this represents a positive policy expression towards rural communities, concerns are equally noted about the perceived cumulative visual impact of inappropriate single house development:

> These growing pressures [of rural housing] present a threat to the open countryside which is a vital resource for sustaining the genuine rural community. The cumulative impacts of this development include: loss of agricultural land and habitats; fields being sold off to house townspeople; increased traffic on

rural roads; the risk of pollution from growing numbers of septic tanks; the increased visual impact of more structures in the landscape; and a weakening of towns and villages (DRD 2001, p. 89).

Following the rural typology, the NSS encouragingly calls for different responses to managing dispersed rural settlement between rural areas under strong urban influences and rural areas that have a strong agricultural base are structurally weak rural areas or possess distinctive settlement patterns, reflecting the contrasting development pressures that exist in the countryside. This is further developed in the Strategy with the distinction the NSS makes between urban and rural generated housing in rural areas. In general, the NSS outlines that development driven by urban areas (including urban-generated rural housing) should take place within built up areas or land identified in the development plan process and that rural-generated housing needs should be accommodated in the areas where they arise. As a more 'sustainable' alternative to dispersed single housing in the countryside, the NSS places considerable emphasis on the role of villages in rural areas.

Interestingly, both the RDS and NSS were careful to avoid detailed policy prescription on rural housing (see Greer and Murray 2003; Scott 2006), and thus avoided additional political controversy at the time of publication. More recently the Republic of Ireland's Department of Environment, Heritage and Local Government have produced Planning Guidelines for Sustainable Rural Housing (2005), ensuring that dispersed rural housing in the countryside remained a high profile issue and a deeply contested feature of the planning policy arena. The Planning Guidelines suggest that the state has shifted to a less restrictive position on housing in the countryside. In summary, the guidelines provide that: (1) people who are part of and contribute to the rural community will get planning permission in all rural areas, including those under strong urban-based pressures, subject to the normal rules in relation to good planning; and (2) anyone wishing to build a house in rural areas suffering persistent and substantial population decline will be accommodated, subject to good planning. In this context, it is worth noting that the term 'good planning' refers to issues surrounding siting, layout and design, rather than planning in a strategic or spatial sense. The sentiments of the new guidelines can be summarized in the following extract from a speech given by the Minister for the Environment, Dick Roche in July 2005:

> Those who would like to prevent homes being built in the countryside attacked me politically. I suggested at the time that planners in our local authorities and critics in some National organizations, all too often did not value the sense of community that exists in rural Ireland. I asked why was it that planners and some national organizations adopted the attitude 'we know best'. I suggested that this exclusivist attitude was wrong: it smacked of arrogance. The sons and daughters of farmers, men and women who were born and were reared in the countryside, people who live in the countryside and work in the countryside – whatever their following in life – have the same right to have a home of their own and a home

in their own place as anybody else. ... All too often planning is seen as a way of preventing people building in and living in their own place (Roche 2005).

In Northern Ireland the Department for Regional Development published Draft Planning Policy Statement 14 'Sustainable Development in the Countryside' in March 2006 as its response to the perceived need to moderate housing pressure. The draft policy was published in the absence of the Northern Ireland Assembly and without local political endorsement. The guidelines, based solely on planning approval evidence, are designed to facilitate public consultation, but are now effectively, following publication, the major material consideration in the determination of new planning applications for development in the countryside. The policy framework applies to all lands outside settlement limits as identified in development plans and imposes a broad presumption against development apart from a number of tightly circumscribed exceptions, for example, farm dwellings, dwellings for retiring farmers, dwellings for non agricultural business enterprises and replacement dwellings. The countryside housing market is narrowly defined, there is no spatial differentiation of the Northern Ireland countryside and in essence all development proposals must demonstrate clear need in order to secure approval. As Murray and Scott (2006) highlight, the result is that below the main town level, rural communities in Northern Ireland face an uncertain, if not bleak, future and the policy framework falls very far short of seeking to understand and provide for the different realities of countryside living in Northern Ireland. The contrast between rural settlement planning policy succession in Northern Ireland and the Republic of Ireland could not be starker.

### Conclusion: Regional planning and contested ruralities

Recent years have witnessed a renewed interest in regional spatial planning within Ireland, and the region has emerged as a key nodal scale for policy intervention. Within this context, regional planning has developed as a key arena for addressing sustainable development conflicts and (as termed by Haughton and Counsell 2004a) 'competing sustainabilities'. An example of this can be found with the rural housing debate. Although rural housing conflicts tend to emerge first on a local scale – the level at which everyday life is most directly impinged upon – recent years have been marked by an 'up-scaling' (as termed by Woods 2005) of rural housing conflicts, as campaigners have been forced to engage in local, regional and national politics in attempts to change policy decisions. This was clearly evident during the formulation of the Northern Ireland Regional Development Strategy and National Spatial Strategy and the subsequent publication of the Planning Guidelines for Sustainable Rural Housing in the Republic of Ireland.

Given that the national and regional tier of policy making is an increasingly important node in establishing rural planning agendas, it is perhaps unsurprising that local actors should come to realize that political decisions taken at higher spatial scales are important in determining the outcomes of their own struggles (Murdoch et al. 2003). In this context, local pro-development interests have begun to build alliances further up the scales of governance. For example, the Irish Rural Dwellers Association (IRDA) has recently emerged as a broad coalition of pro-housing development interests in the Republic of Ireland (including farmers, councillors, community development stakeholders), which has successfully adopted a multi-scaler approach to influence policy outcomes, based on lobbying of elected representatives (both local and national), civil servants and local government officials, as well as forming new alliances with other stakeholders, such as the Royal Institution of Architects Ireland (RIAI). This up-scaling of rural conflicts can be identified in other advanced capitalist societies undergoing rural restructuring processes, as 'rural politics' has been replaced by a new 'politics of the rural' in which the very meaning and regulation of rural space is the defining issue (Woods 2003; 2005).

Adopting a regional approach has also enabled rural policy to be set within a wider spatial context and beyond agricultural sectoral interests. As Marsden (1999) argues, rural space is increasingly playing a key role in the political economy of the modern consumerist state and new demands on rural space are evident not only from agricultural interests, but also rural dwellers unconnected to farming, new rural residents, tourists, environmental groups, and developers. Within this context, there is a clear role for regional planning in managing and regulating rural space in terms of place-making; mediating between conflicting conservation and development goals; and integrating urban and rural dimensions. At present, however, developing holistic rural sustainable development goals remains a deeply contested and fragmented area of policy formulation leading to a disintegrated approach to rural policy.

Furthermore, greater emphasis must be given to addressing the current 'disconnect' between regional policy and rural development. Firstly, this involves an articulation of the role that regional planning can perform for rural areas, particularly for those rural areas beyond urban influence and networks. In relation to the Republic of Ireland, current trends in agricultural restructuring are likely to further reinforce existing regional disparities (Commins et al. 2005), as structurally weak farming activity in the border, midlands and western regions continues to decline. This suggests that policy goals relating to promoting balanced regional development and developing successful gateways and hubs are central to the fortunes of many rural communities. However, key questions remain in relation to the capacity of the selected gateways and hubs to effectively counterbalance the dominance of Ireland's eastern corridor and to disperse the benefits of development to rural areas. As Commins et al. (2005) argue, this requires clear operational programmes for implementing national and regional spatial strategies linked to regional and rural proofing of sectoral programmes.

Although spatial planning has the potential to perform a key role for rural communities, at present planning discourses are dominated by *urban* policy instruments, such as urban capacity studies, the sequential approach to housing location, and urban density tools. In contrast rural planning tools are limited to landscape assessment, which often demonstrate limited affinity with rural community aspirations. Therefore, a key challenge for planners and planning policies is to engage more proactively with understanding rurality and rural place-making, through addressing issues of place and territoriality, identity, attachment to place and community networks. As McDonagh (1998) argues, in this era of what is increasingly being referred to as a 'post-agricultural' society, there is an urgent need to question the understandings of the term 'rural' in Ireland and whether there is a coordinated policy direction for the changing future of rural areas. In this regard, the rural typology developed for the National Spatial Strategy is a significant step in identifying a tailored policy response to diverse rural contexts. However, this process must also be replicated at a local level to develop nuanced policy initiatives for rural areas. Secondly, addressing the current lack of policy coordination at a local level between spatial planning and rural development remains a concern. At present local land-use plans and strategies for social and economic development are poorly integrated in terms of policy formulation processes and delivery (Scott 2004), often leading to a disconnect between land-use and environmental goals and economic and social issues in the local arena. This suggests the need for a more interactive and collaborative style of local policy-making to enable planning officials and rural development stakeholders to explore new 'storylines' of rurality to provide a common departure point for developing an area-based, integrated and holistic approach to rural sustainable development.

## Acknowledgements

The author wishes to thank Stephan Hannon of the School of Geography, Planning and Environmental Policy in UCD for producing Figures 5.1–5.3.

## References

Aalen, F. (1997) 'The Challenge of Change', in F. Aalen, K. Whelan and M. Stout (eds), *Atlas of the Irish Rural Landscape*. Cork: Cork University Press, pp. 145–79.

Albrechts, L., P. Healey, and R. Kunzmann (2003) 'Strategic Spatial Planning and Regional Governance in Europe', *Journal of the American Planning Association*, 69: 113–29.

Bannon, M. (2004) 'Forty Years of Irish Urban Planning: An Overview', *Irish Urban Studies*, 3 (1): 1–18.

Bengs, C. and W. Zonneveld (2002) 'The European Discourse on Urban-Rural Relationships: A New Policy and Research Agenda', *Built Environment*, 28: 278–89.

Breathnach, P. (1998) 'Exploring the 'Celtic Tiger' Phenomenon: Causes and Consequences of Ireland's Economic Miracle', *European Urban and Regional Studies*, 5: 305–16.

CEC (Commission of the European Communities) (1999) *The EU Compendium of Spatial Planning Systems and Policies – Ireland*. Brussels: CEC.

Clinch, P., F. Convery and B. Walsh (2002) *After the Celtic Tiger, Challenges Ahead*. Dublin: O'Brien Press.

Commins, P., J. Walsh and D. Meredith (2005) 'Some Spatial Dimensions: Population and Settlement Patterns', in NUIM/UCD/Teagasc (eds) *Rural Ireland 2025, Foresight Perspectives*. NUIM/UCD/Teagasc: Dublin, pp. 37–45.

Committee for Spatial Planning (1999) *European Spatial Development Perspective: Towards a Balanced and Sustainable Development of the Territory of the EU*. Luxembourg: CEC.

Cooke, P. (2004) 'Regional Knowledge Capabilities, Embeddedness of Firms and Industry Organisation: Bioscience Megacentres and Economic Geography', *European Planning Studies*, 12: 625–41.

DAFRD (Department of Agriculture, Food and Rural Development) (1999) *Ensuring the Future – A Strategy for Rural Development in Ireland*. Dublin: Stationery Office.

DEHLG (Department of Environment, Heritage and Local Government) (2004) *Sustainable Rural Housing, Consultation Draft of Guidelines for Planning Authorities*. Dublin: DOEHLG.

DEHLG (Department of the Environment, Heritage and Local Government) (2005) *Planning Guidelines for Sustainable Rural Housing, Consultation*. Dublin: Stationery Office.

DOELG (Department of Environment and Local Government) (2002) *The National Spatial Strategy 2002–2020, People, Places and Potential*. Dublin: Stationery Office.

DRD (Department of Regional Development) (2001) *Shaping Our Future – Regional Development Strategy for Northern Ireland 2025*. Belfast: DRD.

DRD (Department of Regional Development) (2006) *Draft Planning Policy Statement 14, Sustainable Development in the Countryside*. Belfast: DRD.

Faludi, A. (2000) 'The European Spatial Development Perspective – What Next?', *European Planning Studies*, 8: 237–50.

Faludi, A. (2001) 'The Application of the European Spatial Development Perspective: Evidence from the North-West Metropolitan Area', *European Planning Studies*, 9: 663–75.

Gaffikin, F., M. McEldowney, M. Morrissey and K. Sterrett (2001) 'Northern Ireland: The Development Context', *Local Economy*, 16: 14–25.

Gallent, N., M. Shucksmith and M. Tewdwr-Jones (2003) *Housing in the European Countryside, Rural Pressure and Policy in Western Europe*. London: Routledge.

Giordano, B. and E. Roller (2004) '"Te para todos?" A Comparison of the Processes of Devolution in Spain and the UK', *Environment and Planning A*, 36: 2163–81.

Gkartzios, M. and M. Scott (2005) 'Urban-generated Rural Housing and Evidence of Counterurbanisation in the Dublin City-region', in N. Moore and M. Scott (eds), *Renewing Urban Communities: Environment, Citizenship and Sustainability in Ireland.* Aldershot: Ashgate, pp. 132–58.

Goodwin, M., M. Jones and R. Jones (2005) 'Devolution, Constitutional Change and Economic Development: Explaining and Understanding the New Institutional Geographies of the British State', *Regional Studies*, 39, 421–36.

Greer, J. and M. Murray (1993) 'Rural Ireland – Personality and policy context', in M. Murray and J. Greer (eds), *Rural Development in Ireland, A Challenge for the 1990s.* Aldershot: Avebury, pp. 3–16.

Greer, J. and M. Murray (2003) 'Rethinking Rural Planning and Development in Northern Ireland', in J. Greer and M. Murray (eds), *Rural Planning and Development in Northern Ireland.* Dublin: Institute of Public Administration, pp. 3–38.

Gualini, E. (2004) 'Regionalization as 'Experimental Regionalism': The Rescaling of Territorial Policy-making in Germany', *International Journal of Urban and Regional Research*, 28: 329–53.

Hadjimichalis, C. (2003) 'Imagining Rurality in the New Europe and Dilemmas for Spatial Policy', *European Planning Studies*, 11: 103–13.

Haughton, G. and D. Counsell (2004a) *Regions, Spatial Strategies and Sustainable Development.* London: Routledge.

Haughton, G. and D. Counsell (2004b) 'Regions and sustainable development: regional planning matters', *The Geographical Journal*, 170: 135–45.

Healey, P. (2002) 'Urban-Rural Relationships, Spatial Strategies and Territorial Development', *Built Environment*, 28: 331–9.

Healey, P. (2004) 'The Treatment of Space and Place in the New Strategic Spatial Planning in Europe', *International Journal of Urban and Regional Research*, 28: 45–67.

Healey, P., A. Khakee, A. Motte and B. Needham (eds) (1997) *Making Strategic Spatial Plans: Innovation in Europe.* London: UCL Press.

Jessop, B. (2005) 'The Political Economy of Scale and European Governance', *Tijdschrift voor Economische en Sociale Geografie*, 96: 225–30.

Johnson, J. (1994) *The Human Geography of Ireland.* Chichester: Wiley.

Jones, M., M. Goodwin and R. Jones (2005) 'State Modernisation, Devolution and Economic Governance: An Introduction and Guide to Debate', *Regional Studies*, 39: 397–404.

Kitson, M., R. Martin and P. Tyler (2004) 'Regional Competitiveness: An Elusive Yet Key Concept?', *Regional Studies*, 28, 991–1000.

Marsden, T. (1999) 'Rural Futures: The Consumption Countryside and its Regulation', *Sociologia Ruralis*, 39: 501–20.

McDonagh, J. (1998) 'Rurality and Development in Ireland – The Need for Debate?', *Irish Geography*, 31, 1: 47–54.

McDonagh, J. (2001) *Renegotiating Rural Development in Ireland.* Aldershot: Ashgate.

McEldowney, M. and K. Sterrett (2001) 'Shaping a Regional Vision: The Case of Northern Ireland', *Local Economy*, 16: 38–49.

McGrath, B. (1998) 'Environmental Sustainability and Rural Settlement Growth in Ireland', *Town Planning Review*, 3: 227–90.

Murdoch, J., P. Lowe, N. Ward and T. Marsden (2003) *The Differentiated Countryside*. London: Routledge.

Murray, M. (1993) 'Paradigm Redundancy and Substitution: Rural Planning and Development in Northern Ireland', *Pleanáil – Journal of the Irish Planning Institute*, 9: 195–218.

Murray, M. (2003) 'Strategic Spatial Planning on the Island of Ireland: Towards a New Territorial Logic?', Paper presented to conference: *Linking Development with the Environment – The EU and Accession Countries Perspective*, Bratislava, Slovakia, February.

Murray, M. and J. Greer (2000) *Rural Settlement Patterns and Physical Planning Policy in Northern Ireland*. Cookstown: Rural Community Network.

Murray, M., J. Greer and M. Scott (2003) 'The National Spatial Strategy (2002–2020) for the Republic of Ireland: Implications for Northern Ireland', *Economic Outlook and Business Review*, 18, 63–6.

Murray, M. and M. Scott (2006) *Northern Ireland – Republic of Ireland Perspectives on Rural Development and Rural Planning*, unpublished report for the Northern Ireland Housing Executive's Review of Rural Housing Policy.

Neill, B. and M. Gordon (2001) 'Shaping our Future? The Regional Strategic Framework for Northern Ireland', *Planning Theory and Practice*, 2: 31–52.

NUI Maynooth and Brady Shipman Martin (2000) *Irish Rural Structure and Gaeltacht Areas*. Dublin: DOELG.

NUIM/UCD/Teagasc (eds) (2005) *Rural Ireland 2025, Foresight Perspectives*. NUIM/UCD/Teagasc: Dublin.

Owen, S. (1996) 'Sustainability and Rural Settlement Planning', *Planning Practice and Research*, 11: 37–47.

Porter, M. (2003) 'The Economic Performance of Regions', *Regional Studies*, 37: 549–78.

Richardson, T. (2000) 'Discourses of Rurality in EU Spatial Policy: The European Spatial Development Perspective', *Sociologia Ruralis*, 40: 53–71.

Roche, D. (Minister for Environment) (2005) *Minister Roche Praises Community Effort & Calls for Less 'Dogmatism' in Planning*, 2/08/05, press release available to download from www.environ.ie.

Scott. M. (2004) 'Managing Rural Housing and Contested Meanings of Sustainable Development: Insights from Planning Practice in the Republic of Ireland', paper presented to *Planning and Housing: Policy and Practice, Housing Studies Association Conference*, 9 – 10 September 2004, Belfast.

Scott, M. (2006) 'Strategic Spatial Planning and Contested Ruralities: Insights from the Republic of Ireland', *European Planning Studies*, 14: 811–30.

Shaw, D, P. Roberts and J. Walsh (eds) (2000) *Regional Planning and Development in Europe*. Aldershot: Ashgate.

Spatial Planning Unit (2001) *Rural and Urban Roles – Irish Spatial Perspectives*. Dublin: DOELG.

Sterrett, K. (2003) 'The Countryside Aesthetic and House Design in Northern Ireland', in J. Greer and M. Murray (eds), *Rural Planning and Development in Northern Ireland*. Dublin: IPA, pp. 117–44.

Tewdwr-Jones, M. and R. Williams (2001) *The European Dimension of British Planning*. London: Spon Press.

Turok, I. (2004) 'Cities, Regions and Competitiveness', *Regional Studies*, 38: 1069–84.

Walsh, J. (2000) 'Dynamic Regional Development in the EU Periphery: Ireland in the 1990s', in D. Shaw, P. Roberts and J. Walsh (eds), *Regional Planning and Development in Europe*. Aldershot: Ashgate, pp. 117–37.

Ward, K. and A. Jonas (2004) 'Competitive City-regionalism as Politics of Space: A Critical Reinterpretation of the New Regionalism', *Environment and Planning A*, 36: 2119–39.

## Chapter 6

# Managing Rural Nature: Regulation, Translations and Governance in the Republic of Ireland and Northern Ireland

Hilary Tovey

Applying inappropriate environmental policies may lead to social and economic problems for the people affected, and fail to address the underlying biophysical causes of the problem... Many explanations of environmental degradation... have been constructed without the participation of the affected peoples, and without acknowledging how explanations may reflect social framings (Forsyth 2003, p. 10).

## 'Rural' and the environment

Rural areas are of particular interest to environmental activists and to environmental sociologists. They are the primary site of many of the environmental features which are of concern to regulators: biodiversity, water and other natural resources, landscapes. With the advancing de-agriculturalization of the European countryside, land ownership and land uses are undergoing a period of transformation, with unpredictable environmental consequences. Some formerly agricultural land becomes a site for housing, factories and industrial parks, or infrastructural developments (roads, gas pipelines, mobile phone masts), or for new types of exploitation (forestry, biomass production, golf courses and other recreational and tourist uses); other land is left largely unattended and reverts to a more or less advanced state of dereliction. The largely 'post-productivist' orientation of both state and population towards the rural encourages treating it as a place to be moulded increasingly to urban needs and demands, whether for recreation or for disposal of unwanted urban problems. Official actors see it as a space for prisons, landfill sites or incinerators, unofficial actors as a place to dump unwanted consumer goods and their by-products.

Environmental 'problems' are problems both *by* rural people and *for* rural people. The rural environment is a product of previous as well as current natural resource uses – often sponsored or encouraged by the state – from 'modernized' agriculture to conifer afforestation to mining. On the other hand, much that remains aesthetically valuable in rural areas, as well as valuable from a biodiversity point of view, is a product of previous human uses of the countryside. Very little of the European countryside, particularly of the Irish one, can accurately be categorized as 'wilderness'; the 'unspoilt' places that we still have, such as National Parks, are socio-historical and class-based constructions, often the work of 18th and 19th century landlords following fashionable ideas of the 'sublime' in landscape painting during the Romantic period (Slater 1993). And as rural areas were used in the past to express and represent new intellectual and aesthetic ideas, today they are still often seen as a location in which new ideologies and new ways of living can be experimented with. The 'back to the countryside' movement in France and Germany after the 1968 student revolt, for example, brought many young people dissatisfied with an urban industrial lifestyle into the European countryside to set up alternative forms of living and to find new and ecologically friendly ways of making a livelihood (Willis and Campbell 2004).

The Republic of Ireland (RoI) experienced some aftershocks of this new movement in the 1970s when in-movement of people from a number of 'core' European countries, attracted to Irish rural areas (particularly in the north west and in west Cork) both by the low price of farm land and the perception that RoI was still a place of unspoilt nature, led to a remarkable growth in the numbers and articulacy of the fledgling Irish organic movement of the time. This vision of the countryside as a space for ecological and social experimentation lives on today, not only in alternative farming and food making, as before, but increasingly in the construction of ecological houses and built environments, for example, in the Village Project in North Tipperary, or eco-builders experimenting with new types of one-off housing and with solutions to the problems these pose in sewage disposal, energy use and visual impact, in Louth and in South Tipperary.

The activities, both pro- and anti-environmental, of rural 'natives' are also important to the environmental picture. While rural people, particularly farmers, are still often represented as a particularly obstinate and unenlightened section of the population in terms of their environmental practices, much of the activism in support of the Irish environment has developed within rural settings and within rural civil society. Many of the key environmental struggles of the last three decades have been rural-based: the fight against the exploitation of the countryside for gold mining in Mayo; protests against fish-farming, against chemical factories and other forms of 'inappropriate' industrial development (Allen 2004); and the work done to protect biodiversity by generations of anglers, hunting and shooting clubs. Mobilization by local groups in defence of the rural environment is widespread across rural Europe, often stimulated by state and European Union (EU) regulatory and managerial interventions. However, many other cases against such 'external' actors are motivated by either their inattention to ecological concerns of the local people in promoting or supporting economic development in rural areas, or because of the nature of the managerial regime (science- and expert-controlled, uninterested in local understandings or inappropriately centralized in form) which they seek to impose.

Kousis (1999) argues that in the more 'peripheral' (i.e. rural) countries of Europe – Greece, Spain, Portugal, and we could add, RoI – environmental movements have tended to take a different form and to mobilize themselves around different concerns than in the more industrialized, 'core' areas (see also Tovey 2005). Much local environmental mobilization in RoI is centred on what she calls resistance to 'ecological marginalization', which takes the form of relatively spontaneous and unorganized opposition to the introduction into local areas of state-sponsored industrial and infrastructural projects which threaten to degrade or destroy natural resources on which local livelihoods depend. Because of their informal organization and the place-focussed nature of their concerns, rural mobilizations tend to be left out of accounts of 'the environmental movement' in Europe; their non-recognition then feeds into a picture of environmentalism as a struggle by enlightened core elites against 'backwardness' and 'ignorance' about environmental issues among rural populations (Tovey 1993).

Rural environmental management, then, is an issue which concerns both elites who identify risks to water, natural species and habitats, or landscape aesthetics, and also rural people themselves who face problems in sustaining their livelihoods or their desired quality of life. This chapter investigates how management of the rural environment is understood in Ireland, north and south. Starting from the argument that there are different regulatory discourses surrounding the Irish environment, it asks how rural environmental regulation is constructed both by environmental managers and by actors within rural civil society. Following Latour (1987, 1988), I explore how the notion of environmental management is 'translated' by different groups of institutional actors; in particular, who translates it as 'sustainable development' and what are the effects of the introduction of a discourse of sustainable development into attempts to manage and regulate the use of nature within rural settings, particularly in regard to engagement of civil society actors within projects for environmental 'governance'? I end by arguing that a key and largely ignored issue for achieving environmental and social sustainability is that of different environmental *knowledges*, and the differentiated power, cultural and symbolic capital associated with them.

## 'Translating' environmental regulation

A number of different state or semi-state institutions in the Republic of Ireland have responsibilities for environmental management. The Department of the Environment, Heritage and Local Government, particularly through the National Parks and Wildlife Service (NPWS) is the key player; but we could also include the Environmental Protection Agency, the Heritage Council, the Department of Agriculture and Rural Development, the Department of the Marine and Natural Resources, and the Department of Community and Rural Affairs. From a rural point of view, the Department of Agriculture and Food is an important actor. It can directly influence, through regulations or subventions, the production and waste management practices of 120,000 farmers; and through its control over the agri-environmental programme Rural Environmental Protection Scheme (REPS) and its close connections with LEADER projects it can shape the future of rural nature and biodiversity. The Environmental Protection Agency primarily concerns itself with the management of pollution and waste, and has initially focussed on regulating industrial enterprises and local government activities in these areas, although it is also increasingly bringing large-scale industrial agriculture under its remit (Taylor 2001). In Northern Ireland (NI), the key environmental manager is also the Department of Environment; its large in-house agency, the Environment and Heritage Service, combines most of the activities that in the Republic are divided between the NPWS and the Heritage Council. In the following discussion, however, the focus is on the Department of the Environment in the Republic of Ireland, and its associated actors, the NPWS and the Heritage Council.

How do these actors understand the concept of 'environmental management'? We can identify at least three different ways in which they translate the regulating of rural practices in order to maintain or protect nature: as 'scientific management of nature', as 'heritage conservation', and as 'sustainable development'. Each translation is associated with a different network of institutions and actors, and each gives a different role to 'the public' or civil society as participants in environmental management and protection. Translations of environmental management can be regarded as 'storylines' (Hajer 1995): ideas which have a capacity to enrol a range of different actors, who have different interests and different understandings of what the term means, but who can nevertheless use it as an umbrella to engage in more or less wide-ranging co-operation with each other. Some storylines, however, have a greater capacity to enrol and mobilize actors than others.

*Environmental management as scientific management of nature*

The Departments of Environment in both NI and the Republic of Ireland (RoI) use ideas of both 'heritage' and 'sustainable development' to articulate their vision of 'environmental management'. In Northern Ireland, the objectives set out by the department for itself include: 'To protect, conserve and enhance the natural environment and built heritage', 'To improve the quality of life of people in Northern Ireland in ways which are sustainable and which contribute to creating a better environment', and 'To support a system of local government which meets the needs of residents and ratepayers' (www.doeni.gov.uk). The 'mission statement' of the Department in the RoI similarly commits it 'To promote and improve the quality of life through protection of the environment and heritage, infrastructure provision, balanced regional development and good local government' (www. environ.ie). In the Republic, the Department has more explicit responsibilities to contribute to economic growth: it aggregates within itself the two elements that are generally thought to contribute to 'sustainable development' – nature conservation and resource development for economic growth. However it is notable that in each Department, the leading role in environmental management within their own organizational structures is given to scientific management: in NI, the staff of the Environment and Heritage Service offer 'many different scientific and professional skills and expertise'. In the Republic, the NPWS is also primarily a scientific body, charged with the designation, on scientific bases, of areas of special conservation concern and also with their conservation management.

The discovery that rural Europe both faces and is producing environmental risks can be attributed primarily to natural scientists. Latour argues that every scientific 'discovery' brings new actors into the social world, for whom other actors must 'make room' if the knowledge is to be socially institutionalized. Science 'renegotiates what the world is made up of, who is acting in it, who matters, and who wants what' (Latour 1988, p. 40). Other social actors must reorganize their world to incorporate the new 'actors' made visible by scientific work – landscapes, rivers and lakes, soil micro-organisms, insects, plants, animals and habitats, and their 'natural' processes of establishment and decay. Reconstituting social reality is a momentous task, likely to create strong resistance from many social groups, and new 'scientific facts' cannot achieve it on their own. 'An idea or practice cannot move from A to B solely by the force that A gives it; B must seize it and *move* it' (ibid., p. 15 – original italics). Scientific knowledge is not diffused to passive recipients; it has to be seized and moved by actors who can see interests for themselves in mobilizing the new knowledge. 'Seizing' and 'moving' knowledge inevitably involves re-working it; every 'translation' of knowledge produces a 'drift, betrayal, ambiguity' or 'diversion' of knowledge. Latour (1988, p. 253) argues that translation has a strategic intent: 'It defines a stronghold established in such a way that, whatever people do and wherever they go, they have to pass through the contender's position and help him to further his interests'. By following the translation process we can identify networks of knowledge actors; and it is through these networks that scientific knowledge is able to act on the world. The successful establishment of new scientific knowledge requires a process of 'cognitive convergence' (Lahsen 2004) between scientists and other social actors.

The activities of the NPWS (formerly Dúchas) concentrate primarily on identifying and designating sites of interest to natural scientists across the country. The Irish Wildlife Act of 1976 licensed it to engage in site designation and conservation, and it has subsequently been given the responsibility for implementing EU Bird and Habitat Directives and most recently Natura 2000, which requires all member states to identify and protect ecologically important habitats, species and sites within their territory. The NPWS has carved out a space of considerable autonomy for itself within its parent department, manned by a large group of scientific experts who develop and maintain scientific information on ecological conditions; it controls a number of regional environmental managers (such as the managers of the National Parks) and on-the-ground 'environmental police', in the Park and Wildlife Rangers Service. It has management responsibilities for 6 National Parks, 77 Nature Reserves, 7 'Refuges for flora and fauna' and 68 Wildfowl Sanctuaries in addition to the hundreds of National Heritage Areas, Special Areas of Conservation and Special Protection Areas which it has designated.

The activities in which the NPWS is engaged are potentially (and sometimes actually) controversial and conflictual. Its information and advice can be very unwelcome to those within its own or other Government departments who are promoting new roads, approving sites for quarries and rural factories, or facilitating businesses to exploit natural resources. The Department of Agriculture's project of intensifying and industrializing food production stands in direct challenge to the NPWS's interest in freezing habitats and sites, often on farmed land, against further productive use. Within this context, the NPWS emphasizes its scientific knowledge and credentials, which allow it to position itself as outside political negotiation and bargaining. Its interventions into rural site conservation are justified strictly on a scientific basis: farmers or other landholders who are unhappy at discovering their land has been designated for conservation may appeal the designation, but only if they can produce *scientific* evidence to challenge the designation. NPWS-designated sites are not publicized or advertised, locally or nationally; they are not seen as locations open to the public or resources for public education about nature, but as sites 'owned' by science and properly given over to scientific research. NPWS operations are largely non-transparent; they are accountable for their management of ecological sites not to the general public but to other scientific experts.

The networks of knowledge circulation in which the NPWS are embedded are primarily networks of scientists, or of institutions with a similar self-understanding as scientific or research-based. NPWS scientists co-operate with scientists from the NI Environment and Heritage Service, with scientists in other state departments and agencies in the Republic, and with academic researchers, often those who have trained NPWS scientists or who work in the university department from which they graduated; the National Platform for Biodiversity Research, a grouping of state- and university-employed researchers, is co-sponsored by the NPWS and the Environmental Protection Agency. Through such networks, state scientists have access also to transnational institutions and expertises, such as the European Platform for Biodiversity Research, the European Science Foundation, and a series of global environmental conventions. The NPWS also exchanges knowledge with some environmental non-governmental organizations (NGOs) – those that, such as Birdwatch Ireland, employ scientific researchers on their staff or are receptive to scientific direction of their activities.

The NPWS orientation to environmental management is well captured in a 1997 report on *Principles for Sustainable Development*, produced by Comhar, the agency within the Department which communicates state thinking about the environment to the Irish public:

> Ecological systems are the basis and preconditions of all life. The intrinsic value of diversity of species and habitats should be recognized. Maintenance of biodiversity is the prerequisite for the continuation of all living systems. Loss of biodiversity at global level is a serious problem... Research shows that ... ecological processes operate much more efficiently in species-rich communities but there are many gaps in our knowledge.
>
> In addition to the intrinsic value of a diversity of species and habitats, biological communities have other significant attributes such as protection

of water supplies, providing us with food, plants and sources of novel drugs and horticultural species. From an economic perspective, gene-based science provides opportunities for the development of new crops, drugs and raw materials. Biodiversity and the appreciation of nature in diverse forms are central to the quality of life of humankind: species rich habitats and landscapes are enormously important aesthetic and amenity resources.

*Relevance to Ireland:* Ireland has a rich diversity in habitats and species. To halt the loss experienced and maintain that diversity involves taking action to ensure that a sufficient range and number of sites and species are designated for protection from unsustainable development activity (that is any activity which would undermine the conservation of habitat and species). The greatest threats to biodiversity in Ireland are habitat loss, pollution and introduced species. The absence of adequate data for all plant and animal groups is also a serious problem. ... We need to eliminate all sources of pollution to land, sea and air that undermine the carrying capacity of living systems and ensure that nutrient and pollution loads in watercourses do not impair biological diversity. There is a need to accelerate the process of transparent sustainable management of designated sites and species and to require all development to be consistent with planning guidelines; this would include strictly regulating and controlling drainage and extraction activity to prevent damage to bogs, fens, turloughs and other wetlands as well as coastal habitats. In addition guidelines should be developed for key professions, and ecological education introduced into all types of education and training (Comhar 1997, p. 15).

This is a discourse clearly addressed to fellow scientists, and secondarily to science-funding authorities who might help scientists to make up the gaps in their knowledge; it is addressed to the general public only insofar as they are prepared to act as 'novice scientists'.

The network of scientific knowledge actors is based in domains of power, decision-making and control over nature; scientific knowledge is owned and guarded by an 'epistemic community' of experts, for whom 'the public' are insufficiently educated to participate in decision-making. On this account, we might conclude that the environmental management regime in RoI is profoundly undemocratic. But it may also be largely ineffective: Latour's analysis directs us to ask whether ecological science is able to 'move the world', and achieve the power over the Irish rural environment which would make all other actors 'pass through its stronghold'. There are other actors and networks that also have an interest in environmental management and who can translate, 'divert' or 'betray' it to fit their own circumstances.

## Environmental management as heritage conservation

The Heritage Council, like the NPWS, operates with a scientific understanding of nature conservation, but in its storyline 'the environment' is translated as 'heritage'.

The Council was established as a semi-state agency under the 1995 Heritage Act, to be an independent policy advisor to the Minister for the Environment and his Department; however it is largely funded by the Department of Environment and responsibility for the administration of its funds lies with the NPWS. The Council's brief covers all matters to do with built and natural heritage (biodiversity, wildlife habitats, inland waterways and wetlands, architecture, archaeology and geological features), and it is charged with co-ordinating state, semi-state and NGO actions towards these. It is largely staffed by ecological scientists, but its network also includes many non-scientific actors: politicians (national and local), country councils, LEADER committees, and local or voluntary groups.

Again like the NPWS, Council staff find the task of advising central state departments (including its own Department of the Environment) often problematic. Developing networks of relations above and below the national level can help to manage relations with national Government. This includes interactions with sympathetic experts within the EU Commission, and relations with local government, NGOs, and local groups. Since 2002 the Council have been co-funding Heritage Officer posts with the county councils; now found in 28 local authorities, these advise on planning decisions, draw up Local Heritage Plans, and give educational talks to schoolchildren and other groups within the local authority's area. Under the National Biodiversity Plan (Department of Environment 2002), local authorities are required to draw up a Local Biodiversity Plan for their area, and this has also involved co-operation with Heritage Council staff. The Council also invites applications from local groups to apply for heritage conservation funding, assesses the applications, and assigns staff members to work with applicant groups.

The Heritage Council is an institution which has a clear interest in 'seizing' and 'moving' expert ecological knowledge in support of environmental management. And as Latour (1988) suggests, interesting translations of that knowledge follow. The translation of 'the environment' into 'heritage' opens it up to claims of ownership from the population as a whole; 'heritage' belongs to 'the nation', not to a scientific elite. Council staff express a conservation philosophy that is at odds with that of the NPWS and its scientific networks: heritage should be managed at the lowest level possible. 'The only future for nature conservation and for biodiversity is getting the local landowners and local groups actually involved in it' (interview with staff member, March 2005[1]). The resources which would be needed (of both finance and expert knowledge) are not available to put a top-down approach into practice, and in any case such an approach would bypass a key resource which is available – local knowledge and interest in nature: 'There's a lot of expertise, and there's a huge amount of enthusiasm and goodwill, and

---

1   This interview, and other research material used in this chapter, was collected as part of the CORASON (A Cognitive Approach to Rural Sustainable Development) cross-national research project which is funded by the EU 6th Framework research programme). My thanks to Petra Aigner for her assistance in collecting the data.

I feel that biodiversity should be focusing on tapping into that goodwill, and empowering that and facilitating that... we should be building on that kind of, on that enthusiasm' (ibid.).

If environmental management should include lay actors and expertises, what is the place of scientific expertise? The staff member quoted above 'moved' this issue by distinguishing between what he called 'hard' and 'soft' nature conservation. 'Hard nature conservation... is your designated sites', which must be established and managed scientifically, through the sort of work done by the NPWS; and 'the soft side, which is I think the biodiversity side' is 'all about empowering people and tapping into what's out there and making it into a feel good factor'. For example, Council staff undertook a habitat mapping project that they could have done by using remote-sensing equipment. Instead, they carried out a farm-gate survey: 'We insisted that people go and find the landowners, explain what you're doing, and ask them to walk the land with you, and it's amazing the feedback we got'. He saw farmers, not as 'malevolent' to biodiversity, but as people who have been educated out of appreciating nature through 50 years of policies for agricultural modernization and technological intensification. The strategy is to 'link them back in', so conserving rural nature must go along with conserving the farm population.

Heritage Council staff are thus 'diverting' environmental management in the direction of greater democracy, greater social concern, and a wider enrolment of relevant knowledges. However the Council's continued existence and funding depends on the recognition that it is an expert body capable of giving expert, uncontested advice to Government. This requires that lay knowledge can never be granted equal status with scientific expertise: 'soft' conservation projects are inevitably assessed and measured by the criteria of 'hard', scientific ecology. Applications from local conservation groups for funding are turned down because they are not sufficiently scientific in their approach. In one such case, 'They were hugely committed and they have great energies to achieve what they did, but I felt that they probably didn't see the potential for the site...If they wished to attract wildfowl, for example, you do need quite specific habitat requirements, I mean with a wet grasslands, you do need to know things like water level, what kind of vegetation, does it need to be heavily grazed in winter... It's the whole knowledge bit, the whole scientific thing' (ibid). Heritage conservation institutions appear to rediscover the value of 'local' knowledges, but then use them to reconfirm and reassert the superior value of scientific expertise (Martello and Jasanoff 2004).

*Environmental management as sustainable development*

A third set of actors translate their activities as 'sustainable development' rather than nature conservation. County councils and LEADER committees, for example, 'seize' the ideas of ecological scientists and translate them so that they construct a 'stronghold' for their own institution, extending its networks and enrolling local and voluntary groups. The storyline about sustainable development formulated by one local authority is 'to realize the economic, social and cultural potential of the county in a manner that will not undermine such aims for future generations' (Tipperary South Riding County Council Development Plan 2003, p. 5). Environmental management is incorporated into 'economic, social and cultural' matters: here the local authority shares the Heritage Council's 'diversion' of nature as heritage, but with a greater recognition of its economic and livelihood implications.

Since 2002 it has been state policy in the Republic of Ireland that local authorities must institute 'participatory' procedures for governance. These are being institutionalized in the form of Community Development Forums to which local groups elect representatives to discuss a range of issues – from economic development to land use planning to environmental concerns – and make recommendations to the County Council concerned. Through this structure the local authority enrols voluntary and community groups into the practice of sustainable development. Management of nature thus becomes an integral part of management of local civil society; environmental management is valued as much for its capacity to develop 'community' as to 'conserve' nature. LEADER programmes in Tipperary display similar 'moves'. For example, they have pioneered a project called 'The Golden Mile', in which rural groups compete for a prize for having maintained the best mile of roadway in the county during the previous year. The 'best' road is one which has been cleaned up, litter and weeds (brambles, nettles) removed, the verges maintained to allow movement of wildlife, and the hedges replanted, as necessary, with native tree species.

The criteria for judging the roads were drawn up by a group of people from farming organizations, the community and voluntary sector, tourism interests, and a leading national environmental NGO; the latter were not seen as holders of the only relevant knowledge, because it was equally important to enrol experts in rural development, for example in rural tourism promotion. Tipperary LEADER understands such competitions as 'both community and environmental projects… to encourage communities to become more aware of the rural environment' (interview with LEADER manager, April 2005). The Golden Mile competition is not just a way of raising ecological awareness, however, or of mobilizing community collectivity; it is also a way of bringing more and more people into LEADER's own networks and passing them through its stronghold: 'What is important is getting as many people as is possible involved in what's happening, because if you don't have people behind you, you kind of have nothing really' (ibid.).

While the network of 'scientific managers' of the rural environment occupies the domains of power and symbolic capital, it appears to be the translation into 'sustainable development' that is most successful in 'moving the world'. This draws a wide range of social, political, economic and scientific actors into its network, and recognizes that environmental management policies, if they are to be successful, require the engagement of those affected by them. It may not be the best translation, however, for the protection of nature, particularly where 'development' is given more emphasis than 'sustainability'. As a policy discourse, sustainable development seems to achieve rather little 'reorganization of their world' to incorporate new 'natural' actors by those who endorse it, which probably goes a long way in explaining its broad political and popular appeal. The contradiction between economic growth and the protection of nature has not gone away, even if we now have a discourse which says that it has.

### 'Cognitive justice' in environmental regulation

'Sustainable development' is not only a site of struggle over the relative importance of 'development' versus 'sustainability'; it is also a site of contestation over the issue of *environmental knowledge* and its use in environmental management. As the concept is set out in the Brundtland Report *Our Common Future* (World Commission on Environment and Development [WCED] 1987), achieving sustainable development requires civil actors to subordinate themselves to scientific authority, both in regard to the identification of environmental risks and problems and in regard to their solution. If we ask whose knowledge can tell us what is a 'sustainable' use of resources, the answer given by Brundtland was unequivocally 'science' (Irwin 2001). However in the decade after Rio sustainable development came to be seen as something that requires 'local action' (Local Agenda 21), public participation, and the inclusion of non-state and non-scientific actors in the decision-making process, including all the stages of identifying, interpreting and acting on environmental problems. The focus of concern began to shift, from the condition of nature bequeathed to following generations, to the types of social institutions which we pass on (Redclift 1997). The institution of science itself has come under particular scrutiny, as recognition grows of the need to find 'means by which to diversify and localize environmental science, including greater local determination by people not currently represented in science' (Forsyth 2003, p. 22). An environmental management agenda which is unable to recognize and incorporate a diverse range of knowledges of nature is likely today to be judged both ineffective and undemocratic, and hence itself 'unsustainable'.

Martello and Jasanoff (2004) argue that sustainable development requires a shift from 'environmental management' to 'environmental governance' – a form of regulation in which the exercise of power is oriented towards openness, democratic participation and accountability, as well as effectiveness and coherence. Again this brings the issue of environmental knowledge into particular focus. At a global level, environmental decision-making is increasingly in the hands of bodies that do not conform to normal democratic requirements of representation and accountability, and (given the current underdevelopment of a global civil society) are not routinely held accountable for their decisions by a mobilized public. They increase reliance on scientific knowledge and authority alone, and routinely reproduce the belief that scientific knowledge is superior to all other forms. However the institutionalization of this global environmental regime has 'paradoxically' led to a rediscovery of local knowledge and its significance for environmental governance. Global environmental regulators increasingly confront the fact that environmental solutions are, in practice, applied to specific localities – which challenges generalized expectations about, for example, how climate change is impacting on the world – and equally that global environmental programmes cannot be implemented without the participation of local actors. Even the World Bank, they note, has begun to argue that local, indigenous and traditional knowledges should be enlisted within programmes for sustainable development; science can no longer be regarded as the only cognitive resource useful for managing ecological problems.

But making room for local knowledges within an environmental governance regime is not easily achieved (Leach, Scoones and Wynne 2005). 'Unconventional forms of expertise cannot be accommodated within global environmental regimes without renegotiating basic rules of decision-making' (Martello and Jasanoff 2004, p. 12); significant 'procedural innovations' are required, not just in political processes but also in science itself. Expert and non-expert knowledges must be enabled to interact with one another; 'constant translation back and forth across relatively well-articulated... knowledge-power formations' (ibid., p. 5) is needed. In practice, despite the various experiments in 'public participation' that have begun to accrue around environmental governance (particularly within the EU), regulating institutions still operate in such a way as to 'invoke and thus reinforce' a boundary between science and other forms of knowledge; still constitute scientific knowledge as 'universal knowledge' and relegate non-universal knowledges to the inferior category of 'local'. Thus the 'paradoxical' rediscovery of the value of local knowledge accompanies a recreation of the symbolic domination of science as the legitimate way to know and address environmental risks and crises.

Problems of cognitive justice arise at a number of points in environmental management in the Republic of Ireland. The scientific management regime of the NPWS appears to be an example of 'orthodox' or 'unreconstructed science' (Forsyth 2003), which displays little reflexive interest in either the social bases of its own assumptions or the social effects of their implementation. The approach of the Heritage Council, while moving towards a participatory form, fits well with Martello and Jasanoff's (2004) description of how 'global' institutions rediscover 'local' knowledges and then use them to reassert and reconfirm the superior value of scientific expertise. Local groups in return may refuse to pass through the stronghold of 'bottom-up but science-based nature conservation' which the Council is seeking to establish. Lay approaches to conservation do not seek to 'implement specific management regimes', because they do not have a purely scientific goal in sight: reconstructing local wetlands to attract migratory birds, for instance, is inextricable from a set of other goals such as providing leisure facilities for the population of the adjacent town, increasing tourism, enrolling the interest of members of the local gun club, and so on.

'Lay knowledge', being place-based, resists standardization into a set of precepts abstracted from the particularities of the local site that can then be applied universally to wetland habitats designed to attract certain sorts of birds. But the 'sustainable development' translation, which appears at first sight to offer more scope for the development of environmental governance and the enrolment of local actors, also appears to be implemented in ways which prioritise scientific, professional and managerial over local and lay forms of knowledge. The 'constant translation' between expert and lay knowledges which Martello and Jasanoff (2004) identify as a necessary part of 'just' governance does not occur. While welcoming local participation, and even defining it as participation in decision-making, actors using the sustainable development storyline do not appear to have taken hold of the possibility that institutionalising 'lay' voices should have any impact on the status of 'expert' knowledge. Members of the new Community Forums are offered assistance to grasp and familiarise themselves with the discourses of experts, but few if any provisions are made the other way round.

To live in the countryside, and particularly to make a livelihood from rural resources, requires practices using nature and hence the development of knowledges about nature. Rural civil society is both shaped by and mobilized around practices and knowledges about nature which blend 'lay' and 'expert' forms, often in unpredictable but effective ways. An environmental regulatory regime which devalues local and lay knowledges makes rural environmental 'governance' almost impossible to achieve.

## References

Allen, R. (2004) *No Global (The People of Ireland Versus the Multinationals.* London: Pluto Press.

Comhar (1997) *Principles for Sustainable Development.* Dublin: Government Publications Office.

Department of Environment Heritage and Local Government (2002) *National Biodiversity Plan (Ireland).*

Forsyth, T. (2003) *Critical Political Ecology (The Politics of Environmental Science).* London: Routledge.

Hajer, M. (1995) *The Politics of Environmental Discourse.* Oxford: Clarendon Press.

Irwin, A. (2001) *Sociology and the Environment (A Critical Introduction to Society, Nature, and Knowledge).* Cambridge: Polity Press.

Jasanoff, S. and M. L. Martello (eds) (2004) *Earthly Politics (Local and Global in Environmental Governance).* Cambridge, Mass and London: MIT Press.

Kousis, M. (1999) 'Sustaining Local Environmental Mobilisations: Groups, Actions and Claims in Southern Europe', in Chris Rootes (ed.), *Environmental Movements: Local National and Global.* London: Frank Cass, pp. 172–98.

Lahsen, M. (2004) 'Transnational locals: Brazilian experiences of the climate regime', in Jasanoff and Martello (eds): *op. cit.,* pp. 151–72.

Latour, B. (1987) *Science in Action: How to Follow Scientists and Engineers Through Society.* Milton Keynes: Open University Press.

Latour, B. (1988) *The Pasteurisation of France*. Translated by Alan Sheridan and John Law. Cambridge, Mass: Harvard University Press.

Leach, M., I. Scoones, and B. Wynne (eds) (2005) *Science and Citizens (Globalisation and the Challenge of Engagement)*. London: Zed Books.

Martello, M. L. and S. Jasanoff (2004) 'Introduction: Globalisation and Environmental Governance', in Jasanoff and Martello (eds), *op. cit.*, pp. 1–30.

Redclift, M. (1997) 'Frontiers of Consumption – Sustainable Rural Economies and Societies in the Next Century?', in Henk de Haan et al. (eds), *Sustainable Rural Development*. Aldershot: Ashgate, pp. 35–50.

Slater, E. (1993) 'Contested Terrain: Differing Interpretations of Co. Wicklow's Landscape', *Irish Journal of Sociology*, 3: 23–55.

Taylor, G. (2001) *Conserving the Emerald Tiger*. Galway: Arlen House.

*Tipperary South Riding County Development Plan 2003* http://www.southtippco co.ie.

Tovey, H. (1993) 'Environmentalism in Ireland: Two Versions of Development and Modernity', *International Sociology*, 8, 4 (December): 413–30.

Tovey, H. (2005) *The Environmental Movement in Ireland and the Making of Environmental Activists*. Report delivered to the EPA from the research programme on Environmental Attitudes, Values and Behaviour in Ireland.

Willis, S. and H. Campbell (2004) 'The Chestnut Economy: The Praxis of Neopeasantry in Rural France', *Sociologia Ruralis*, 44 (3): 317–31.

World Commission on Environment and Development (1987) *Our Common Future* (Brundtland Report). Oxford: Oxford University Press.

# PART II
# Primary Production
# and Sustainability

Chapter 7

# Agriculture and Multifunctionality in Ireland

John Feehan and Deirdre O'Connor

## Introduction

In any of its multifarious definitions farming is always about making a living from the land. In earlier and simpler economies the emphasis is on direct utilization of the resources immediately available in order to provide for the needs of family and community. With the evolution of greater complexity in societies specialization and geographical integration develop apace, making the community and the greater social construct of which it is now part more vulnerable to the consequences of change in one or more of the variables necessary to the maintenance of the broader culture. Even without such evolution of greater social complexity, even the most isolated agricultural community cannot be isolated entirely. Almost inevitably there is a slow influx and outflow or know-how and resources – of improved cultivars or techniques developed in nearby communities: and of people. There are very few Easter Islands.

Farm families in Ireland operate in a complex and rapidly-changing market and policy environment. These developments have evoked a wide-ranging set of responses at farm household level as people attempt to construct their livelihoods against such a backdrop. The changing agricultural and rural policy landscape is clearly a key factor in these decisions. Of specific relevance is the emergence of concepts such as the Living Countryside and the European Model of Agriculture which are underpinned by the notion of a multifunctional agriculture, with its broadened set of functions for the farming sector and the wider society. This chapter provides an overview of the nature and extent of the adaptive practices undertaken by Irish farm households and analyses the policy context which shapes them. From a starting point which explores how and why the need for such adaptive practices arose, the chapter then explores the emergence of the 'multifunctionality' policy context within which these activities take place. This is followed by an exploration of the extent to which multifunctional-type activities have developed in Ireland in recent years. It concludes with some remarks concerning the challenges posed by a multifunctional approach to agriculture in terms of the development of appropriate policy and institutional support mechanisms.

**Charting the decline of on-farm sustainability**

In the Irish situation our location had a profound effect on the nature of self-sufficiency. We are an island on the edge of Europe, distanced from innovation originating in mainland Europe in both time and space: remote but not removed, and all these innovations did eventually find their way here, if often in an attenuated form of an echo or a trickle, or modified and adapted to suit our more Atlantic conditions.

In one form or the other, farming has always involved *agri-cultura*: cultivation of the soil and/or management of a sward to produce food or other necessities or luxuries, either for local consumption or export into the wider market economy. What those products might be was constrained in the first place by geography, climate etc.; and constrained secondarily by the market – what the market wanted and would pay for; and thirdly by competition. In most parts of the world a sophisticated agronomy – often developed over millennia – enabled communities to make the maximum use of their resources within the limits imposed. Until recently the challenge of maintaining fertility was a major constraint (Feehan 2003). The arrival of cheap imported fertiliser, organic at first and later (with the development of industrial chemistry) synthetic, broke through this restraining bottleneck, paving the way towards an apparently unlimited horizon of productivity, but at a price.

The challenge to so well understand the possibilities and constraints of local soils and landscapes as to be able to maximize their potential to produce in a way and at a level that does not compromise the future – sustainably, that is – and which does not compromise other values and functions of land (and it is of course often the case that these are not always foreseen, and at times cannot be foreseen, either because the knowledge is not yet there to enable understanding, or because they develop at a later stage of social awareness) is lessened where particular key aspects of the need to understand and manage on the basis of that understanding are short circuited. A good example is the steadily increasing availability of imported nutrients in post-Famine Ireland: organic nutrients mined from the sub-fossil guano deposits of South America initially, later supplemented and in time replaced by inorganic fertiliser ('artificials') supplied by the nascent chemical industry, especially when the steep decline in demand for their use after the First World War made a concerted move on agriculture necessary to maintain profit.

The increasing efficiency and scale of global transport in the 19th century made this substitution possible in the first place. In fact, the scarcity of manure had always been a big problem in Irish farming, but the possibility of a sustainable solution was provided in principle by the development of alternate husbandry during the agricultural revolution. That solution was demonstrated in practice by the best of Victorian farming, reaching its supreme achievement perhaps in the systems of Robert Elliot at Clifton Park and his work on laying down land to grass (Elliot 1908). Its impact in the Irish situation was limited for two reasons. One was the inadequate resourcing available to the majority of small farms, and the other the insufficiency of training and the lack of a tradition in farm management.

Fertiliser use remained limited however, because it was still an expense not easily incurred on most small farms, which consequently retained a significant level of nutrient sustainability as well as self-sufficiency until the middle of the 20th century. This changed in precipitous fashion after accession to the EEC in 1973. Farmers welcomed the change almost unreservedly because it held out the prospect of greater prosperity and less labour. The cost of this greatly increased fertility and productivity was not however simply the price of fertilizer and other external inputs at the farm gate – modest in the beginning but escalating steeply with the ending of the era of cheap oil. It also did away with basic self-sufficiency and the need for a more nuanced understanding of what it required to farm well.

## Moving towards multifunctionality

The ability to support one's enterprise from local resources – always a challenge, always dependent on intelligent management learned over ages – has been lost. The inability to support a family today by 'traditional' agriculture means that for the majority of farm families, if they want to stay on the land, alternative land-utilizing enterprises need to be found, or employment found off-farm by one or more members of the farm household. Such *pluriactivity* has now become the norm for a broad sector of the rural community. It depends to a considerable extent on an increasingly sophisticated awareness of the *multifunctional* nature of the countryside, and it will be useful to trace the evolution of the latter concept before turning to an examination of the practices and policies around which farm households in Ireland construct their livelihoods today.

Because the pattern of farming that has evolved over the past half century no longer provides a sufficient income for many small farming families, they must look in other directions to make up the shortfall. There are two overall directions where solutions are sought: *re-evaluation of the possibilities* presented by the farm itself and its resources to generate income through the identification and development of alternative enterprise; and *employment off the farm* by one or more members of the farm household. Van der Ploeg, Long and Banks (2002) conceptualize this development based on the distinguishing of *broadening, deepening* and *regrounding* activities. *Broadening* activities refer to the diversification of the 'products' of the farm, taking advantage of new market opportunities in areas such as tourism, heritage and landscape management. *Deepening* activities are those which add value to farm products via different forms of production or alternative supply chains – such as organic production or farmhouse food production. *Re-grounding* activities refer to the reorganization of household assets such as labour and capital through engagement in off-farm employment or cost-reduction strategies on-farm.

The concept of multifunctional agriculture has gained prominence in the recent past as the basis underpinning the Agenda 2000 proposals for CAP reform and the subsequent shift to the single farm payment instrument under the Mid-Term

Review agreement of 2003. However, earlier echoes can be found in the Green Paper of the European Commission on the Future of Rural Society (1988) and in EU Directive 286/75 on Mountain and Hill Farming in Less Favoured Areas (LFAs), which argued the case for subsidizing agricultural production in areas where it could not be competitive but nevertheless fulfilled important economic, social and environmental functions (Buller 2003). In Ireland, as in many other EU member states, these payments came to be regarded as central for maintaining viability in sparsely populated areas of the country (Dunne and O'Connell 2003). In the wider international arena, the concept of a broadened set of roles and functions for agriculture appeared in the Brundtland Report (1987) and was carried forward into the Rio Convention of 1992. It has been the subject of lengthy consideration from the OECD (OECD 2001) and more recently has emerged as a bone of contention within the context of the WTO negotiations on agriculture. This debate has been ongoing between the 'friends of multifunctionality', who regard it as a key vehicle for safeguarding the multiple functions of agriculture, and its opponents who regard it as disguised protectionism (Burrell 2001; Thomson 2001; Mahe 2001; Potter and Burney 2002; Losch 2004). Arguing that multifunctionality within the EU context emerged from the debate over agri-environmental policy, Buller (2003) claims that 'green coupling' is an integral component of the European model of agriculture in which the rural landscape, biodiversity and countryside access cannot be considered as public goods independent of the process and practices of agricultural production.

Marsden and Sonnino (2005) propose three main interpretations of multifunctionality that correspond to the different agricultural paradigms which have shaped European agricultural policy in recent decades. Within the context of the *agro-industrial paradigm*, where agriculture is characterized by economies of scale, concentration and specialisation of production, multifunctionality can be interpreted as a palliative to the productivist cost-price squeeze which emphasizes the role of pluriactivity, often viewed as an 'unwanted' economic adaptation strategy, enabling less competitive producers to survive in an increasingly hostile market environment. Viewed through the lens of a *post-productivist* paradigm, multifunctional agriculture can be seen as the spatial regulation of the consumption countryside, where nature is conceived chiefly in terms of landscape value as a consumption good. This paradigm also underpins the depiction of multifunctionality as a set of social demands on agriculture and the expectations or requirements of the society of which it is part (MULTAGRI 2005).

The *sustainable rural development paradigm* reasserts the socio-environmental role of agriculture in sustaining rural economies and cultures, re-emphasizes food production and agro-ecology and, according to Knickel and Renting (2000), repositions farmers as the 'centre of gravity' in the rural development process. Buller (2003) argues that multifunctionality can be seen as a 'proactive form' of farm-based integrated rural development with the emphasis primarily upon agriculture and agriculture enterprises. According to Marsden and Sonnino (2005), it is within this context that multifunctionality finds its most comprehensive expression, where it emerges as a vehicle for engagement in a variety of activities and functions for the farm and for the wider society. Such contributions include, *inter alia*, its role in food supply chains; its ability and potential to fulfil new societal goals; its contribution to rural employment and its role in the maintenance of rural population in less favoured areas (MULTAGRI 2005).

## The current context for multifunctionality

A recurring theme in agricultural and rural policy analysis is the fundamental contradiction which runs through it, which Buller (2003) ascribes to the 'curious blend' of post-war interventionism and late-20th century liberalism that characterizes the CAP as a policy mechanism. van Huylenbroeck and Durand (2003) note the ambivalent stance of European agricultural and rural policy which exhorts farmers to meet the growing demands from society for 'non-productive' functions of agriculture while simultaneously becoming competitive in an increasingly liberalized and globalized market. Marsden and Sonnino (2005) refer to the 'bifurcated' nature of policy recommendations for the future of the UK agri-food system, which urges re-engagement with the conventional agri-food system, while at the same time advocating the pursuit of opportunities to develop the alternative, speciality and niche food sectors. In the Republic of Ireland (RoI) context, the most recent policy statement on the future of the Irish agri-food sector shows evidence of a similarly contradictory approach. It suggests a set of principles which favour the 'market-driven' development of the sector, while simultaneously arguing that the production of certain public goods (rural landscape, culture and heritage and biodiversity), intrinsically associated with agricultural production, constitutes an important rationale for the state's continuing role in agriculture (Agri-Vision 2015 Committee 2004). This policy statement is notable for the explicit use of the concept of multifunctionality as a justification for on-going intervention in the sector. In a similar vein, policy statements from Northern Ireland (NI) emphasize the importance of subsidizing farmers both for the purposes of supplying consumers with high quality food and as a mechanism for ensuring the provision of a 'living countryside' (Department of Agriculture and Rural Development 2005). Furthermore, as Commins (2005) notes, the drivers of change in the rural economy – typically characterized by values that espouse the free market (downward pressure on farmgate prices and the impetus to increase scale in production and processing) – are also being challenged by counter-trends in which other concerns are paramount. These other concerns embrace physical environmental quality, social and environmental sustainability, safety, authenticity and traceability in food as well as shorter food supply chains.

Gorman (2004) argues that while traditionally rural development initiatives in Ireland have been characterized as 'marginal activities for marginal people', this perspective is changing as more and more farm families negotiate a range of complex adjustment strategies as a means of constructing viable livelihoods. This is borne out by O' Connor et al. (2006) who note the increasing prevalence of farm-based rural development initiatives among more 'commercial' operators in the RoI in recent years. A similar trend is evident in NI whereby a higher proportion of 'larger' farms are involved in diversification activities compared to the regional average (Departmernt of Agriculture and Rural Development 2006).

Within the context of a larger project, an attempt has been made previously to analyse the socio-economic impact of rural development policies in the RoI, using 1998 as a base year. This project, known as IMPACT, made a comparative analysis of the existence and potential of rural development activities in seven European countries using the broadening, deepening and regrounding framework discussed earlier.[1] In the RoI the most important and common strategies employed by farm households were those of regrounding, in the take-up of off-farm employment or by cost-reduction strategies such as reduced expenditure on external inputs/ hired labour and reduced borrowings for investment. Broadening activities were the second most important pattern of rural development in Ireland. Within this category participation in nature and landscape management schemes was the most widely adopted while agri-tourism also emerged as a significant activity. Deepening activities were the least important in terms of the numbers of households involved and were lower for the RoI than for many of its European counterparts involved in the IMPACT study (O' Connor et al. 2006). The most recent Census of Agriculture (2001) conducted for the RoI showed that approximately 5 per cent of farm households were engaged in diversification activities, the most important of which were forestry and farm tourism (Central Statistics Office 2004). In Northern Ireland, the latest estimates show that almost 9 per cent of farm households have diversified activities (Department of Agriculture and Rural Development 2007). However, the most common among these, namely agricultural contracting/haulage, is not classified as a diversification activity in many definitions on the basis that it is part of the agricultural industry with payments made from one farmer to another.[2] If this category is excluded, than approximately 6 per cent of Northern Ireland farms are diversified with the most common activities being farm tourism and farmhouse food production/sales.

### The specificities of multifunctionality

Consequently, it appears there are a number of specific rural development activities which are important dimensions of multifunctionality in Ireland. These include forestry, agritourism, environmental quality, off-farm activity and artisanal food production. These elements are considered in more detail below.

---

1  FAIR-CT-4288 'The Socio-Economic Impact of Rural Development Policies: Realities and Potential'. The seven countries studied were the Netherlands, Italy, Spain, Germany, the United Kingdom, Ireland and France.

2  See the discussion on diversification in http://statistics.defra.gov.uk/esg/reports/ divagri.pdf.

*Forestry as a dimension of multifunctionality*

Forestry accounts for just over 9 per cent of the land area in the RoI significantly below the EU average for forestry cover of 30 per cent (Agri-Vision 2015 Committee 2004). The Government's National Forestry Strategy was published in 1996 and set a target of planting 20,000 ha per annum until 2030 with a view to doubling forestry cover to approximately 17 per cent of the land area. However, a subsequent review of that Strategy concluded that many of its objectives were not being achieved and suggested a range of alternative measures for the future development of the sector (Bacon and Associates 2004). This is the latest phase in Ireland's campaign to reduce our dependence on imported timber (especially in the RoI) and in recent years there has been a concerted effort to increase the proportion of hardwoods. Until a few decades ago almost all planting was Sitka spruce and lodgepole pine. On the one hand this may be seen as an attempt to restore to the country the percentage forest cover of an earlier time. But it stops short of the recovery of an earlier tradition in woodcraft in which the management of trees was seen as an integral part of farming itself (Feehan 2005a).

Within the context of the most recent round of CAP reform, the provision that allows farmers, in certain circumstances, to plant forest while still receiving their full single farm payment (SFP) will be an important factor in the future development of this sector. Several studies, including those by Convery and Roberts (2000), Behan and McQuinn (2002) and FAPRI-Ireland (2003), note the potential positive impact from a combination of more extensive agricultural production and increased forest planting in enabling Ireland to meets its commitments under the Kyoto protocol.

In Northern Ireland the area under plantation forest is 6 per cent, much lower than in the RoI and only half the area in Great Britain. The Government's new Forest Strategy seeks to double the area of forest over the long term, and in the shorter term secure a modest increase in combined public and private forest by 1500ha by 2008 (at a rate of 500ha a year) (Forest Service of NI 2006). As in the RoI, this will be achieved through the further conversion of farmland to forest, the plantation of bogland being no longer considered acceptable on environmental grounds.

*Agritourism as a dimension of multifunctionality*

A key element of the broadened set of functions associated with multifunctional agriculture is recreational access. The term 'recreation' is used here in a broader, more literal sense to articulate the importance of time spent in the countryside for human well-being (Feehan 2005b). It is among the deep psychological roots which this sort of discussion attempts to explore that the ultimate explanation is to be found for why farming people will attempt to 'stay on the land' at any cost, and in order to do so are prepared to explore ways of making a living that may take them far from the familiar routine of a traditional *agricultura*. It is why the city worker will make his home in the country even if it means spending perhaps a thousand wasted hours driving to and from work every year; it is what the tele-worker linked cybernetically with the city office is in search of. It is the experience the rural tourist is in search of during the precious days or weeks of holiday. Heneghan (2002) suggests that rural tourism has the potential to be a 'serious instrument' of rural development, with important income generating potential. Its synergistic potential with other rural development activities, such as artisanal food production, culture, heritage and environmental quality, is also one of its defining characteristics (Gorman et al. 2002; Agri-Vision 2015 2004). Such was the widespread take-up rate of grants for rural tourism projects in the first LEADER programme in the RoI that concerns were expressed in policy circles about market saturation; and subsequent LEADER funding was directed towards marketing initiatives as distinct from the development of new tourism products. As noted previously, it remains one of the most common diversification strategies adopted by farm families on the island of Ireland. Notwithstanding the above, many commentators have argued that the sector effectively operates in a policy vacuum without a cohesive strategy and is currently characterized by a plethora of fragmented small-scale initiatives at local level (Commins 2005). While visitor numbers to Ireland continue to grow, the benefits have not been evenly distributed across the regions. Urban centres have been the principal beneficiaries and rural tourism remains under pressure (Fáilte Ireland 2008).

*Environmental quality as a dimension of multifunctionality*

In the RoI, the Rural Environment Protection Scheme (REPS), introduced in 1994, is the mechanism used to implement EU Regulation 2078/92. Under this regulation and its direct descendant, EU Regulation 1257/99, over 59,000 farmers participated in 2006 with approximately 40 per cent of the utilizable agricultural area (UAA) being farmed under the scheme (Department of Agriculture, Fisheries and Food 2007). In Northern Ireland, approximately 33 per cent of the land area was being farmed under agri-environmental schemes in 2005, which is comparable to that for the United Kingdom overall (Department of Agriculture and Rural Development 2007). In terms of delivering environmental quality, the REPS scheme in the RoI scheme has produced mixed reviews. Feehan (2004) and Harte and O'Connell (2003) note that it offers very limited scope for farmers to be innovative in how REPS plans are developed or implemented.

Organic farming is another important component of the link between environmental quality and multifunctionality. A growing minority in the 'developed' world has reached a level of awareness, sufficiently supported by their favourable economic situation, to be able to significantly influence agricultural practice so that it reflects their concern for food to be produced in a 'healthy' way under 'more ethical' conditions. The growth of the 'organic' movement has its roots in this development. In many respects the growth of organic farming represents a return to earlier and more sustainable ways of land management, but constrained by the need to conform to one or other of a variety of modern sets of standards and the need for certification. Between 1985 and 1999 the percentage of organic farming areas as a proportion of total agricultural area in the EEA18 rose from almost nothing to 2.5 per cent. More recent statistics show that at EU-25 level, organic and in-conversion area amounts to 3.6 per cent of UAA (European Commission 2005). In the case of Ireland the increase was from a just over 5,000 ha in 1993 to 38,000 ha in 2006 which equates to approximately 0.7 per cent of available UAA (Department of Agriculture, Fisheries and Food 2007). In Northern Ireland, the sector has expanded from 0.02 per cent of UAA in 1998 to 0.87 per cent in 2005, significantly below the level for the UK which stood at 3.3 per cent of UAA in the same year (Department of Agriculture and Rural Development 2007a). The expansion of organic farming is seen by the EU as an important indicator of environmental quality, and as one important factor in alleviating the environmental problems associated with modern farming, though it is not seen as the whole answer, needing to be supplemented by more general adoption of low input farming, integrated crop management and integrated pest control.

*Off-farm activity as a dimension of multifunctionality*

Maintaining the maximum number of rural households and especially family farms is a specific objective of the most recent major relevant policy statement in the RoI, the White Paper on Rural Development (1999). While many constructions of pluriactivity are framed as expressions of poverty or 'deficient agriculture' (van der Ploeg et al. 2002), the reality is that farm households increasingly have multiple sources of income associated with the transfer of resources from the urban to the rural economy, which bridge the farm/non-farm divide in a substantial way (Kinsella et al. 2000; Frawley and Phelan 2002; Frawley et al. 2005).

Wilson, Mannion and Kinsella (2002) argue that part time farming is neither a state of transition into full-time farming nor a movement out of farming altogether, but a structural phenomenon which will be central to future developments in rural Ireland. Clearly such developments are impacted upon by a range of non-agricultural policy drivers and the wider macro-economic situation. In this context, it appears that while all regions of the RoI benefited from the strong period of economic growth in Ireland since the early 1990s, many enterprise and employment initiatives supported by State agencies have been predominantly concentrated in the larger urban centres with foreign direct investment playing a key role. Initiatives such as LEADER have proven to be an important counterpoint to this trend, given their focus on local indigenous resources and rural areas (Commins 2005). Off-farm employment has grown rapidly in importance to farm household livelihoods in recent years. Recent statistics for the RoI for 2005 suggest that on 55 per cent of farms, the holder and/or spouse had an off-farm job, while on 81 per cent of farms, either the farmer or spouse had some source of off-farm income, from employment, pensions or social welfare (Department of Agriculture, Fisheries and Food 2007). Similar evidence of dependence on off-farm income is evident in data for NI for 2005, which shows that on 52 per cent of farms, either the farmer or spouse or both had other work in 2005 (Department of Agriculture and Rural Development 2007b)

*Artisanal food production as a dimension of multifunctionality*

As van Huylenbroeck and Durand (2003) note, artisanal or typical food products can be an efficient channel for the promotion of multifunctionality in that they frequently draw on traditional and/or non-conventional farming systems which can contribute to landscape, biodiversity, cultural heritage and environmental quality, among others. The positive contribution of speciality food production to rural development has been noted in the studies of McDonagh and Commins (1999), O' Reilly (2001), Sage (2002) and O' Connor and Gorman (forthcoming). The synergy effects with rural tourism, environmental quality, culture and heritage are also noteworthy features. In recent policy statements, there has been explicit acknowledgement of the actual and potential contribution of the speciality food sector to rural development objectives (Agri-Vision 2015 Committee 2004) and there is evidence of increased institutional support for the establishment of farmers' markets and other alternative food networks. Another indicator of its rising profile is that farmhouse food-related activities (processing and direct sales) constitute the second most important form of diversification activity in NI currently (Department of Agriculture and Rural Development 2007).

**Multifunctionality in a changing agricultural context**

A comparison of the results of the Census of Agriculture for the RoI taken in 1991 with the most recent estimates available provides evidence of the extent of structural change which has taken place in Irish farming. Data for 1991 show that there were approximately 170,000 farms in Ireland with an average farm size of 26 ha while in 2005 there were 131,000 farms with an average farm size of 33.4 ha (Department of Agriculture, Fisheries and Food 2007). This decline in farm numbers looks set to continue with projections for 105,000 farms in Ireland by the year 2015 (Agri-Vision 2015 Committee 2004). Much of the decline to date has occurred amongst the smaller holdings, with the numbers of farms with less than 20 ha falling by 39 per cent between 1991 and 2005, and the numbers of farms over 30 ha increasing by 9 per cent over the same period (Department of Agriculture, Fisheries and Food 2007). Significant structural adjustment is also evident in NI where the number of farms has declined by 30 per cent over the twenty-year period between 1985 and 2005 and has become more specialized, with a marked decline in pig and cereal production and an increase in cattle and sheep enterprises (Department of Agriculture and Rural Development 2006).

   Agriculture is more important to the RoI economy than it is too many other EU member states. This is so despite a decline in the contribution that it makes to Gross Domestic Product, which has fallen from 17 per cent in 1973 (when Ireland joined the EU) to just over 5 per cent in 1998 and to 2.3 per cent in 2006 (Department of Agriculture, Fisheries and Food 2007). The corresponding contribution of agriculture to the NI economy is lower at 1.9 per cent, but higher than that for the average of the United Kingdom overall (Department of Agriculture and Rural Development 2007a).

Farm income has depended enormously on EU support in recent decades. In the years between 2001–2005, direct payments equated to approximately 80 per cent of average family farm income (FFI) while for certain enterprises (such as drystock farming), they amounted to over 140 per cent of FFI in the RoI. The picture over the same period in NI is even more pronounced, with the value of direct payments representing twice the value of Net Farm Income across all types of agriculture (Department of Agriculture and Rural Development 2007b). The threshold for what is considered an acceptable income has risen enormously in recent decades. A full-time dairy farmer today would need an output of at least 70,000 gallons of milk, or the equivalent in other enterprises, to be viable on these terms. It is expected that by 2010 the pressures outlined above will reduce the number of full-time farmers in the RoI to 20,000, with 60,000 part-time and a further 20,000 in transitional groups (Agri-Vision 2015 Committee 2004).

Academic soothsayers are still engaged in the process of trying to predict the fallout from the introduction of the single farm payment (SFP). In general terms we can foresee a greater concentration of inputs and effort on the better land, allowing more marginal areas to revert to scrub. The enforcement of good farm practice in a policy future increasingly concerned and stringent about environmental health should ensure that this increased intensification is not accompanied by deterioration in such key indicators as water quality and food safety; and the withdrawal of production concern for the more marginal land will greatly benefit biodiversity and environmental quality in general.

All these outcomes can also be sketched in terms of new resource opportunity. Two considerations are especially worth considering. Firstly, the withdrawal of intensive agriculture to a more productive centre means an increase in the area of *recreational* land in the sense in which this term has been used earlier. A re-assessment of the resource potential of this land in the light of the earlier discussion of community tourism, at farm and at community level, would be a useful and potentially profitable exercise. Secondly, an opportunity will have been lost if all such land is allowed uncritically to slip back to waste, in the way much marginal arable land, reclaimed from the wild and managed at such human cost was allowed to revert to brake and heath in the decades following the Great Famine (Feehan 2003).

## Concluding remarks

The foregoing analysis has attempted to chart some of the main mechanisms through which Irish farm households are attempting to construct their livelihoods in a complex and rapidly-changing market and policy environment. Against this backdrop, the development of appropriate public policy and institutional support measures represent formidable challenges. At the heart of this challenge is the fact that Irish agriculture exhibits the 'competitive dualism' alluded to earlier, characterized by the co-existence of a sector with sufficient capacity to withstand and adapt to radically changing market conditions, alongside a less competitive sector which has limited response capacity, but one which is potentially viable if its supply of public goods is remunerated However, a final policy challenge is the identification of public preferences for such goods. In Ireland, farming interests have dominated the debates on agricultural and rural policy and the consumer and citizen perspective has thus far been neglected in policy development, planning and research. The issue of institutional capacity to elicit information on public demand is a key question in this regard, requiring a more transdisciplinary mix of approaches of methodologies and instruments to address these issues.

## References

Agri-Vision 2015 Committee (2004) *Report of the Agri-Vision 2015 Committee.* Dublin: Stationery Office.

Bacon, P. and Associates (2004) *A Review and Appraisal of Ireland's Forestry Development Strategy, Final Report.* Dublin: Stationery Office.

Behan, J. and K. McQuinn (2002) *Projections of Agricultural Land Use and the Consequent Environmental Implications – End of Project Report No. 4822.* Dublin: Teagasc.

Brundtland, G. (1987) *Our Common Future: The World Commission on Environment and Development.* Oxford: Oxford University Press.

Buller, H. (2003) *Changing Needs, Opportunities and Threats – the Challenge to EU Funding of Land use and Rural Development Policies: The Background To Reform.* Paper presented to the Land Use Policy Group Conference on 'Future Policies for Rural Europe – 2006 and Beyond'. Brussels: March.

Burrell, A. (2001) *Multifunctionality and Agricultural Trade Liberalization.* Paper presented to 77th EAAE Seminar/NJF Seminar No. 325. Helsinki: 17–18 August.

Commins, P. (2004) 'Poverty and Social Exclusion in Rural Areas: Characteristics, Processes and Research' *Sociologia Ruralis*, 44, 1: 60–75.

Commins, P. (2005) 'The Broader Rural Economy', in NUI Maynooth, UCD and Teagasc (eds), *Rural Ireland 2025 Foresight Perspectives.* Dublin: COFORD, pp. 37–44.

Convery, F. and S. Roberts (2000) *Farming, Climate and the Environment in Europe.* Environmental Studies Research Series Working Paper. Dublin: University College Dublin.

Department of Agriculture and Rural Development (2005) *A Study on Rural Policy.* Available from www.dardni.gov.uk/pwc_study_march_05-2.pdf.

Department of Agriculture and Rural Development (2006) *EU Farm Structure Survey 2005: Northern Ireland Report on Agricultural Labour Force, Farm Diversification And Contractor Use.* Available from http://www.dardni.gov. uk/euss2005-2.pdf

Department of Agriculture and Rural Development (2007a) *Statistical Review of Northern Ireland Agriculture 2006.* Available from http://www.dardni.gov.uk/ statistical_review_of_ni_agriculture_2006.pdf.

Department of Agriculture and Rural Development (2007b) *Farm Incomes in Northern Ireland 2005/2006.* Available from http://www.dardni.gov.uk/farm-incomes-2005-06.pdf.

Department of Agriculture, Fisheries and Food (2007) *Annual Review and Outlook for Agriculture.* Dublin: Stationery Office.

Department of Agriculture, Food and Rural Development (1999) *Ensuring the Future. A Strategy for Rural Development in Ireland.* Dublin: Stationery Office.

Dunne, W. and J. J. O'Connell (2003) 'Evolving EU Food Production Policy: Implications for Ecolabeling', in W. Lockeretz (ed.), *Ecolabels and the Greening of the Food Market.* Boston: Tufts University, pp. 1–10.

Elliot, R. H. (1908) *The Clifton Park System of Farming.* London and Kelso (4th edition).

European Commission (2005) *Organic Farming in the EU: Facts and Figures.* Available from http://ec.europa.eu/agriculture/qual/organic/facts_en.pdf.

Fáilte Ireland (2008) *End of Year Review and Outlook for 2008.* Available from http://www.failteireland.ie/About-Us/News-and-Events/Ireland-welcomes-record-7-8m-visitors-in-2007.

FAPRI-Ireland Partnership (2003) *The Luxembourg CAP Reform Agreement: Analysis of the Impact on EU and Irish Agriculture.* Dublin: Teagasc.

Feehan, J. (2003) *Farming in Ireland: History, Heritage and Environment.* Dublin: UCD Faculty of Agriculture.

Feehan, J. (2004) *Enhancing Biodiversity: The Challenge and Opportunity of REPS 3.* Paper presented to the National REPS Conference 2004. Dublin: Teagasc.

Feehan, J. (2005a) 'The Woodland Vegetation of Ireland, Past, Present and Future', *Forest Perspectives: Irish Forestry*, 62.

Feehan, J. (2005b) 'Community Development: the Spiritual Dimension', *Perspectives on Community Development in Ireland*, 1, 1: 63–74.

Forest Service of Northern Ireland (2006) *Northern Ireland Forestry. A Strategy for Sustainability and Growth.* Forest Service of Northern Ireland.

Frawley, J., D. O'Meara and J. Whiriskey (2005) *County Galway Rural Resource Study.* Dublin: Teagasc.

Frawley, J. and G. Phelan (2002) *Changing Agriculture: Impact on Rural Development.* Paper presented to the Teagasc Rural Development Conference 2002. Tullamore. 14 March.

Gorman, M. (2004) *Socio-Economic Impact of Rural Development: Livelihood Realities and Prospects for Irish Farm Families.* Unpublished PhD Thesis, University College Dublin.

Gorman, M., J. Mannion and J. Kinsella (2002) 'Agri-tourism in Ireland: Ballyhoura in South West Ireland', in J. D. van der Ploeg, A. Long and J. Banks (eds), *Living Countrysides – Rural Development Processes in Europe: State of the Art*. Doetinchem: Elsevier, pp. 100–105.

Harte, L. and J. O'Connell (2003) 'How Well do Agri-Environmental Payments Conform with Multifunctionality', *EuroChoices*, 2, 1: 36–40.

Heneghan, M. (2002) *Structures and Processes in Rural Tourism*. Paper Read to Teagasc Rural Development Conference 2002, Tullamore, 14 March.

Kinsella, J., S. Wilson, F. de Jong and H. Renting (2000) 'Pluriactivity as a Livelihood Strategy in Irish Farm Households and its Role in Rural Development', *Sociologia Ruralis*, 40, 4: 350–70.

Knickel, K. and H. Renting (2000) Methodological and Conceptual issues in the Study of Multifunctionality and Rural Development, *Sociologia Ruralis*, 40, 4: 512–28.

Leavy, A. and S. McCarthy (2004) *Economics of Forestry as a Farm Enterprise in a Rural Development Context*. Paper Read to Teagasc Rural Development Conference 2002. Tullamore. 16 March.

Losch, B. (2004) 'Debating the Multifunctionality of Agriculture: From Trade Negotiation to Development Policies by the South', *Journal of Agrarian Change*, 4, 3: 336–60.

Mahe, L. (2001) 'Can the European Model be Negotiable in the WTO?', *EuroChoices*, Spring: 10–15.

Mannion, J., M. Gorman and J. Kinsella (2001) 'Connecting Farming, the Environment and Society: A Living Countryside Perspective', *Tearmann*, 1: 11–18.

Marsden, T. and R. Sonnino (2005) *Setting Up and Management of Public Policies with Multifunctional Purpose: Connecting Agriculture with New Markets and Services and Rural SMEs*, UK National Report. Available from www.multagri. net. Accessed 8 November, 2005.

McDonagh, P. and P. Commins (1999) 'Food Chains, Small Scale Food Enterprises and Rural Development: Illustrations from Ireland', *International Planning Studies*, 4, 3: 350–71.

MULTAGRI (2005) *Concept Oriented Research Clusters: Application to the Multifunctionality Concept*. Available from www.multagri.net. Accessed 8th November, 2005.

O'Connor, D., M. Gorman, H. Renting and J. Kinsella (eds) (2006) *Driving Rural Development: Policy and Practice in Seven EU Countries*. Assen: Van Gorcum.

O'Connor, D. and M. Gorman (forthcoming) 'Regional Quality Food Production and Rural Development in Ireland', in H. Renting, K. de Roest and N. Parrott (eds), *Relocalising Food: Quality Food Production in Europe and its Role in Rural Development*. Assen: Van Gorcum.

OECD (2001) *Multifunctionality: Towards an Analytical Framework*. Paris: OECD.

O'Reilly, S. (2001) *The Fuchsia Brands Ltd. Case Study.* Unpublished Mimeo. Department of Food Business and Development, University College Cork.

Potter, C. and J. Burney (2002) 'Agricultural Multifunctionality in the WTO: Legitimate Non-Trade Concern or Disguised Protectionism?', *Journal of Rural Studies*, 18: 35–47.

Sage, C. (2002) 'Social Embeddedness and Relations of Regard: Alternative "Good Food" Networks in South-West Ireland', *Sociologia Ruralis*, 19, 47–60.

Thomson, K. J. (2001) 'Agricultural Economics and Rural Development: Marriage or Divorce?', *Journal of Agricultural Economics*, 52, 3: 1–10.

Van der Ploeg, J. D., A. Long and J. Banks (eds) (2002) *Living Countrysides: Rural Development Processes in Europe – The State of the Art.* Doetinchem: Elsevier.

van Huylenbroeck, G. and G. Durand (2003) *Multifunctional Agriculture – A New Paradigm for European Agriculture and Rural Development.* Aldershot: Ashgate.

Wilson, S., J. Mannion and J. Kinsella (2002) 'The Contribution of Part-Time Farming to Living Countrysides in Ireland', in J. D. van der Ploeg, A. Long and J. Banks (eds), *Living Countrysides – Rural Development Processes in Europe: The State of the Art.* Doetinchem: Elsevier, pp. 164–174.

# Chapter 8

# Sustainable Forestry in Northern Ireland and the Republic of Ireland

Roy W. Tomlinson and John Fennessy

## Introduction

The sustainable management of forests is concerned with delivery of benefits for the present generation whilst protecting the environment and resources for future generations to enjoy. Forestry is a complex activity – a renewable resource with a minimum 40-year cycle, an alternative land-use, an agent of landscape change, a provider of wildlife habitats, an environment for recreation, a carbon store and – not least – the source of raw material for timber-based industries that enable provision of the wider benefits of forests. A brief history of forestry on the island of Ireland, placing current forest strategies and policies in context, is followed by sections that discuss issues surrounding forestry and sustainability. These include: economic and social issues; effects on the environment; measures taken to meet these concerns and to sustain the environment; and forestry as part of the wider landscape, including its role in carbon storage in relation to global warming.

## Development of forestry in Northern Ireland and the Republic of Ireland

In 2004, approximately 10 per cent of the Republic of Ireland (RoI) and 6.3 per cent of Northern Ireland (NI) (Forestry Commission 2004; Food and Agriculture Organization of the United Nations [FAO] 2005) was wooded compared with a European Union (EU) average of around 33 per cent. The majority of woodland and forest is of non-native conifers; most broadleaved woodland dates from the nineteenth century or later.

Woodland developed from about 10,000 years ago as environmental conditions improved after the last ice age, but clearance from the Mesolithic time period reduced woodland cover. Population growth and development of settlements, creation of farmland, and commercial exploitation contributed to woodland loss (Neeson 1991). By the 1600s, woodland may have occupied about 12 per cent of the country (McCracken 1971), or as little as 2.1 per cent (Rackham 1995) and by the 1700s, Ireland was importing timber to sustain shipbuilding and construction. Woods continued to be used for firewood, charcoal iron smelting, making glass, and for leather tanning (McCracken 1971; Carey 2005). These commercial interests may have diminished woodland area but, alternatively, they may have managed some woods sustainably to ensure survival of the resource.

Research on woodland history in Ireland is meagre, but extensive use of wattle in Cork and Dublin implies coppice management, and Watts (1984), Jones (1986) and Carey (2005) have shown a tradition of coppice wood management in parts of Ireland into the eighteenth and nineteenth centuries. Generally, these woodland management skills appear to have been lost as Anglo-Irish landowners adopted plantations earlier and more completely than landowners in England (Rackham 1986). Only around 10 per cent of woodland present in the mid-1600s remained when the first '6-inch' Ordnance Survey maps appeared (1836–44); most had been converted to farmland.

Today, therefore, ancient woodland (that existing from before 1600 AD) is rare in Ireland. It is often of high biodiversity and may include rare species; so one aspect of present sustainable forest management must be maintenance of ancient woodland and, where planted over, its restoration. Current management policies also seek to maintain later broadleaved woodland and to encourage its planting, thereby sustaining landscape value and biodiversity – often regarded as greater than that of conifer forests.

The eighteenth and nineteenth centuries saw planting of trees on estates, but woodland area again declined in the late nineteenth century with the Land Acts that transferred ownership to former tenant farmers. By 1905, woodland was just over 122,000 hectares (ha) (including 16,800 ha in the future NI), or 1.4 per cent of the land area (O'Carroll 2004). A strong awareness of the crisis in forestry, in the context of land reform, grew among public and private notables in Ireland and led to the 1908 report of the Departmental Committee on Irish Forestry. This recognized private owners' inability to carry out afforestation programmes and recommended that public authorities should plant significant areas of forest. Although this strategy marks the beginning of modern Irish forestry policy, purchase of lands by public funds for afforestation proceeded modestly up to the First World War when forest resources in Britain and Ireland came under further pressure, with extensive felling of older forests. The 1919 Forestry Act and establishment of the Forestry Commission sought to expand forestry, but the political status of Ireland was about to change.

A Forestry Act, introduced in the Free State in 1928, transposed most previous legislation to Irish law and introduced forestry grants and felling controls. The Act of 1946 enabled the state to purchase land and gave ministers powers to promote forestry through education and research. A State-planting target of 10,000 ha per annum was set, to achieve 400,000 ha within 40 years.

In 1950, the Irish Government invited the Food and Agriculture Organization (FAO) to advise on this planting policy, which would result in extensive planting of 'rough mountain grazing'. The FAO recommended division into two programmes:

- a commercial programme designed to meet minimum requirements for sawn softwood in times of emergency; and

- a social programme for soil conservation, stabilization of employment in congested areas and reclamation of 'idle land'.

The social programme was never explicitly accepted as policy, although emphasis on planting in western counties, as shown in Ministers' Reports from 1960 onward, suggests it was not dismissed (O'Carroll 2004). Like earlier reports, the FAO accepted private landowners' inability to carry out afforestation programmes and again recommended public authorities should plant significant areas of forest. Private afforestation remained at a low level until introduction of a forestry scheme in 1980 under the European Economic Community (EEC) Regional Policy Programme (Fennessy 1986) (popularly known as the 'Western Package'). Private afforestation has since expanded rapidly and now comprises almost the entire planting programme. Not only has state planting almost ceased, but the national estate is managed by an independent State-owned company – Coillte Teoranta (Coillte) – established by the 1988 Forestry Act. Current forest strategy was set out in 1996 (Department of Agriculture, Food and Forestry 1996). Around 82 per cent of present woodland and forest is of conifers.

Post-partition, forestry in NI, under the influence of the Forestry Act (NI) 1953 and the White Paper on Forestry in NI 1970 (Kilpatrick 1987), followed a similar path to that in the RoI. A social/employment element existed alongside the strategic need to produce timber. The area under trees increased from 18,500 ha in 1940 to 85,700 ha in 2004 (Forest Service [NI] 2004). From the 1950s to the 1980s expansion of public forests was on upland peat bogs and wet mineral soils; but from 1987 greater emphasis was placed on the private sector, which by 2004 accounted for around 28 per cent of forest area. Forest expansion in NI is currently around 700 ha each year, principally from agricultural land and aided by grant schemes. In consequence, 75 per cent of private woodlands are broadleaves or broadleaf-conifer mixtures; Forest Service species are 91 per cent conifers (Forest Service [NI] 2004).

## Economic and social issues

The current strategic plan for the forestry sector in the Republic of Ireland aims to develop forestry to a scale and in a manner that maximizes its contribution to national and economic well being on an environmentally sustainable basis. A critical mass of timber production of 10 million cubic metres is deemed necessary to achieve these aims. To meet this production, a target was set of 1.2 million hectares of productive forest by 2030 with annual afforestation of 25,000 ha to 2000 and 20,000 ha from 2001 to 2030. A reforestation programme was to maintain the forestry estate after clear-felling. Yield class 18 was to be the national average.[1]

---

1   'Yield class' is based on average annual volume of wood produced by a forest over the rotation.

The public to private afforestation ratio was to be 30:70, with emphasis on farmer participation (which required improved compatibility between incentives for forestry and other farm support).

Planting targets have not been met (around 14,000 ha per annum was achieved from 1996 to 2003). Moreover, the prospective range of log sizes and species is inadequate to meet market demands, and information to guide the private sector is deficient (Bacon 2004). In 2004, the Irish Government commissioned Peter Bacon & Associates to review the forestry programme and to identify reforms required to achieve further progress.

Bacon estimated around 3,780 people were employed in forest establishment and harvesting in the RoI, a further 6,000 were engaged in timber processing and every five jobs in forestry supported an additional three in the Irish economy. In total, forestry supported approximately 16,000 jobs (Bacon 2004). Particularly with regard to forest planting and harvesting, this employment benefit is significant in the disadvantaged west of RoI, where most private planting has occurred. Kearney and O'Connor (1993) showed forestry may have potential to ease *under*-employment and aid in stabilizing populations, pointing out that declining agricultural populations are inevitable regardless of whether there is forestry development. However, local perceptions do not always support national analysis. In South Leitrim, despite recent arrival of a pulp-wood factory, Papageorgiou et al. (2000) found local respondents to a questionnaire believed forestry had not delivered the number of jobs envisaged. Meanwhile, saw-mill owners complained of difficulty in employing workers because better paid jobs were available.

Recognizing that much current planting is by farmers, Bacon (2004) noted that although creating €100 in grower income costs €121.60 in support payments, the comparable support figure for competing agricultural land uses is between €140 and €147. Although farmers have shown reluctance to plant forests because of long-term commitment of land and perceived lack of adequate return for risks involved, Bacon suggested that Common Agricultural Policy (CAP) reform will change farmers' perceptions and that they will plant sufficient land to meet the 20,000 ha per annum target. In South Leitrim, farmers nearing retirement were satisfied with their annual payment for planting their land. Indeed, there is little doubt that those who have planted benefited financially from tax-free grants and premia (Papageorgiou et al. 2000). Nevertheless, many people, not necessarily farmers, saw forestry as a competitor for land (O'Leary et al. 1999) and questioned the quality of timber produced, because of the rapid growth rate of Sitka spruce and lack of management on small farm plantations (Papageorgiou et al. 2000). Similarly, on the Mayo/Roscommon border, Kearney and O'Connor (1993) found opposition to forestry, which was seen as inimical to agricultural development and a cause of depopulation – although this poorly forested area has a long history of rural depopulation.

Bacon (2004) concluded that the combined benefits of forestry and wood processing exceeded costs by €571 million in Net Present Value terms (discount rate of 5.5 per cent). However, there are considerable difficulties in establishing costs and benefits of non-timber goods such as landscape, environmental goods and carbon accumulation, goods not recognized sufficiently under the 1996 *Strategy* (Bacon 2004).

In NI there is insufficient investment in forestry by the private sector, largely because of difficulties farmers and landowners have in investing in long-term projects with no prospect of a return for many years (Forest Service [NI] 2003). In consequence, opportunities for regional development in rural areas, for public access to the countryside, and for protection and conservation of the countryside are insufficient (ibid.). Following a review of options, the base case, which maintains the existing range and balance of outputs within the limits of public finance currently available, gave a Net Present Value of forestry and wood processing of £144 million (discount rate 6 per cent) (c. €100M). Forestry and wood processing currently account for around 950 jobs; combined 'the annual subvention cost per job is £7,800 (€5,500). The annual value added per job is £19,500 (€13,500)' (ibid., p. 10).

This economic appraisal of forest policy in NI, and public consultation on options for forestry (Forest Service [NI] 2004), informed the recent NI forest strategy (Forest Service [NI] 2006). This emphasizes the need to expand forest area (eventually to double it), but at a modest rate. It is believed this expansion will be achieved because, as the Minister wrote in the foreword, reform of the CAP and introduction of the Single Farm Payment will provide additional confidence that forestry is a credible option for land use – a view paralleled in Bacon (2004). Funds available under the Woodland Grant Scheme and the NI Rural Development Regulation Plan will be reviewed.

Although the strategy aims to meet the needs for timber production, much of it relates to environmental goods, including perceived needs of the public for access and recreation. As the published strategy is not accompanied by an economic analysis and there is no articulation with options in the economic appraisal (Forest Service [NI] 2003), its sustainability is difficult to assess.

### Environmental impacts of forestry

*Impacts on habitats and biodiversity*

Much forest expansion in RoI over recent decades was on blanket peatland; this was more easily obtained than lowland farmland and farmers were prepared to plant it. Consequently, peatland habitats and species were lost. Blanket peatlands, confined to northwest margins of Europe and classed as a 'Priority Habitat' under the EU Habitats Directive, have plants that cannot grow elsewhere and are adapted to waterlogged, nutrient-poor and acidic conditions (however, some do occur on lowland raised bogs).

Forest planting on peat requires large drains to lower water tables and produce drier rooting zones. Whether large drains within forests affect peatland outside them is difficult to determine. Some conservationists argue there is general draw-down of the water table with change in plant species composition and loss of micro-habitats. However, this conflicts with findings that water movement in peat is limited largely to the acrotelm (upper 10–20cm of the bog) (Tomlinson 1979). Aerial application of fertilizers to forests, so as to counter poor nutrient supply, may drift onto surrounding bog and, with any effective drainage, cause change in species composition. The specialized plants cannot compete with species suited to drier, more nutrient-available conditions.

Blanket bogs are usually extensive tracts of open country that provide habitats for predator birds requiring large territories (e.g. the hen harrier, one of Europe's rarest birds of prey). Young plantations may increase populations of species preyed upon, provide nesting habitats for the hen harrier and lead to an increase in its population (Coillte no date), but as forests become denser and no longer suitable for foraging and nesting, the hen harrier moves to new plantations. The increased population of the hen harrier in NI is similarly explained by expansion of young forest (Environment and Heritage Service 2005), but uniquely, a small number are regularly recorded as nesting in trees of mature conifer forest (Scott 2000). Forest plantations also reduce habitats for migratory and wetland birds – including curlew, dunlin and golden plover, particularly as damp lowlands have been drained.

Blanket peatlands are of archaeological significance; anaerobic conditions sustain evidence of past landscapes (e.g. the pollen record of past land covers). Further, peat holds 53 per cent of the soil carbon stock in the RoI and 42 per cent in NI (Cruickshank, Tomlinson, Devine and Milne 1998; Tomlinson 2005), which, if released, would add to atmospheric carbon and to global warming.

Peatland was considered traditionally as wasteland, suitable for low productivity grazing and harvesting of peat for fuel, and into which forest could be planted rather than into more valuable farmland. Realization of the need to sustain species and habitat diversity, and recognition of blanket peat as a unique habitat is a recent phenomenon.

In NI, the Forest Service has not planted new areas of oligotrophic or dystrophic peat since 1993 (Forest Service [NI] 1993) and the total area of new private plantations on peat is small. In the RoI, the amount of forest planted on peatland between 1990 and 2000 is disputed, with 70 per cent of the area afforested according to the Minister for Agriculture and Food,[2] and afforestation of up to 84 per cent of the area estimated by the European Environment Agency (Spatial Analysis Group, EEA 2004). The different estimates arise largely from different survey methodologies, but even 70 per cent represents significant peatland loss. It has been recommended recently that planting should avoid blanket and raised bogs (O'Halloran et al. 2002), as well as other priority conservation areas (Hickie et al. 1993); indeed, attention has been given to rehabilitation of peatland towards conditions prior to plantation.

---

2   Reply to Parliamentary Question 225 of 27 April 2004.

*Acidification*

Enhanced acidification of soils, streams and lakes is an environmental impact of forestry (Allott et al. 1998). Most forest planting in the RoI has been of exotic conifers in regions of high precipitation and frequent low cloud. Conifers have considerable capacity to intercept moisture and pollutants it contains, and dust particles carried by dry winds. Conifer forests thereby are efficient 'scavengers' of acid pollutants and acid-precursors. Eventually, through leaf-drip and stem-flow, intercepted acid pollutants reach forest soils and may increase soil acidification, especially where soil parent materials have low buffering capacities (as on acid rocks). Additionally, preferential uptake of base cations by tree roots may increase soil acidity. Normally, plant litter returns base cations to the soil, but felling and removal of trees also removes this potential return, resulting in net soil acidification. Production of organic acids in the forest floor may also increase soil acidification.

Sampling of soft-water streams in the Wicklow Mountains shows that in catchments with extensive mature forest cover, streams tend to be more acidic than comparable streams in moorland-dominated catchments (Kelly-Quinn et al. 1996). The majority of streams investigated had naturally high levels of dissolved organic carbon. Where catchments were heavily forested, streams had abrupt and prolonged increases in acidity, largely explained by dissolved organic matter. Intense rainfall events increased stream discharge because forest drains channelled water into streams; contact time with soils was reduced and thereby any buffering reaction. Also, during periods of easterly airflow (from urbanized and industrialized Great Britain) inputs of nitrates and sulphates increased stream acidity in forested catchments, probably due to 'scavenging' by conifers.

Increased stream acidity can adversely affect their biodiversity and the sustainability of populations of fauna and flora. Crustaceans, molluscs and many insect larvae are unable to survive and riverine birds, including wagtails and dipper, may be affected (Reynolds 1998). Dipper populations may be reduced because of absence in acidic waters of mayfly nymphs and some caddis larvae, which are important for feeding nestlings (Ormerod et al. 1991). Kelly-Quinn et al. (1997) found that streams in the Wicklow Mountains flowing over granite were less diverse in macroinvertebrates than streams flowing over Ordovician or Silurian materials, and within a rock type, forested streams were less diverse than adjacent non-forested streams.

Rock type is an important influence. Whereas forested streams in western RoI also have increased acidification (Farrell et al. 1997), studies in parts of southern Ireland where the country rock has a higher buffering capacity, have shown little increased stream acidification in forested catchments (Giller and O'Halloran 2004). In consequence, forests appear to have limited influence on stream invertebrates in Munster catchments (ibid.). Similarly, dipper populations there were not reduced by afforestation. Clenaghan et al. (1998) showed that macroinvertebrate communities in conifer-afforested sites were not impoverished, but differed from those above and below the plantation. Here local ecological factors were of primary importance.

Some streams in forested catchments have increased concentration of aluminium, up to and beyond that at which it becomes toxic to salmonids. In the Wicklow Mountains, afforested catchments with a combination of low pH and high inorganic aluminium concentration had either much reduced trout populations or were devoid of fish (Kelly-Quinn et al. 1996). As with invertebrates, forest cover had no effect on trout populations in Munster; local habitat here was more important than amount of forest cover (Lehane et al. 2000).

The effects of forests on acidity and on aluminium concentration in streams are of major concern, not only for sustainability of species *per se* but because of the recreational and financial importance of fish. Angling is an important part of tourism and recreation industries in the RoI. Whelan and Marsh (1988) estimated an annual domestic expenditure of IR£15.6m (over €19m) and a foreign tourist angler expenditure of upwards of IR£12m (over €16m); both supported nearly 2,000 full-time jobs with a IR£15m (over €18.5m) tax revenue. More recently, angling was estimated to be worth IR£78m (over €97m) to the Irish economy (Western Regional Fisheries Board 2004); and Ireland's reputation for the quality of its angling, and particularly for wild salmon (Curtis 2002) and trout, has spread world-wide.

*Eutrophication*

Concern about water quality in Ireland has increased in recent years, especially loss of phosphorous from land to water. Introduction of the Phosphorous Regulation in the RoI (Anon 1998) placed pressure on all economic sectors to protect and improve water quality, and pressure increased further with the European Community (EC) Nitrate Directive (91/676/EC). For successful growth, forests planted on peatlands require fertilizer application at planting and subsequently (Joyce and O'Carroll 2002). These applications can affect streams, lakes and habitats surrounding forests.

Increases in phosphates have been recorded in upland streams that consequently may experience eutrophication and wildlife changes (Giller and O'Halloran 2004). Cummins and Farrell (2003) reported increased phosphorus levels in forest drains and small streams in blanket peatland as a result of clear-felling, reforestation and fertilizing. Increase in phosphorus could be related to fertilizer treatment alone, but effects of felling, reforestation and fertilizing could not be separated to explain increased concentrations of nitrates, ammonium and potassium. Implications of the results for downstream river-water quality were unclear. Giller and O'Halloran (2004) suggest that interactions between harvesting and water quality may be catchment-specific, with a lack of generalized patterns, and related to management practices during the operations. Lakes enriched by forestry tend to be localized and eutrophication, as of rivers, may owe more to intensive agriculture (Allott et al. 1998). Harvesting around Lettercrafoe Lake in 2004 was alleged to have released excessive phosphate into watercourses. Coillte received two District

Court summonses from the Western Regional Fisheries Board (WRFB) because of the importance of local watercourses for salmonids and freshwater pearl mussel (for which the RoI is its stronghold in Europe). The case was dismissed because Coillte had complied with all felling licence requirements, had consulted with relevant authorities and shown cooperation with the WRFB. However, debate continues, in particular concerning acceptable concentrations of phosphorous and sensitive species such as the freshwater pearl mussel (Forest Network Newsletter [FNN] 2006).

## Sustainable forestry

In the 1990s, the importance of environmental and social dimensions of forestry grew, mainly in response to the 'Earth Summit' at Rio de Janeiro in 1992, the Kyoto Protocol, the support for the principles of sustainable forest management and changing societal views on forests and the practice of forestry (Fennessy 2005). The UK and Irish Governments produced policies to meet concerns about forestry's effects on the environment. Both Governments agreed to Helsinki and Lisbon guidelines (*Ministerial Conferences on the Protection of Forests in Europe*, Helsinki 1993 and Lisbon 1998), adopted the 'Pan-European Criteria' (part of the Lisbon meeting), and produced National Forest Standards and accompanying Guidelines of Best Forest Practice (Forest Service [RoI] 2000a, 2000b; Forestry Commission and Forest Service [NI] 2004).

*Policy responses to the negative impacts of forestry on water quality*

The UK Forestry Commission guidelines (Forestry Commission 2003) (which apply in NI) include actions in relation to acidification and eutrophication at catchment and site levels. It is noted that enhanced capture of acidic pollutants by forests (scavenging) could delay recovery of acidified waters or even lead to further acidification in sensitive areas despite the general decline in acid deposition. Therefore, new planting of catchments must assess possible effects using the critical loads concept – i.e. the maximum level of pollutants that a given ecosystem can tolerate without adverse change. Catchments in areas above the critical load threshold should not be planted.

As the critical loads concept applies to relatively large expanses of land, there must be detailed consideration of factors affecting, and the consequences of, acidification of more local streams and water bodies. In existing forests species mix may need to be widened and include more broadleaves to reduce the scavenging effect. In areas above 300m (where scavenging may be stronger), selective deforestation may be necessary. Similar 'Water Quality Guidelines' in the RoI enable acidification-sensitive areas to be identified (Forest Service [RoI] 2000c); application for planting these areas involves consultation with regional fisheries boards and local authorities.

Forestry and Water Quality Guidelines include measures to limit eutrophication of water courses and lakes. For example, drains leading from a forested site must taper into buffer zones, allowing discharged water to fan out before entering streams; buffer zones are a filter, reducing sediment and preventing nutrient-enriched water entering streams. Rates, timing and methods of fertilizer application are included in the guidelines and vary with stage of forestry. At planting application is manual, but for established forests with nutrient deficiency, application may be aerial and tight controls are necessary, including wider buffer zones, non-application in windy conditions or during or after prolonged rain (Forest Service [RoI] 2001a). Harvesting guidelines include ensuring that run-off from extraction routes does not enter streams (Forestry Commission 2003). Sediment in run-off could enhance nutrient status of streams and be harmful to salmonid populations, destroying reeds and reducing feeding potential.

*Forests and sustaining biodiversity*

Broadleaves, particularly if native, generally have higher biodiversity than conifers; a greater mix of broadleaves may increase biodiversity. However, planting broadleaves in peatlands, where most expansion has occurred, may not be possible. Biodiversity may be enhanced by planting two or more species of the non-native trees, by diversity of tree ages and by areas of biodiversity enhancement. The last applies to all forests in the RoI and should account for c. 15 per cent of the area (Forest Service [RoI] 2000d). These areas would include open spaces (5–10 per cent of the forest) and retained habitats. The percentage in open space, however, includes land required for forestry operations (e.g. roads, turning bays) whose suitability for enhancing biodiversity is questionable.

Retained habitats, which should account for 5–10 per cent of site area, aim to conserve and enhance habitats, flora and fauna throughout rotations of the forest. The guidelines stress careful design to avoid disturbance, suggest sustainability of retained habitats may be enhanced by a 3 m buffer zone, and indicate some habitats may require proactive management (Forest Service [RoI] 2000d).

Other evidence of concern for sustaining biodiversity includes the 'Biodiversity Action Plan for the Hen Harrier' adopted by Coillte (Coillte, no date) and engagement of the Forest Service (NI) in the NI Species Action Plan for that species (EHS 2005). The PAWS scheme (restoration of Plantations on Ancient Woodland Sites) in NI aims to return about 200 ha of selected PAWS to native woodland; those sites selected from woodland existing in 1830 and thought to be semi-natural (some may be rare 'ancient woodland') (Forest Service [NI], no date). The Native Woodland Scheme, launched in the RoI in 2001, similarly aims to encourage proactive protection and expansion of native woodland and associated biodiversity, using 'close-to-nature' silviculture (Forest Service [RoI] 2001b).

Work by Coillte in Midland bogs also exemplifies attempts to increase biodiversity of forest land. Between 2004 and 2008 Coillte aimed to restore around 570 ha of raised bog habitat on its property. This is the largest single raised bog restoration project to be undertaken in Ireland, accounting for over 5 per cent of the RoI area of raised bog conserved in Special Areas of Conservation – a significant contribution to conservation of a European Priority Habitat.

## Forests and the wider landscape

### *Forests and landscape quality*

Coniferous plantations are criticized because they do not blend with surrounding countryside, tending to smother the landscape (Tomlinson 1997). Such views may need to be tempered. Research on public opinion of forestry in NI showed that 47 per cent of respondents supported forestry 'to improve the countryside landscape' and 57 per cent because it 'provides places to walk in' (Forestry Commission and Forest Service [NI] 2005). Three-quarters of respondents wanted more woodland in their local area. Woodland type was not specified, but because most woodland with which people are familiar is coniferous, attitudes to conifers may be less adverse than previously thought. As with views on economic benefits of forestry, perceptions may differ with locality. O'Leary et al. (1999) found different attitudes to forestry between sample populations in Co. Wicklow and Co. Leitrim. In Wicklow people were generally positively disposed towards forestry; in Leitrim they were generally negative. In Wicklow, longer experience of forestry, greater cover of forests, higher employment in forestry and more forest parks, may explain the findings. Additionally, Wicklow had greater proportions of residents that were urban and had higher educational levels. Negative attitudes in Leitrim were not restricted to the farming population, but forests were perceived to be taking land that should be devoted to agriculture. Furthermore, many interviewees thought afforestation would pollute rivers and lakes (important constituents of Leitrim's landscapes). There appears to have been a lack of perception of the role of forests in a landscape; the aesthetics of landscape were confused with functional attributes.

Forestry guidelines recommend design criteria for different landscape types (Forest Service [RoI] 2000e). Factors considered include scale and size of planting, and shape, pattern, edge effects, textures and colours of planting. Improvement in design often involves use of broadleaves, for example in a mix of species around forest edges, but they are not always appropriate. Few broadleaved species grow on blanket bog (e.g. birch on drier parts) and where hill slopes are jagged, the conical and stark shape of conifers may be more appropriate than rounded crowns of broadleaves, which are more suitable for drumlin topography or mature farmland.

The number of visitors illustrates amenity importance of forests – although estimating numbers is difficult given the dispersed distribution and extensive perimeters of forests. In 2004–05, NI forests had over 508,000 paying visitors, which is probably a major under-estimate of their amenity value because most forests are freely accessible to pedestrians and entrance charges (vehicles) are levied only for the nine Forest Parks. Around 9,500 educational visitors were escorted in 2004–05 and there were teacher-led visits throughout the year (Forest Service [NI] 2005).

Coillte's forests attract an estimated 8 million visitors each year (Clinch 1999) and deliver an annual recreational value of around €16 million (Coillte 2005), based on an average 'willingness to pay figure' of €1.87 per person. More recently, users of forests and trails typically placed a value of €5.40 on the benefit to them of a single visit (Fitzpatrick Associates 2005). Placing monetary value on amenity is fraught with difficulty (Coillte 2005), but considerable numbers of people cherish opportunities to walk and enjoy quiet recreation in forests. Sustaining these opportunities is of major significance in present and future forest management.

*Forests as possible carbon sinks*

Plants take carbon from the atmosphere and convert it to plant tissue. Trees have a long life-span and have greater volume than vegetation replaced. During their life, forests therefore provide a carbon store and may be a possible carbon sink (i.e. the amount taken from the atmosphere and stored in forests is greater than that lost through respiration).

The carbon density estimated for forest trees in 2000 was around ten times that for peatland vegetation (3.0 t C/ha) (Tomlinson 2004). Recent expansion of forest onto blanket peat offers an opportunity to enhance vegetation carbon stock (M tonnes C); this gain in stock can be used against increased carbon emissions arising from recent economic development in Ireland. Current estimates give a net annual increase in forest carbon stock of 0.11 Mt C from 1990 to 2000 (Gallagher, Hendrick and Byrne 2005), but estimates vary widely depending on estimates of forest areas, yield class and volume of trees, and conversion factors including biomass expansion factors, specific density and carbon content (Kilbride et al. 1999; Gallagher et al. 2005).

Changes in forest carbon stocks (and the possibility of using gains to offset other emissions) should include forest soils because planting, growth and harvesting of trees may affect soil carbon stocks. For example, planting on peat entails drainage to lower water tables, leading to greater aeration, breakdown of peat and thereby possible release of carbon to the atmosphere. Research in Great Britain suggests that carbon losses from forest planted on peat may be less than previously thought and that throughout most of the 20th century, afforested peatlands in the UK will have been a net Carbon sink (Hargreaves et al. 2003).

In the RoI there has been little research on effects of forestry on soil carbon content, particularly in relation to peat, but Byrne and Farrell (2005) concluded that blanket peat forests in Ireland were also likely to be net carbon sinks. However, carbon accumulation alone is not a justification for planting peatland; the effects on biodiversity, including loss of peatland habitat, and on landscape, must be considered.

## Conclusions

Forestry has been shown to be a complex activity. The resource has been created by state and private involvement from meagre beginnings to a significant economic, environmental and social asset. As suggested by forest strategies in both the RoI and NI, continued support is essential to meet economic, social and environmental aims of forestry. That support may change over time; forests are dynamic systems managed in accordance with the different values they have for society.

An example of changing views is the increasing emphasis given to 'farm woodlands', but not only to extend hectarage of productive forest. Farm woods (in addition to larger forests) are seen as a means to offset some of the increase in national carbon emissions by storing carbon and by using wood as a renewable energy resource (EU countries are committed to include renewable energy in their energy mix). A recent extension of this is the 'Short Rotation Coppice Energy Crop', which involves growth of high yielding trees such as willow at close spacing and with harvesting about every three years. As of late 2005, 310 ha were planted in NI and 105 ha in the RoI (Gilliland 2005).

Forests supply extensive, though poorly quantified, benefits for society in recreation, health and well-being. Governments are encouraging an urbanized population to adopt moderate exercise and there is increased demand for access to forests and their expansion in the landscape. Rapid changes in farming, partly as a result of globalization, pose threats to rural environments, including loss of employment and services, but forestry offers opportunities to maintain population, economic activity and rural services. Although elements of forest strategies in the RoI and NI require further analysis, the balance of existing analysis suggests that objectives to increase forest cover are valid economically, socially and environmentally.

## References

Allott, N., G. Free, K. Irvine, P. Mills, T. Mullins, J. Bowman, W. Champ, K. Clabby, and M. McGarrigle (1998) 'Land use and aquatic systems in the Republic of Ireland', in P. Giller (ed.), *Studies in Irish Limnology*. Dublin: The Marine Institute, pp. 1–18.

Anon. (1998) *Statutory Instrument No.258 of 1998.* Local Government (Water Pollution) Act, 1977 (Water Quality Standards for Phosphorus) Regulation, 1998.

Bacon, P. (2004) *A Review and Appraisal of Ireland's Forestry Development Strategy. Final Report*. Killinick: Peter Bacon & Associates.

Byrne, K. and E. Farrell (2005) 'The effect of afforestation on soil carbon dioxide emissions in blanket peatland in Ireland', *Forestry*, 78: 217–27.

Carey, M. (2005) *The Native Woodland Business in County Wicklow from the 17th Century*. Dublin: Ireland's Native Woodlands Conference Proceedings, Galway, 8–11 September 2004. Dublin.

Clenaghan, C., P. Giller, J. O'Halloran, and R. Hernan (1998) 'Stream macroinvertebrate communities in a conifer-afforested catchment in Ireland: relationships to physico-chemical and biotic factors', *Freshwater Biology*, 40: 175–93.

Clinch, J. (1999) *Economics of Irish Forestry*. Dublin: COFORD.

Coillte (no date) *Biodiversity Action Plan for the Hen Harrier.* www.coillte.ie/ managing_our_forests/bio/Hen_Harrier2.htm.

Coillte (2005) *Recreation Policy – Healthy Forest, Healthy Nation.* Dublin: Coillte.

Cruickshank, M., R. Tomlinson, P. Devine and R. Milne (1998) 'Carbon in the vegetation and soils of Northern Ireland', *Biology and Environment: Proceedings of the Royal Irish Academy*, 98: 9–21.

Cummins, T. and E. Farrell (2003) 'Biogeochemical impacts of clearfelling and reforestation on blanket peatland streams 1: Phosphorous', *Forest Ecology and Management*, 180: 545–55.

Curtis, J. (2002) 'Estimating the demand for salmon fishing in Ireland', *The Economic and Social Review*, 33: 319–32.

Department of Agriculture, Food and Forestry (1996) *Growing for the Future: A Strategic Plan for the Forestry Sector in Ireland.* Dublin: Stationery Office.

Environment and Heritage Service (EHS) (2005) *Northern Ireland Species Action Plan: Hen Harrier.* http://www.ehsni.gov.uk/pubs/publications/henharrier_pdf.pdf.

FAO (Food and Agriculture Organisation of the United Nations) (2005) *Global Forest Resources Assessment 2005.* Country Report 085 – Ireland. Rome: FAO.

Farrell, E., G. Boyle and T. Cummins (1997) *A Study of the Effects of Stream Hydrology and Water Chemistry in Forested Catchments of Fish and Macroinvertebrates.* AQUAFOR Report 1. Chemistry of Precipitation, Throughfall and Soil Water, Cork, Wicklow and Galway regions. Dublin: COFORD.

Fennessy, J. (1986) *A Critical Review of the EEC Forestry Development Scheme – Regulation No. 1820/80.* Submitted as part of BA in Public Administration (Unpublished).

Fennessy, J. (2005) 'Foresight Report of the Forestry Sector in Ireland', in *Rural Ireland 2025 – Foresight Perspective.* Dublin: COFORD.

Fitzpatrick Associates (2005) *Economic Value of Trails and Forest Recreation in the Republic of Ireland: Final Report.* Dublin: Fitzpatrick Associates, Irish Sports Council and Coillte.

FNN (2006) *Forest Network Newsletter.* Issue 117, 14 September 2006.

Forest Service (NI) (no date) *Restoration of Native Woodland on Plantations on Ancient Woodland Sites.* http://www.forestserviceni.gov.uk/environment/ ancient_woodland.htm.

Forest Service (NI) (1993) *Afforestation – The DANI Statement of Environmental Policy.* Belfast: Department of Agriculture for Northern Ireland.

Forest Service (NI) (2003) *Economic Appraisal of Forest Policy.* http://www. forestserviceni.gov.uk/Priv_woodands/publications/misc/Economic%20Appr aisal%20of%20Forest%20Policy.pdf.

Forest Service (NI) (2004) *Options for Forestry: Consultation Paper.* Belfast: Department of Agriculture and Rural Development.

Forest Service (NI) (2005) *Annual Report 2004–05.* Belfast: Department of Agriculture and Rural Development.

Forest Service (NI) (2006) *Northern Ireland Forestry: A Strategy for Sustainability and Growth.* Belfast: Department of Agriculture and Rural Development.

Forest Service (RoI) (2000a) *Irish National Forest Standard.* Dublin: Department of the Marine and Natural Resources.

Forest Service (RoI) (2000b) *Code of Best Forest Practice – Ireland.* Dublin: Department of the Marine and Natural Resources.

Forest Service (RoI) (2000c) *Forestry and Water Quality Guidelines.* Dublin: Department of the Marine and Natural Resources.

Forest Service (RoI) (2000d) *Forest Biodiversity Guidelines.* Dublin: Department of the Marine and Natural Resources.

Forest Service (RoI) (2000e) *Forestry and the Landscape Guidelines.* Dublin: Department of the Marine and Natural Resources.

Forest Service (RoI) (2001a) *Forestry and Aerial Fertilisation Guidelines.* Dublin: Department of the Marine and Natural Resources.

Forest Service (RoI) (2001b) *The Native Woodland Scheme.* Dublin: Department of the Marine and Natural Resources.

Forestry Commission (2003) *Forest & Water Guidelines.* 4th edn. Edinburgh: Forestry Commission.

Forestry Commission (2004) *Forestry Statistics 2004.* Forestry Commission, Edinburgh. http://www.forestserviceni.gov.uk/publications/publications/misc/ fcfs004.pdf.

Forestry Commission and Forest Service (NI) (2004) *The UK Forestry Standard.* Edinburgh: Forestry Commission.

Forestry Commission and Forest Service (NI) (2005) *Public Opinion of Forestry 2005: Northern Ireland.* http://www.forestserviceni.gov.uk/publications/ publications/misc/ni_pub_op_for_2005.pdf.

Gallagher, G., E. Hendrick, and K. Byrne (2005) 'Preliminary estimates of carbon stock changes in managed forests in the Republic of Ireland 1990–2000', Appendix G in M. McGettigan, P. Duffy and N. Connolly, *Ireland National Inventory Report 2005: Greenhouse Gas Emissions 1990–2003 reported to the United Nations Framework Convention on Climate Change.* Wexford: Environmental Protection Agency.

Giller, P. and J. O'Halloran (2004) 'Forestry and the aquatic environment: Studies in an Irish context', *Hydrology and Earth System Sciences*, 8: 314–26.

Gilliland, J. (2005) *Short-rotation Coppice – The Irish Experience to Date.* http:// www.coford.ie/iopen24/pub/pub/Seminars/2005/Gilliland.pdf.

Hargreaves, K., R. Milne and M. Cannell (2003) 'Carbon balance of afforested peatland in Scotland', *Forestry*, 76: 299–317.

Hickie, D., R. Turner, C. Mellon and J. Coveney (1993) *Ireland's Forested Future: A Plan for Forestry and the Environment.* Belfast: The Royal Society for the Protection of Birds.

Jones, M. (1986) 'Coppice wood management in the eighteenth century: An example from County Wicklow', *Irish Forestry*, 43: 15–31.

Joyce, P. and N. O'Carroll (2002) *Sitka Spruce in Ireland.* Dublin: COFORD.

Kearney, B. and R. O'Connor (1993) *The Impact of Forestry on Rural Communities.* Dublin: Economic and Social Research Institute.

Kelly-Quinn, M., J. Bracken, D. Tierney and S. Coyle (1997) *A Study of the Effects of Stream Hydrology and Water Chemistry in Forested Catchments of Fish and Macroinvertebrates.* AQUAFOR Report 3. Stream Chemistry, Hydrology and Biota, Wicklow region. Dublin: COFORD.

Kelly-Quinn, M., D. Tierney, C. Coyle and J. Bracken (1996) 'Factors affecting the susceptibility of Irish soft-water streams to forest-mediated acidification', *Fisheries Management and Ecology*, 3: 287–301.

Kilbride, C., K. Byrne and J. Gardiner (1999) *Carbon Sequestration & Irish Forests.* Dublin: COFORD.

Kilpatrick, C. S. (1987) *Northern Ireland Forest Service – a History.* Belfast: Department of Agriculture for Northern Ireland.

Lehane, B., P. Giller, J. O'Halloran and P. Walsh (2000) 'Conifer forest location and fish populations in southwest Ireland', *Verh. Internat. Verein. Limnol.*, 27: 1116–21.

McCracken, E. (1971) *The Irish Woods since Tudor Times: Distribution and Exploitation.* Newton Abbot: David and Charles.

Neeson, E. (1991) *A History of Irish Forestry.* Dublin: The Lilliput Press.

O'Carroll, N. (2004) *Forestry in Ireland – A Concise History.* Dublin: COFORD.

O'Halloran, J., P. M. Walsh, P. S. Giller and T. C. Kelly (2002) *Forestry and Bird Diversity in Ireland: A Management and Planning Guide.* Dublin. COFORD.

O'Leary, T., A. McCormack and J. Clinch (1999) *Afforestation in Ireland: Regional Differences in Attitude.* Dublin: Department of Environmental Studies, University College, Dublin.

Ormerod, S., G. Rutt, N. Weatherley and K. Wade (1991) 'Detecting and managing the influence of forestry on river systems in Wales: Results from surveys, experiments and models', in M. Steer (ed.), *Irish Rivers: Biology and Management.* Dublin: Royal Irish Academy, pp. 163–84.

Papageorgiou, K., B. Elands, K. Kassioumis and T. O'Leary (2000) *Local Perspectives on European Afforestation.* www.dow.wau.nl/docs/presentations/ Aberdeen_2000.pdf.

Rackham, O. (1986) *The History of the Countryside.* London: Dent.

Rackham, O. (1995) 'Looking for Ancient Woodlands in Ireland', in J. R. Pilcher and S. S Mac an tSaoir (eds), *Wood, Trees and Forests in Ireland.* Dublin: Royal Irish Academy, pp. 1–12.

Reynolds, J. (1998) 'Human impacts on freshwaters', in *Ireland's Freshwaters*. Dublin: The Marine Institute, pp. 82–95.

Scott, D. (2000) 'Marking a decade of tree nesting by hen harriers in Northern Ireland, 1991–2000', *Irish Birds*, 6, BirdWatch Ireland.

Spatial Analysis Group, EEA (2004) *Revision of the Assessment of Forest Creation and Afforestation in Ireland.* EEA/AIR/AIR3/JLW. http:// friendsoftheirishenvironment.net/pdf/peatrevision.pdf.

Tomlinson, R. (1979) 'Water levels in peatlands and some implications for runoff and erosional processes', in A. Pitty (ed.), *Geographical Approaches to Fluvial Processes.* Norwich: Geo Abstracts, pp. 149–62.

Tomlinson, R. (1997) 'Forests and woodlands', in F. Aalen, K. Whelan and M. Stout (eds), *Atlas of the Irish Rural Environment.* Cork: Cork University Press, pp. 122–33.

Tomlinson, R. (2004) *Impact of Land Use and Land Use Change on Carbon Emission/Fixation.* Report to the Environmental Protection Agency on Project 2000-LS-5.1.2-M2.

Tomlinson, R. (2005) 'Soil carbon stocks and changes in the Republic of Ireland', *Journal of Environmental Management*, 76: 77-93.

Watts, W. (1984) 'Contemporary accounts of the Killarney woods 1580–1870', *Irish Geography*, 17: 1–13.

Western Regional Fisheries Board (2004) 'About Us'. www.rrfb.ie/aboutus/ boards_responsibilities.php.

Whelan, B. and G. Marsh (1988) *An Economic Appraisal of Irish Angling.* Report for the Central Fisheries Board. Dublin: Economic and Social Research Institute.

Chapter 9

# Governance and Sustainability: Impacts of the Common Fisheries Policy in Northern Ireland and the Republic of Ireland

David Meredith and Joan McGinley

**Introduction**

In 2002 the United Nations (UN) Food and Agriculture Organisation (FAO) estimated that 70 per cent of global commercial fish stocks were overexploited or in danger of being depleted through fishing activities (FAO 2002). At the World Summit on Sustainable Development (WSSD) held in Johannesburg later that year, world leaders reached a consensus acknowledging the significant contribution of marine fisheries to economic and food security and to biodiversity in general. A number of commitments were entered into at the WSSD designed to achieve sustainable fisheries. These included maintaining or restoring fish stocks to levels that can produce maximum sustainable yields (MSYs) through the management of fishing capacity (International Union for the Conservation of Nature [IUCN] 2003). From a European Union (EU) perspective, these are long-held policy objectives. In addition to regulating the EU fish market and negotiating access arrangements with non-member states, the Common Fisheries Policy (CFP) contains regulatory frameworks concerned with the allocation of fishing quotas on the basis of MSYs and legislative instruments aimed at regulating the size and structure of member states' fishing fleets. Despite applying these common laws since 1983, the EU Commission estimated that, in 2003–2004, of 43 fish stocks for which data is available, 81 per cent were over-fished and a further eight (18.6 per cent) were at their MSYs (COM 2006).

These data raise serious questions regarding the likely success of the strategy agreed at the WSSD in 2002. If the EU, with its significant resources and exclusive control over a large, contiguous maritime area, cannot implement an effective fisheries management system is it feasible to expect developing countries to succeed where many developed countries have not?

Though we do not address the latter issue directly, this chapter examines the socio-economic consequences of resource allocation and fleet control measures for fishing-dependent communities. A brief introduction to fisheries management theory and how the concept of sustainability is incorporated into conventional management systems precedes an overview of the structure and development of the CFP. We then focus on examining, with reference to Northern Ireland (NI) and the Republic of Ireland (RoI), the implementation and impacts of these policy measures.

## Fisheries management and sustainable development

'Exploitation of marine resources is generated by complex and polymorphous societies with multiple but real dynamics acting on several spatial and temporal scales' which have found their ability to respond to global changes threatened by the increasing constraints of integration and external control to which almost all productive activities are now submitted (Delbos and Premel 1996, p. 129). It is this response that state and supra-state policies seek to manage. Fisheries management is a relatively new area of study that, since its inception, has been dominated by biologists and more recently by economists. Despite limited availability of data and only a very basic understanding of oceanic processes, individuals such as Petersen sought to outline the relationship between fish stocks and human activities as early as 1894 (Symes 1996, p. 6). Initially only partial theories existed to explain how fish stocks (determined by the relationship between reproduction, growth, natural mortality and fishing-induced mortality) interacted with their environment and human activities. Following the Second World War the development of complex fish stock population models became feasible with the collection of considerable quantities of oceanic data. Nonetheless these remained of limited practical value as they concentrated on biological aspects of fishing and did not take into account the role of fishers and their possible impact upon resources (Holm 1996, p. 180). This prompted economists to contribute to fisheries management theory by introducing assumptions regarding fishers' behaviour into scientific principles to create bio-economic fisheries management theory. Eventually an integrated model was developed – complex enough to incorporate fishing-induced fluctuations whilst also allowing the prediction of future fish yields at given levels of fishing effort.

Figure 9.1 depicts the primary concepts incorporated into the classical bio-economic fisheries model. Combined within this model are the relationships between the number of fishing vessels, cost of fishing, and the return from a stock at a given level of effort. Sustainability, as will be seen, has long been a core component of fisheries management systems; notwithstanding this, the limited conception of what constitutes sustainability within fisheries must be considered one of the primary contributors to the emergence of unsustainable development practices.

In Figure 9.1 $E^1$ represents the maximum economic yield (MEY) – the point at which the least investment of capital results in the greatest economic value return from the fishery. Amongst economists and administrators concerned with economic efficiency, this point is perceived as the primary objective of any management

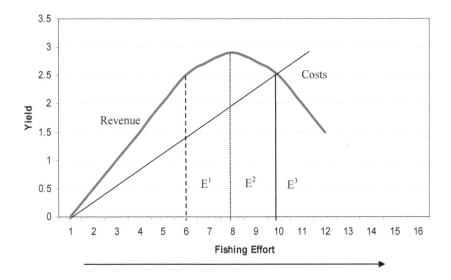

**Figure 9.1    Gordon-Schaefer Bio-economic Fisheries Model**
*Source*: Based on Grafton, Hill, Adamowicz, Dupont, Renzetti and Nelson, 2004, p. 108

regime 'as it is in the best interests of the owner that resource rents be maximized because these rents are the return from ownership of the resource' (Coull 1972, p. 13; Hartwick 1986, p. 253). However, as more fishers enter the fishery and more fish are caught, the MEY is soon surpassed. This is allowed to happen for a number of reasons, the most significant of which relates to the 'open access' nature of fisheries. As the term implies, access to fish stocks is construed as being largely unlimited and/or unregulated. In many states, the MEY is not considered the primary objective as social and rural/regional development policy concerns commonly perceive fish stocks as a means of supporting economic development activities in areas with few alternative resources. Notwithstanding this, once the MEY has been exceeded the return on investment for all fishers declines with the introduction of each new vessel. This process is commonly referred to as 'The Tragedy of the Commons', a term coined by Garret Hardin in 1968.

Under open access conditions the individual fisher has no incentive to do what benefits the group as a whole (Scott 1996). The resource is open to anyone who purchases a vessel and/or gear and hence there is unlimited competition for fish. In these circumstances 'no one fisherman is personally motivated to conserve the resource, for any fish returned to the water to grow larger in size will likely end up in the nets of a rival fisherman'; by the same token, 'where no individual is able to recoup an investment made in the fish stock, everyone will personally incline to neglect the future of the resource' (Copes 1981, p. 113). Based on such reasoning fishers willingly partake in the destruction of fish stocks, as benefits for acting otherwise are not returned to them or their communities under open access conditions. Fishing at this level is deemed to be beyond the maximum sustainable yield of the stock in question ($E^2$ in the diagram), and is defined by fishery biologists to be the point where reproduction equals natural and fishing-induced mortality, the MSY. $E^3$ represents the stage at which the fishery is depleted and fails to return any significant economic benefits to an economy.

At a theoretical level this model explains fish-stock dynamics and their interactions with fishers in terms of simple inputs (vessels) and outputs (catches). When applied to real-world problems and political decision-making processes such as those common to the EU, the model fails for a number of reasons that are considered below.

In formulating the most commonly-used theories, as in the model outlined above, assumptions had to be made so as to enable the modelling of intricate oceanic processes and environments. With continued declines in global fish populations, however, many are questioning the validity of those assumptions that can be shown to be a '…simplistic image of marine ecosystems, and a faith in the human capacity to predict and control them' (Holm 1996, p. 179). Take the following three assumptions: fishing constitutes an important determinant of the state of a fish stock; stocks are inherently stable, behave predictably at appropriate levels of exploitation and tend towards an equilibrium state; and fishers are rational individuals with a low capacity for collective action. It is currently accepted that the first of these assumptions is broadly correct with evidence from all fisheries of declining landings corresponding to increased fishing effort (Barinaga 1995). However 'the conclusion that fishing is the only, or even the primary determinant of the state of the stocks, is under increasing pressure' (Holm 1996, p. 183). There are indications that, within some fisheries, other factors are as important (if not more so) than those impacts induced solely through fishing effort (Grima and Berkes 1989). Whether regulation can help restore a fish stock is certainly a contentious issue. Bio-economic models not only fail to take account of the tendency towards instability within ocean environments, but also simplify the behavioural and interactive characteristics of fish species (Acheson and Wilson 1996). In the North Atlantic the ecosystem is characterized by being basically unstable; '…indeed this is one of the main reasons why this ecosystem can yield such enormous quantities of fish' (Sandberg 1996, p. 36). Furthermore, persistent and prolonged fishing effort can fundamentally alter the ecosystem as fishing gear, particularly the towed varieties, destroy plant life and reshape the seabed. In these circumstances the habitat of specific species may be altered to an extent that precludes recovery of fish stocks in the short or medium term.

In relation to the final assumption, case studies by social scientists indicate that fishers have collectively managed local fisheries in the past. It is only with the increasing influence of modern market economics and management systems that local co-operative/co-management structures have disintegrated (McGoodwin 1990; Pinkerton 1993). Many fisheries considered to be open access in the past were in fact regulated by complex socio-cultural based management institutions (Kalland 1996). These had the power to exclude, harvest, manage and control access to resources. With the subversion of these rights by state bodies through the imposition of top-down management regimes, local practices and structures were destroyed and, having no constraining moral obligations, fishers reacted by extracting the largest return from the fishery as promoted by market-based reasoning (Alegret 1996).

Despite intensive implementation of bio-economic models within open access fisheries over the past 50 years, resource depletion has become endemic to many. These models are, in theory and practice, intrinsically bound to science and the accompanying worldview 'of humans being apart from and above the natural world order' (Gadgil and Berkes 1991, p. 151). The dominance of this paradigm has contributed greatly to the success of maximizing exploitation of fish stocks and, to some extent, the value of fish taken from the world's oceans annually. However, modern fisheries administration models have failed largely due to their success in treating the MSY as a target within a highly competitive resource exploitation environment. In terrestrial-based economic activities 'that are not limited by some resource that cannot be augmented and where production units can be replicated anywhere without negative repercussions; competition will then result in increased productivity' (Hannessan 1993, p. 3).

Fisheries, however, prove to be a striking exception to traditional economic models precisely because the resource cannot be augmented and also because production units cannot be replicated anywhere. Nowhere are the shortcomings of science's 'world view' more obvious than when confronted by complex regional and local social and ecological systems such as those common to fisheries. These tend to vary spatially and temporally, rendering useless the assumptions and generalizations of modern scientific fisheries management models which are appropriate to conventional, rather than sustainable, resource development (Gadgil and Berkes 1991). This being true raises the question of how fisheries can be managed sustainably.

The first point to be made is that, ostensibly, the key problem is not fisheries science but rather the way research and recommendations have been interpreted, and frequently abused, by those responsible for the management of fisheries: the European Council of Ministers in the EU. In defence of fisheries science it can be said that fisheries present a particular difficulty to sustainable development, as it is generally perceived, because key factors controlling or strongly influencing stock productivity are not amenable to regulation (e.g. water temperature and weather conditions). Indeed it is only in recent years that core elements of oceanographic and climate sciences were incorporated into fisheries models.

Critically, however, sustainable development has largely developed around the premise that presupposes the negative affects of production can be mitigated through the adoption of new technologies or practices. Examples from Newfoundland show that this is not the case, as a total cessation of all commercial fishing activities has not resulted in the recovery of some fish stocks. Furthermore, unlike most terrestrial-based activities, the management of inputs and outputs is not practical and, in some instances, not possible. Simply put, when a fisher – particularly those exploiting demersal – or seabed-dwelling fish – casts a net, they cannot know with any degree of certainty the quantity or composition of their catch. As a result of these difficulties existing fisheries management regimes based on the regulation of inputs and outputs are considered by many to have failed.

Despite the considerable difficulties confronting the sustainable development of fish stocks, the concept has an important role to play in bringing about the conditions whereby available resources may be exploited in a rational manner. The significance of the concept lies in the holistic perspective it advocates when considering the management of practice and process. Increasingly it is incorporated into new or existing regulatory policies in an effort to secure the future of fish stocks. In many respects this is a simple continuation of previous management initiatives as the focus remains on fish stocks with relatively little consideration given to the socio-economic sustainability of fishing communities. However, there are moves amongst those charged with fisheries management to incorporate co-management approaches into fishery management policies (Dubbink and van Vliet 1996; Couper and Smith 1997). The EU has recently developed institutional structures empowering stakeholders at the regional level to engage directly with fisheries management processes.

Although there is no widely-accepted definition of co-management, the term is generally accepted to imply the integration at various levels of local, national and, in the case of the EU, supranational management systems involving 'genuine power sharing between community based managers and government agencies, so that each can check the potential excesses of the other' (Pinkerton 1993, p. 37; Gadgil and Berkes 1991, p. 12). This approach recognizes that supranational fishery management policies applied at local levels are rarely successful, as communities perceive them to be inequitable and ineffective given their specific social, cultural and ecological experiences. Effective co-management systems harness these experiences and establish a partnership between the stakeholders and top-down administrators. As fishers have a broader contextual knowledge of the ecosystem upon which they depend and interact with on a regular basis, they are ideally positioned to monitor those changes that affect fish stocks. Fishers are also positioned to react to or, in some cases, anticipate changes to the fishery through modification of their activities. It is the objective of co-management governance bodies to observe these actions and ensure that they are indeed in the best interests of the fishery.

As will be demonstrated in the next section, the Common Fisheries Policy has evolved over a long period to incorporate sustainable development as a core objective, and in recent years has devolved much of the responsibility for achieving sustainability to national and regional levels through the development of co-management structures. The next section examines this evolution in detail with a description of the structure and evolution of the CFP from 1970–2002.

## The Common Fisheries Policy, 1970–2002

The CFP is one of the few fully-fledged European Union policies with common rules and regulations governing all aspects of member states' commercial sea fisheries. Moves to instigate the development of a common fisheries policy were first made by France and Italy in 1967 following trade liberalization within the European Economic Community (EEC) common market area and subsequent decline of their fishing sectors. The mandate for a CFP is contained in the 1957 Treaty of Rome. Article 38 (4) of this treaty provides that: 'The function and development of the Common Market in respect to agricultural [fish] products shall be accompanied by the establishment of a common agricultural [fisheries] policy among the member states' (EEC 1957). Article 39 further develops this aim by providing basic objectives for the policy, which include: increasing productivity; raising income levels amongst fishers; stabilizing markets; and guaranteeing regular supplies and ensuring reasonable prices for the consumer, whilst also recognizing and taking into consideration the importance of fishers' activities within peripheral areas (EEC 1957).

Utilizing the opportunity presented by the French and Italian request, the European Commission undertook a comprehensive review of fishing in each of the then six member states. Focusing on fleet structure, supporting infrastructure, market organization, external trade and finally social aspects of fishery operations, the study provided a framework from which it was possible to draft a Common Fisheries Policy. Nevertheless, it was not until 1969 when Denmark, Norway, the RoI and the United Kingdom applied for membership of the EEC that any further action was taken. All four applicants had major fishery interests or resources, on which the six members depended to varying degrees for fish supplies. The possibility of gaining permanent access to these resources provided considerable motivation for the existing members to begin talks with the RoI, and also with Norway, Denmark and the United Kingdom (Wise 1984; McGinley 1991). Before accession negotiations could begin, however, it was vital that a common policy be agreed amongst the six as failure to do so would enable newly joined members to influence any future development of the policy – particularly those regulations relating to conditions of access to resources (Holden and Gorrod 1994). Although a consensus was arrived at only a few hours before accession negotiations commenced, the policy became part of the *acquis communitaire* that the applicants had to accept.

## The Common Fisheries Policy, 1970–1973

The policy agreed to by France, Belgium, Luxembourg, the Netherlands, Germany and Italy consisted of two separate and distinct bodies of legislation. Structural matters were considered in a body of regulations that sought to promote fleet modernization and reconstruction whilst market-related issues formed the second sub-policy. Although mutually exclusive these sub-policies were complementary by virtue of the deficit in fish supplies experienced by the six member states at this time.

Originally the structural policy aimed to promote the introduction of new, more efficient vessels, technologies and gear. It focused on increasing the capacity of member states to catch greater quantities of fish, thereby reducing import requirements (OJEC No. L236/1, p. 704). Achieved through provision of financial subsidies, this led to growth in the EU's fishing capacity and can be considered a significant contributor to contemporary problems of overcapacity. Long-term impacts of increased fishing effort and subsequent increased exploitation of stocks, many of which are considered in retrospect to have been at or even to have surpassed maximum sustainable yields, were not considered. The reason for this is that there was little scientific expertise within the Commission during this period (Holden and Gorrod 1994).

During the 1980s and 1990s the structural policy was altered to take into account the growing problems of overcapacity and continued depletion of the EU's most important fish stocks. These changes came about as a result of the growing awareness and understanding within the Commission of the consequences of overfishing and also an increased political willingness to occasionally take appropriate, unpopular action. By the late 1990s the structural policy was largely focused on restructuring the EU's catching sector so as to balance fishing capacity with available resources. Grant aid for the construction and modernization of fishing boats was severely curtailed, whilst a series of Multi-Annual Guidance Programs (MAGPs) were introduced designed to regulate member states' fishing capacities. These programs were elevated into a distinct Fleet Policy following the 2002 review of the CFP.

The Markets Policy, although having a number of objectives – including the establishment of marketing standards, stabilization of market prices, avoidance of surplus formation, supporting producers' incomes and protecting consumers' interests – had as its primary focus the provision of a guaranteed market for fish as a means of promoting greater volumes of landings (OJEC No. L236/1). As greater expertise was introduced to the section responsible for fisheries policy at the EU level (DGXIV or DG FISH as it is now known) and the fish market became better understood, the policy has progressively evolved. Overall, however, this section of the CFP has remained relatively stable by virtue of the fact that the '...Community has an increasing deficit in supplies of most fish species' associated with increasing demand and decreasing supply (Holden and Gorrod 1994, p. 34). This statement, unfortunately, remains valid at the time of writing.

**The Common Fisheries Policy, 1976**

Events at the global scale prompted the next phase of the CFP's development. The Community was required to react following general acceptance that 200 nautical mile exclusive fishery zones (EFZs) could be successfully established unilaterally

(Glassner 1993). Coastal states immediately set about extending their territorial limits, thereby excluding foreign fishers from what were now national waters. The European Council of Ministers followed international trends and set a deadline of 1 January 1977 for adoption of exclusive economic zones by all Atlantic member states. Enlargement was necessary for a number of reasons, foremost among which was the decision by Canada, Iceland and the United States to create enlarged North Atlantic EFZs. Henceforth fishing by other states in these waters would be illegal. Although only seriously affecting the distant water fleets of the United Kingdom, Denmark and (Western) Germany, the EC could not afford, biologically or economically, to entertain fishing capacity redirected to the waters of member states (DGXIV 1996). Consequently, the external policy has two objectives:

- to regularise external fisheries relations between member and non-member states, and
- to maintain as many distant-water vessels on fishing grounds outside of the EU as possible (Holden and Gorrod 1994, p. 35).

These regulations act as a common foreign fisheries policy that was drafted in such a manner as to enable one body to regulate all access agreements with third countries on behalf of the EC/EU. Under this system, member states are unable to compete amongst themselves for rights to fish in the waters of non-member states. The policy has assumed greater Irish significance in recent years with the introduction of the Atlantic Dawn, the world's largest fishing vessel, to the Irish fleet register. This vessel avails of the EU's agreement with Mauritania to fish off the west coast of Africa.

**The Common Fisheries Policy, 1983**

The ratification of the external policy and extension of Atlantic member states' EEZ's to 200 nautical miles, thereby creating a common fisheries area, resulted in the need for the development of a common fisheries management system. The regime would apply to all EC Atlantic and Baltic waters as a consequence of:

> ...the perspicacity of the Commission officials who had drafted Article 2 of Regulation 2141/70 on equal access, the existence of this article meant that, following the agreement of all Member States to extend their fishery limits to 200 miles...a community sea to which a common conservation policy could be applied was automatically created as from 1 January 1977 (Holden and Gorrod 1994, p. 40).

The objectives of this regime were set out in a 1976 Commission document that anticipated the ratification of the external policy (COM (76) 80, 18.2.76). It proposed the adoption of total allowable catches (TACs) on the basis of the following: Maximum Sustainable Yields; allocation of quotas by species to each member; establishment of control and enforcement systems; special provision for fishery-dependent communities in the RoI, Northern Britain and Greenland; and the development of 'conservation' measures. Consequently this body of legislation became known as 'the conservation policy'.

Notwithstanding the efforts of the Commission in establishing detailed goals for this policy in 1976, it took six years of negotiations before all members ratified the regulations in question. The policy ratified in February 1983 was in many respects the same as that which had been proposed by the Commission in 1976. Overall, the policy's goals remained virtually unchanged. Six years had been spent essentially refining regulatory details and the fine print associated with EU legislation, resulting in a plethora of articles, paragraphs and sub-paragraphs, some of which conflicted with each other or with other elements of the CFP. From the perspective of policy makers and industry stakeholders, the most important elements of this policy related to resource allocation. What few had considered were the potential implications of fleet management measures.

### EC/EU fleet management initiatives

Having failed to sustain resources through the control of outputs by imposing catch limitations (TACs and quotas), the EC responded by introducing legislation that strengthened regulations contained in the 1970 CFP, limiting the catching capacities of member states' fleets. These restrictions are implemented within the structure of MAGPs; national development plans were prepared whose acceptance by the Fisheries Commission 'translates the institution's commitments with regard to the proposals made by each member state into its sectorial development's plans' (DGXIV 1996, p. 20). Implicit in all this were commitments to reduce the capacity of each state's fleet. The underlying conceptual framework of MAGPs rested on the knowledge that biological limits had been reached for many of the more important fisheries and that further expansion of catching capacity would not be countenanced. MAGPs were to establish broad limits within which each state was given the freedom to choose how their fleets would be structured. From a rural socio-economic development perspective this was of considerable importance given the ability of the fishing industry to support considerable numbers of people through direct and indirect employment in areas with few if any alternative indigenous industries. A member state could decide to encourage many small enterprises, a few large enterprises or a combination of both. This point is critical as member states were responsible for the national/regional institutional framework, setting goals for their industry and their approach to (sustainable) development. Ultimately this national/regional governance system shaped and determined the social and economic impacts of common EC/EU policies – in this case the CFP. And it is those (political) decisions which privilege one form of sustainability (e.g. biological rather than social) that determine the subsequent impact of the CFP at local levels.

When first proposed in 1970, MAGPs were seen as a means of co-ordinating aspects of the structural policy and enabling financial assistance to be better planned (OJEC No. L236, p. 704). Their aim was redefined in 1983 as achieving 'a satisfactory balance between the fishing capacity... and the stocks, which are expected to be available during the period of validity of the programs' (OJEC L290, p. 2). Before ratifying and implementing this legislation it was necessary for members to adopt a means of quantifying a fleet's cumulative fishing capacity. Following considerable debate, gross registered tonnage[1] (GRT) and engine power (kWs) were accepted as appropriate indicators of a vessel's capacity to catch fish. GRTs and kWs are, however, arbitrary measures that do not reflect actual catching capacity, as 'the capacity of a fishing fleet may be increased significantly in different ways, including the incorporation of more effective fishing equipment or the installation of electronic equipment to facilitate the detection of resources' which may not have any discernible impact on either GRT or kWs (OJEC No. C180, p. 7). Reflecting only the potential of a boat to catch fish, this system made any significant reduction in the level of overfishing improbable (Holden and Gorrod 1994). That such a system was adopted reflects the prevailing attitude of the time: that fish were plentiful and consequently members should be allowed continue their fleet expansion (Wise 1984). However, as problems of overfishing became more apparent throughout the 1980s and 1990s supporting legislation for MAGPs and capacity restrictions was progressively strengthened.

By and large member states treated the first MAGP (1983–1986) as a paper exercise, with little consideration given to the plans submitted and agreed with the Commission not to increase the amount of fishing effort. MAGP II, which operated between 1987 and 1991, was more tightly monitored with EU grant-aid linked to attainment of the agreed targets. The third MAGP was to be adopted by 1 January 1992 and was to conclude on 31 December 1996. However, the negotiation of this MAGP proved difficult and an early agreement was not forthcoming. The difficulties stemmed from the inability of many member states to achieve the objectives of MAGP II, the impact reductions had had on fishery-dependant regions, the ongoing economic crisis brought on by overfishing, and the scale of restrictions being proposed by the Commission in the third programme. Consequently, one-year interim measures were adopted and these filled a legal vacuum in the absence of a full MAGP. Prior to commencement of MAGP III negotiations, the Commission tasked a group of independent experts, led by Professor Gulland, to assess the biological status of the more important fish stocks. The Gulland report (1990) recommended an immediate minimum fishing effort reduction of at least 40 per cent.

Following fraught debate, the Commission proposed 30 per cent cuts for those fleets involved in demersal fisheries, 20 per cent for crustaceous flatfish,

---

1    Subsequently GRT was replaced by Gross Tonnage (GT) in an effort by the EU Commission to agree a common means of measuring capacity between member states.

and no reductions for those exploiting pelagic species.[2] Eventually agreement was reached with the member states that involved accepting a 20 per cent decrease in demersal fleets, 15 per cent in those exploiting crustaceous and flatfish and no decrease in fleets exploiting pelagic stocks. When it came to negotiating the fourth MAGP there were a number of difficulties, most of which derived from member states' reluctance to agree to the MAGP III capacity cuts. Although scheduled to commence on 1 January 1997, it was not formally ratified by the Council of Ministers until 3 July 1997. The main proposals contained in this programme focused on the continuing need to bring fishing capacity into line with catching opportunities through reduction of the fishing effort.

In order to prepare MAGP IV, a group of independent experts was once again tasked with reviewing the state of the stocks and commenting on the reductions in fishing effort that were needed. The conclusions of this group, similar to those of the Gulland Report, formed the scientific basis for the Commission's proposal for a Council decision to fix the guidelines for the MAGP IV. In the event the Council of Fisheries Ministers refused to accept the scale of the effort reductions proposed by the Commission – approximately 30 per cent of the total fleet (COM 2000, 272, p. 3). Under a compromise proposal developed by the Irish presidency, adopted in June 1997, the reduction rates applied to the segments were weighted according to the proportion in total catches of depletion risk and over-fished stocks. This made the MAGP IV objectives very much less ambitious than those proposed by the Commission, the global objectives for the Community fleet representing a reduction of approximately 5 per cent over the 5-year period. This is about half the reduction achieved by the MAGP III. Moreover, six member states (France, Germany, RoI, the Netherlands, Sweden and the United Kingdom) opted to achieve their objectives in certain segments by adjusting activity as well as capacity. This means that the objectives of MAGP IV represented a reduction of between two and three per cent in terms of capacity (COM 2000, 272, p. 10).

## The Common Fisheries Policy, 2002

Twenty-five years after the CFP was finalized a major review was undertaken to amend regulations and introduce new policy structures and instruments that reflected contemporary circumstances. These included an acceptance within DGFISH that management measures based on total allowable catches and quota allocations, agreed in 1983, had failed and that significant changes were now required. After much negotiation amongst member states it was agreed that MAGPs would be replaced by a Fleet Policy focused on managing fishing vessels and fishing capacity. Changes to the quota allocation system saw a move towards

---

2   Normally pelagic species are p found at the surface and include mackerel, herring, tuna etc. Demersal species, in contrast, are those that live on or near the seabed and include cod, haddock, hake, monkfish, etc.

the introduction of multiannual quotas within either Stock Recovery Plans (SRP) (where fisheries were depleted through overfishing) or Stock Management Plans (SMP) (where the fishery was under the MSY). These can be construed as co-management structures with the state responsible for drafting and implementing SRPs and SMPs that have been agreed with the EU Commission. New regional co-management structures were also introduced in 2002 as a means of providing a forum for industry, policy and scientific stakeholders that would transcend national boundaries and bring together those with a relevant interest in key fisheries.

These developments stem from a belated acceptance within the EU Commission that many of the contributing factors to overfishing related to the unwillingness of states to implement measures that would have a detrimental impact on their own fishing communities. In effect, the introduction of these governance structures removes the EU Commission from the management equation and places full responsibility for the sustainable development of fisheries with member states and industry stakeholders. In both the Republic of Ireland and Northern Ireland these developments have led to the introduction of new legislation setting out a national governance framework for fisheries in which individual fishers and others within the fishing industry are subject to financial penalties and criminal charges for failure to comply with the law. Strange as this statement may seem to those unfamiliar with fisheries and fisheries management, there has long been a reluctance to enforce strict management legislation – particularly in the Republic of Ireland. The general response of fishers and the industry has been to withdraw co-operation from co-management bodies. As an aside, it is worth noting a similar response amongst organisations representing farming interests to the introduction of environmental legislation that restricted certain activities and applied financial and criminal penalties for non-compliance.

## Impact of fleet management policies in the Republic of Ireland and Northern Ireland

Before commencing MAGP II in 1986, the Republic of Ireland proposed to the Commission (OJEC, L67/27) that the fleet would be cut by 9,124 GRT by the end of 1991. While these proposals were being put to the Commission, three new pelagic vessels with a combined tonnage of 7,724 GRT were registered, increasing the capacity reduction target from just over 9,000 to 16,484 GTR. In 1982 the 24 largest vessels in the fleet had a lesser-combined GRT than the three new vessels permitted to enter the fleet. Since capacity was now subject to limits under the MAGP, state approval of these three vessels provides a clear indication of support for and encouragement of capacity accumulation in the fleet.

During the late 1980s the whitefish sector began to recover from severe financial difficulties that had affected it for over a decade. This revival was reflected in the 260 applications for new whitefish vessel licences, which rested with the Department of the Marine (DOM). Due to the necessity of achieving cuts of 16,484 GRT no new licences were approved, and in May 1989 a formal embargo was imposed (DOM 1998, Oral Communication). Eventually the ban was lifted in May 1990 with the introduction of a new policy stipulating that the introduction of new or second-hand boats into the country required the removal of equivalent capacity from the fleet register. This has since become known as the 'equivalent tonnage' policy and ultimately gave rise to the trade in tonnage. Under the new policy, capacity can be bought and sold on an 'open market', thereby allowing the market to value fishing tonnage.

In taking these actions the state made the right to fish a capital asset as the vessel capacity is separate from the actual fishing boat, the equipment, gear and licence. In reality it is a newly created, limited resource that was very much in demand. However, the pressure generated by the introduction of the three large pelagic vessels in the mid 1980s meant that not only was further fleet expansion impossible for the whitefish sector, but whatever capacity reductions were required in future MAGPs would have to come from this sector. With the introduction of a trade-in capacity, the whitefish sector was further disadvantaged given the lack of capital traditionally available to demersal fishers. Conversely the pelagic sector, being modern and therefore seen as more progressive by the state and financial institutions, had greater access to capital, thus allowing it a comparative advantage in the capacity marketplace.

Following the licensing embargo, and decisions allowing the introduction of pelagic vessels at the expense of the renewal of the whitefish fleet, there was considerable anger among demersal fishers. This anger eventually manifested itself in the actions of a group of fishermen from Dingle. These individuals had purchased boats from other countries in 1990, and proceeded to fish without either a licence or having purchased the required tonnage. Within the industry it was presumed that these fishermen had been given unofficial sanction by the then Taoiseach of the country who had a holiday home not far from Dingle. Throughout 1991, a number of these vessels were arrested, some more than once. Pressure, from the boat operators and from the sector in general, mounted for the situation to be regularized. Exacerbating the climate of ill will within the industry was the seeming unwillingness of the state to take action against these illegal operators. Indeed, despite having been arrested and detained on several occasions none of these boats was ever prosecuted. On 18 July 1991, the Minister for Marine announced a special whitefish scheme whereby up to 20 licences were to be issued, ostensibly for the introduction of 20 all-weather whitefish boats into the fleet. By this stage 11 such vessels were in the country fishing illegally. When the licence allocations were announced at the close of 1991, vessels that had up to then not complied with the capacity legislation received licences. Many other applicants, who had abided by the law and were awaiting permission from the Department prior to purchasing new vessels, were denied access to the special scheme.

Two sets of conditions were issued with the licences – one for those who were non-compliant and another for those who had yet to purchase their vessels. The common element to both however, was the phased introduction of replacement tonnage:

- 5 per cent of relevant tonnage would be removed prior to granting of the initial licence for those yet to obtain their vessel and within 28 days of receipt of the approval for those with vessels already fishing;
- A further 15 per cent before the end of the first year of the licence; and
- 20 per cent before the end of each subsequent year until 100 per cent was removed.[3]

By the autumn of 1992, a number of the 20 vessels were experiencing difficulties and the five-year limit for introducing the replacement tonnage was extended to seven. In 1993 further difficulties were encountered and the scheme was extended to 8.5 years. With operational problems persisting through January and February of 1994 some banks refused to provide any further finance to several of the licensees. After much discussion, the DOM announced a settlement through implementation of a Special Whitefish Licence Scheme in December 1994. The main thrust of this scheme was:

- The combined owners had to make a once-off payment of at least £100,000;
- Purchasing not less than 150 tonnes; and
- 'Vessels could not be sold within the Irish fleet before 1/1/98 without the full balance of the replacement tonnage being withdrawn' (Department of the Marine, correspondence 25/7/1994).

Arising out of this apparent 'favouritism' towards the demersal sector, pressure was brought to bear by those lobbying on behalf of pelagic vessel owners, many of whom are based in Killybegs, Co. Donegal. This ultimately resulted in the introduction of a special pelagic vessel renewal scheme which allowed further modernization and accumulation of capacity in the pelagic sector. The essential components of this agreement were:

- The pelagic sector could purchase replacement tonnage from the demersal sector – a practice which until then had not been permitted; and
- Those introducing new pelagic boats were granted permission to buy tonnage over a phased period of two to three years.

These developments took place in the Republic of Ireland within a tightening fleet regulatory environment resulting in, as will be explored below, a sharp decline in the number of fishing vessels. The impacts of this process were diverse, with some fishers and ports benefiting from new investments whilst other areas and communities experienced significant declines. Overall there was a significant concentration of

---

3   This information is taken from the formal agreement between the DOM and a recipient of one of the licences that was made available by the recipient to the authors for the purposes of this research.

**Table 9.1     Evolution of Northern Ireland's Fishing Fleet, 1994–2002**

|                  | 1994 | 1996 | 2002 |
|------------------|------|------|------|
| **Under 10 Metres** | 274  | 224  | 167  |
| **10–18 Metres**    | 104  | 94   | 62   |
| **18–24 Metres**    | 108  | 78   | 74   |
| **Over 25 Metres**  | 15   | 15   | 16   |

*Source*: Department of Agriculture for Northern Ireland – Fisheries Division

fishing capacity in a limited number of ports and communities including Killybegs, Castletownbere, Greencastle, Rossaveal and Dunmore East/New Ross.

In contrast to the Republic of Ireland experience, the impacts in Northern Ireland were substantially different. Whilst the fishing fleet in the region did decline in terms of the number of vessels, there was no significant concentration of remaining capacity. This development is largely explained with reference to the fact that restructuring did not lead to the expansion of the number of larger vessels. Fleet changes did, however, result in substantial declines in the number of boats under 24 meters in length (Table 9.1).

The different outcome to common EC/EU fishery regulations in the Republic of Ireland and Northern Ireland is interesting as it points to the role of the state, rather than to that of the EC/EU, in shaping the impact of regulations on fishing communities. This issue is explored in greater detail below.

**EU fleet management strategies: Impacts in Northern Ireland and the Republic of Ireland**

Between 1993 and 2004 the number of fishing boats decreased by 13 per cent in Northern Ireland and by over 32 per cent in the Republic of Ireland (Tingley 2006; CSO 2007). Analysis of the structure of the fishing fleets indicates that, in both jurisdictions, the number of vessels between 12 and 24 meters declined to a greater extent than smaller and larger vessels (Meredith 2001; Tingley 2006). In many respects the decline in mid-sized vessels is unsurprising given that they were faced with increasing pressure from both larger and smaller vessels.

Differences in the scale of fleet restructuring reflect global processes and national and/or regional fleet management objectives. The global processes in question include increases in economies of scale driven by reduced catching opportunities as well as increasing fuel and other input prices, resulting in greater competition within fisheries. Though global processes are of significance in understanding the drivers of change affecting fishing fleets, it is clear from the analysis presented thus far that if one is to understand the differences in and the spatial impacts of fleet restructuring in the Republic of Ireland and Northern Ireland, then we must consider the state's approach to fleet management.

As was outlined above, the state plays a key role in determining the evolution of fishing fleets through policy objectives and the implementation of strategies designed to restructure the scale and composition of fleets. The two most common strategies include fleet renewal and capacity reduction programmes. So as to encourage capacity reduction, states can avail of EU funding to incentivize fishers to decommission fishing vessels. In the period under consideration (1991–2004) both the Republic of Ireland and Northern Ireland implemented decommissioning schemes. The EU provides support for these schemes as a means of encouraging and assisting states to achieve agreed MAGP capacity-reduction targets.

Comparing the operation of these schemes in the Republic of Ireland and Northern Ireland highlights significant differences in their design and implementation. In a review of the structure and safety of the Irish fishing fleet the DOM stated that the objective of decommissioning was the removal of older, less efficient vessels by the RoI (DoMNR 1996). This practice of targeting older vessels was continued in more recent (2005 and 2006) decommissioning schemes operated by the Republic of Ireland with the aim of achieving '…a more modern and efficient fleet' (BIM 2006, p. 25). In contrast, the decommissioning schemes operating in Northern Ireland, focused on younger (a minimum of 10 years old) and larger (at least 10 metres) vessels targeting specific species, especially cod (Stationery Office Limited 1997).

The differences in approach to decommissioning largely reflect differences in fleet management objectives. That the entire reduction in the number of Northern Ireland fishing vessels between 1993 and 2004 is accounted for by decommissioning schemes indicates that achieving MAGP capacity reduction targets was the primary goal of these fleet-reduction measures. In contrast, within the Republic of Ireland, only 9 per cent of the decline in the number of fishing vessels is accounted for by the 1996 decommissioning scheme (DoMNR 1996; BIM 2000; 2004). This was the only fleet-reduction programme operated in the RoI during the 1993–2004 period.

It is clear that there are additional processes at work within the Republic of Ireland fleet that account for the bulk of restructuring activity. Assessment of the 1991–2004 period highlights the emergence of a policy of reducing fleet size and concentrating the remaining vessels in fewer ports. Approximately €200 million was provided by the state for investment in new demersal fishing vessels during the 1998–2003 period. These subsidies were given despite advice from the Marine Institute, the Republic of Ireland's fishery science agency, that there was insufficient quota or indeed fish available to support the introduction of new, larger vessels. Under national regulations, capacity equivalent to that of the vessel being introduced has to be removed from the national fishing fleet before the new vessel can commence fishing activities. This ensures that the overall fleet size cannot be increased.[4]

---

4    These rules were breached when the Atlantic Dawn, which had first been registered as a mercantile vessel, was allowed to enter onto the Irish Fishing Fleet Register.

The introduction of over 33 new, larger and more powerful demersal vessels resulted in the capacity of smaller vessels being purchased and the fishing capacity transferred to the new boats. The smaller boats were then scraped or sold outside of the EU. In affect this led to the concentration of fishing capacity into fewer vessels and ultimately fewer ports. It is this process that accounts for the large fall in the number of vessels on the Irish Fishing Fleet Register. At this point it is worth noting that, since 2004, a series of decommissioning schemes have been introduced to remove 25 per cent of the total capacity of the whitefish fleet. The primary aim of these schemes is to remove 'excess' fishing capacity from fisheries where the stock is considered to be at risk or already depleted.

Though centralization of capacity is economically and socially destructive for smaller communities, and ecologically unsustainable owing to the need for larger boats to catch more fish in order to remain viable, it corresponds to the stated vision of future Irish fisheries. Mr T. Carroll, secretary of the Department of the Marine, stated at the 1990 Sherkin Island Conference, that he foresaw 'the centralization and amalgamation of the catching sector into one or two ports'. Contrasting this position is a statement by Mr. Gerry Lavery, the then Fisheries Secretary at the Department of Agriculture in NI:

> The Department is committed to retaining employment and vessels in the three ports. There is no question of us doing anything that would lead to the ports being ostracized or starved of resources. If we were to reach a point where we had such small landings and such a small fleet that it was no longer viable to have three ports, that would have been signalled well in advance to the industry, to the councils and so on and we would be having very wide ranging consultations. We are not signalling that. We want to see the traditional pattern of employment in fishing and fish processing maintained. That is a very important commitment (Lavery 1996).

## Conclusion

The manner in which EU fisheries have been managed over the past 25 years provides 'a telling commentary on the idea of sustainable development and the way this concept has been appropriated and applied by policy makers' (Drummond and Symes 1996, p. 152). Despite being the specified goal of the Common Fisheries Policy, the industry is currently enduring a crisis largely induced by a lack of resources. To date the EC/EU has stressed the sustainability of fish stocks and fishing activities in a manner that assumes 'that to speak of natural resource sustainability is to speak at one and the same time of the sustainability of communities' (Ommer 1996, p. 3). Yet this assumption is not borne out in the national-level implementation of structural, conservation, external, market or fleet policies that often see inherent contradictions resulting in the disenfranchizing of communities and their ability to act sustainably. Overfishing and accumulation of capacity, to mention but two outcomes of current policies, whilst not sustainable, are the logical reaction by fishers who find their livelihoods, culture and communities under threat.

Within the Republic of Ireland, the state supported and fostered these developments through a number of decisions taken to restructure both the pelagic and whitefish fleets during the 1990s. EU-supported decommissioning schemes were implemented in a manner designed to remove older, less-efficient vessels rather than targeting, as happened in Northern Ireland, vessels with the capacity to undermine fish stocks. These results demonstrate that the state, rather than the EU, has considerable influence over the outcome of common policies. With this in mind it is worth noting that the responsibility for the development and implementation of fishery management plans was allocated to member states following the 2002 review of the CFP. The (re)empowerment of states potentially offers significant scope to change the development trajectory of the fishing industry and fishing-dependent communities. The data presented above highlights the differentiated outcomes that occur within alternative regulatory systems. The issue for states will increasingly be one of choosing how they wish to approach the issues of attaining sustainability. Is it to be on the basis of a few large boats, a mix of larger and smaller vessels or a greater number of more geographically dispersed, small fishing vessels?

## References

Acheson, J. and J. Wilson (1996) 'Order out of Chaos. The Case for Parametric', Fisheries Management', *American Anthropologist*. Vol. 93, 2: 579–94.

Alegret, J. (1996) 'Ancient Institutions Confronting Change: The Catalan Fishermen's Cofradias', in K. Crean and D. Symes (eds), *Fisheries Management in Crisis*. Oxford: Fishing News Books, pp. 92–8.

Barinaga, M. (1995) 'New Study Provides Some Good News for Fisheries', *Science*. Vol. 269, 1043.

BIM (2000) *Annual Report*. Dublin: Bord Iascaigh Mhara.

BIM (2001) *Annual Report*. Dublin: Bord Iascaigh Mhara.

BIM (2003) *Annual Report*. Dublin: Bord Iascaigh Mhara.

BIM (2004) *Annual Report*. Dublin: Bord Iascaigh Mhara.

BIM (2006) *Building a Sustainable Future for Ireland's Fishing Fleet*. Dublin: Bord Iascaigh Mhara.

COM (2000) 272 final. *Report on the Preparation for a mid term review of the Multi-annual Guidance Programmes (MAGP)*. Brussels: EU Commission.

Commission Decisions 92/588/EEC to 92/598/EEC of 21 December 1992 on a *MultiAnnual Guidance Programme for the fishing fleet for the period 1993 to 1996 pursuant to Council Regulation (EEC) No 4028/86. (OJEC No L 401 1992)*. Brussels: EU Commission.

Commission Decisions 94/180 of 15th June 1994 on laying down guidelines for *Global grants or integrated operational programs for which member states are invited to submit applications for assistance within the framework of a*

*Community initiative concerning the restructuring of the fisheries sector (PESCA). (OJEC No C 180 1992).* Brussels: EU Commission.

Copes, P. (1981) 'Rational Resource Management and Institutional Constraints: The Case of the Fishery', in A. Butlin (ed.), *Economics and Resource Policy*. London: Longman, pp. 113–29.

Coull, J. (1972) *The Fisheries of Europe: An Economic Geography*. London: Bell & Sons.

Council Directive (EEC) 83/515 of 4th October 1983 concerning *Certain measures to adjust capacity in the fisheries sector. (OJEC No 290 1983).* Brussels: EC Council.

Council Regulation (EEC) No 2908/83 of 4th October 1983 on a *Common measure for restructuring, modernising and developing the fishing industry and for developing aquaculture. (OJEC No 290 1983).* Brussels: EU Council.

Council Regulation (EEC) No 2909/83 of 4th October 1983 on *Measures to encourage exploratory fishing and co-operation through joint ventures in the fishing sector. (OJEC No 290 1983).* Brussels: EU Council.

Council Regulation (EEC) No 4028/86 of 18 December 1986 on *Community measures to improve and adapt structures in the fisheries and aquaculture sector as last amended by Regulation (EEC) No 3946/92 of 19 December 1992 – (OJEC No L 401 1992).* Brussels: EU Council.

Couper, A. D. and H. D. Smith (1997) 'The Development of Fishermen-Based Policies', *Marine Policy*. Vol. 21: 111–19.

CSO (2007) *Fishing Fleet Structure by State, Statistic and Year*. Dublin: Stationery Office.

Delbos, G. and G. Premel (1996) 'The Breton Fishing Crisis in the 1990s: Local Society in the Throes of Enforced Change', in K. Crean and D. Symes (eds), *Fisheries Management in Crisis*. Oxford: Fishing News Books, pp. 129–40.

Department of the Marine and Natural Resources (1996) *Report of the Fishing Safety Review Group, May 1996*. Dublin: Department of the Marine and Natural Resources.

DGXIV (1995) *Structural Policy to Assist Fisheries and Aquaculture*. Brussels: European Commission.

DGXIV (1996) 'Fact Sheet 1–6', in *The Common Fisheries Policy: Information File*. Brussels: European Commission Publications.

Drummond, I. and D. Symes (1996). 'Rethinking Sustainable Fisheries: The Realist Paradigm', *Sociologia Ruralis*. Vol. 36, 2: 152–62.

Dubbink, W. and M. van Vliet (1996) 'Market Regulation Versus Co-Management?' *Marine Policy*. 20, 6: 499–516.

European Economic Community (1957) *Treaty Establishing the European Communities*. Rome: EEC.

Food and Agriculture Organization of the United Nations (2002) *The State of World Fisheries and Aquaculture*. Rome: Food and Agriculture Organisation.

Gadgil, M. and F. Berkes (1991) 'Traditional Resource Management Systems', *Resource Management and Optimization*. Vol. 18, 3–4: 127–41.

Glassner, M. (1993) *Political Geography*. New York: John Wiley & Sons.

Grafton, Q., Hill, R., Adamowicz, W., Dupont, D., Renzetti, S. and H. Nelson (2004) *The Economics of the Environment and Natural Resources*. London: Blackwell.

Grima, A. and F. Berkes (1989) 'Natural Resources: Access, Rights to Use and Management', in F. Berkes (ed.), *Common Property Resources*. London: Belhaven Press, pp. 33–54.

Hannessan, R. (1993) *Bio-economic Analysis of Fisheries*. Oxford: Blackwell Scientific Publications.

Hardin, G. (1968) 'The Tragedy of the Commons: The Population Problem has no Technical Solutions; It requires a Fundamental Extension in Morality', *Science*. Vol. 162, 1243–8.

Hartwick, J. (1986) *The Economics of Natural Resource Use*. New York: Harper & Row Publishers.

Holden, M., and D. Gorrod (1994) *The Common Fisheries Policy*. Oxford: Fishing News Books.

Holm, P. (1996) 'Fisheries Management and the Domestication of Nature', *Sociologia Ruralis*. Vol. 36, 2: 177–88.

International Union for the Conservation of Nature (2003) *Statement to the second committee concerning sustainable ocean development. UN General Assembly, 58th Session, 24 November, 2003*. Gland: International Union for the Conservation of Nature and Natural Resources.

Kalland, A. (1996) 'Marine Management in Coastal Japan', in K. Crean and D. Symes (eds), *Fisheries Management in Crisis*. Oxford: Fishing News Books, pp. 71–86.

Lavery, G. (1996) *Review of Fisheries Issues in Northern Ireland by Standing Committee D (Agriculture and Fisheries Issues)*. Belfast: The Secretariat, Northern Ireland Forum for Political Dialogue.

McGinley, J. (1991) *Irelands Fisheries Policy?* Teelin: Croughlin Press.

McGoodwin, J. (1990) *Crisis in the Worlds Fisheries: People, Problems, and Policies*. California: Stanford University Press.

Meredith, D. (2001) *Collection of Socio-Economic Data on the Irish Catching and Processing Sector in Support of the National Strategy Review Group on the Common Fisheries Policy 2002*. Dublin: Marine Institute.

Ommer, R. (1996) *Sustainable Development and Coastal Communities*, Conference Paper presented at Halifax, Nova Scotia, Canada, August 1996.

Pinkerton, E. (1993) 'Co-management Efforts as Social Movements; The Tin Wis Coalition and the Drive for Forest Practice Legislation in British Columbia', *Alternatives*. Vol. 19, 3: 33–8.

Sandberg, A. (1996) 'Community Fishing or Fishing Communities?', in K. Crean and D. Symes (eds), *Fisheries Management in Crisis*. Oxford: Fishing News Books, pp. 34–44.

Scott, A. (1996) 'The ITQ as a Property Right: Where it Came From, How it Works, and Where is it Going', in B. Crowley (ed.), *Taking Ownership. Property*

*Rights and Fishing Management on the Atlantic Coast.* Halifax: Atlantic Institute for Market Studies, pp. 31–98.

Stationery Office (1997) *The Fishing Vessels (Decommissioning) Scheme 1997, Statutory Instrument 1997 No. 1924.* London: The Stationery Office.

Symes, D. (1996) 'Fishing in Troubled Waters', in K. Crean and D. Symes (eds), *Fisheries Management in Crisis.* Oxford: Fishing News Books, pp. 3–18.

Tingley, D. (2006) *Northern Ireland Fleet Futures Analysis (2004–2013) – Methodology and Results.* Portsmouth: CEMARE.

Wise, M. (1984) *The Common Fisheries Policy of the European Community.* London: Methuen.

Chapter 10

# Production, Markets and the Coastal Environment: Exploring the Social Sustainability of Irish Aquaculture

John Phyne

## Introduction

More than 20 years have passed since Wilson's (1984) ethnographic review argued that Irish rural studies, in moving from 'Clare' to the 'Common Market', had shifted away from functionalist ethnographies of local communities towards a wider political economy perspective that linked communities to the historical development of Irish capitalism. For Wilson (1984), rural Ireland was ridden with social divisions and becoming ever more subject to the policy regimes of the European Community.

Leading on from Wilson, my discussion of Irish aquaculture will similarly go beyond the local 'community' to highlight wider political economy considerations.[1] The forces that have given shape to Irish aquaculture as an export-driven industry require analysis. Raising salmon provides employment for native Irish speakers in the Republic of Ireland (RoI) (see Phyne 1999), and shellfish cultivation provides employment throughout the island, but the stability and prospects of the industry are influenced by wider factors that range from globalized market forces to the politics of user conflicts in coastal waters.

Drawing from the literatures on global commodity chains and environmental risk, I apply a political economy perspective in considering a number of issues that bear on the sustainability of Irish aquaculture. I propose to address the role of capital and labour in the social organization of Irish finfish and shellfish production, the impacts of 'buyer-driven' European Union (EU) markets (especially France), the influence of EU policy on the price and quality of Irish farmed fish production, and the attempts of local, national and EU actors to monitor and regulate the environmental impacts of Irish aquaculture. Based on the overview I provide a research agenda for developing an integrated sociology of aquaculture, capable of covering both the social organization of production and the impact of 'buyer-driven' food chains and environmental risk.

---

1    In this chapter, the term 'Irish aquaculture' refers to aquaculture throughout the island of Ireland – both Northern Ireland (NI) and the Republic of Ireland (RoI).

The overall social sustainability of Irish aquaculture is conditioned by the degree to which industry actors 'internalize' demands emanating from Europe, and succeed in providing stable employment as well as an inclusive consultative role for residents of Irish coastal communities – especially in matters relating to the environmental impacts of aquaculture. The emergence of 'buyer-driven' global seafood chains has coincided since the 1980s with greater concerns for local-level environmental protection. This process, known as the Coordinated Local Aquaculture Management System, exists on an embayment-by-embayment basis. This is an Irish version of risk management that does not abandon the commitment to economic growth. While the Coordinated Local Aquaculture Management System was developed in the Republic, it is being used as a template for management in the cross-border areas in the wake of the Northern Ireland Peace Agreement (NIPA) of 1998. Yet, this may be tempered by some controversy over the recent move to devolve aquaculture licensing in the two cross-border loughs (Foyle and Carlingford) to the Foyle, Carlingford and Irish Lights Commission (FCILC), a North/South administrative body. On top of all of this, the RoI, NI and cross-border areas are embracing sustainable development within the context of an evolving, yet poorly defined framework of integrated coastal zone management linked to the EU.

## Global food chains and Irish aquaculture

Global commodity chain (GCC) analysis uses commodity-by-commodity case studies to discern the social and geographical dimensions of a given supply chain (Bair and Gereffi 2001; Gereffi 1999 and 1994; Gibbon 2001; Humphrey and Schmitz 2001; Kaplinsky and Morris 2002; Phyne and Mansilla 2003; Ponte 2001). When GCCs are 'producer-driven' (Gereffi 1994) power is concentrated in the hands of the actual producers of a given commodity. In 'buyer-driven' chains, design companies (apparels) and retail firms (apparels and food) act as the 'lead drivers' in determining the quality and price of the products they purchase (see Humphrey and Schmitz 2001; Ponte 2001; Kaplinsky and Morris 2002). In the apparels and food industries, production is decentralized whereas the points of distribution are centralized.

The food industry has shifted in the post-Fordist era from a situation where processors had the greatest influence to one where import and export agents and giant supermarket firms have the decisive influence over both prices and quality (Winson 1993; Atkins and Bowler 2001; Gibbon 2001; Ponte 2001; Phyne and Mansilla 2003). Retailers and supranational agencies (such as the World Trade Organization [WTO]) are powerful players. In addition to price levels, food safety and quality demands are passed downstream to food processors and producers (Busch and Bain 2004). Following scares over 'mad cow disease' and 'dioxins' in the food supply, the EU developed its Food Safety Authority to institutionalize full traceability in food supply chains for all food being sold on EU markets (European Commission 2002 and 2000).

In the event of producers and processors being unable to comply with new requirements, food producer's associations can assume an important role (Humphrey and Schmitz 2001; Kaplinsky and Morris 2002; Schmitz 2004). In Chile, for instance, the food quality requirements set by the United States market forced the producers' association to fund the 'upgrading' of standards by salmon farming firms in order to maintain access to the market (Phyne and Mansilla 2003). To meet the traceability demands of the EU market, the RoI's aquaculture industry depends upon the Irish Sea Fisheries Board (An Bord Iascaigh Mhara [BIM]) and the fish farming section of the Irish Farmer's Association. In addition, BIM and Northern Ireland Seafood (NIS) are promoting quality issues in NI and in cross-border areas. On the basis of the available evidence, however, the support for upgrading appears to be weak. To further the 'upgrading process', the Republic's Seafood Industry Strategy Review Group (SISRG) is calling for improved quality provisions and improved supply chain integration for a restructured seafood sector that includes expanded production from aquaculture (BIM 2006).

GCC analysis is often biased in favour of the horizontal (and unequal) relations between food producers and buyers in global food chains. A fuller analysis needs to capture vertical social relations at the point of production. This also involves considering the issues of gender in aquaculture firms, household-firm relations and the socio-cultural situation of Irish coastal communities. Obviously, such an analysis requires an ethnographic approach guided by the issues and evidence presented in this chapter.

## Production: The social and regional organization of Irish aquaculture

### Capital and labour in the finfish and shellfish industries

Finfish aquaculture provides more value at the level of production, but a trait of shellfish aquaculture is the greater employment it disperses along the coastline (Ruddy and Varley 1991; O'Connor, Whelan, Crutchfield and O'Sullivan 1992; Phyne 1999). This section provides data on production, value and employment in the Irish aquaculture industry and brief considerations of the social organization of aquaculture in the Republic, NI and cross-border areas. A major difference between the RoI and the other two areas is the presence of a much larger industry and the pivotal role of salmon aquaculture along the western seaboard. Despite some differences, shellfish aquaculture is dispersed along the Irish coastline and characterized by lower economies of scale than its salmon counterpart.

Table 10.1 provides data on production and value in the Irish aquaculture industry. The largest sector (in terms of value) is salmon farming in the RoI. In 2004, 14,067 tonnes were produced with an ex-farm gate value of €51,289 million. Greater tonnage is present in the bottom mussel industry (28,560 tonnes), a trend that is indicative of the historical development of both sectors (Parsons 2005). The bottom mussel industry is the largest industry in NI (6,500 tonnes) and in the

**Table 10.1    Production (Tonnes) and Value (000 € [RoI] and 000 £ [NI and Cross-Border]) in the Irish Aquaculture Industry**

| Species (Production [P] and Value [V]) | RoI (2004 data) | NI (2005 data)[a] | Cross-Border[b] (2002 data) |
|---|---|---|---|
| Salmon | | | |
| P | 14,067 | 750 | 1,700 |
| V | 51,289 | 1,500 | 3,200 |
| Bottom Mussels | | | |
| P | 28,560 | 7,000 | 3,300 |
| V | 21,014 | ? | 941 |
| Rope Mussels | | | |
| P | 8,755 | ? | 35 |
| V | 6,871 | ? | 15 |
| Pacific Oysters | | | |
| P | 5,103 | 280 | 400 |
| V | 12,204 | ? | 334 |
| Native Oysters | | | |
| P | 390 | ? | 70 |
| V | 1,636 | ? | 152 |
| Total | | | |
| P | 56,875 | 8,030 | 5,505 |
| V | 93,014 | 4,500 | 4,642 |

*Notes:*

a.  The data for salmon in this column includes 'salmon and trout' (Department of Agriculture and Rural Development [DARD] 2006). The data for mussels is most likely for bottom mussel production. Intrafish News (2005) reports 6,500 tonnes for Belfast Lough alone in 2004. Hence, I reported all mussel (bottom and rope) data here. Although no 2005 data are available for Native oysters in NI, in 2002 there was production of 12 tonnes of Native oysters in Strangford Lough totalling £103,000 (Roberts, et. al. 2004). Finally, the overall value for shellfish production (all species) in NI is £3 million; this combined with the value of finfish production gives the total reported above.

b.  The cross-border data for Carlingford Lough and Lough Foyle includes production in both NI and the RoI. Hence, some of the data in columns 2 and 3 will be present in column 4. Only a survey of producers in both loughs would facilitate precise knowledge of production in the loughs for both the RoI and NI. The data are for 2002.

*Sources*: For the Republic – Parsons (2005); NI – DARD (2006), Intrafish News (2005) and Roberts et al. (2004); cross-border areas – BIM (2003)

cross-border areas (3,300 tonnes). The former statistic is based upon cultivation in Belfast Lough whereas the latter pertains to production in Lough Foyle and Carlingford Lough. While no specific data are provided, two employees with the Cross-Border Aquaculture Initiative (CBAIT) reported to the United Kingdom's Fisheries Project that cross-border aquaculture production is concentrated in Carlingford Lough (Cabinet Office 2003). The latter also specialises in Pacific

**Table 10.2    Firms and Employment Levels in the Irish Aquaculture Industry (percentages in brackets)**[†]

| Species | No. of Firms | Employment (Full-Time Equivalents – FTE) | Average FTE/ Firm |
|---|---|---|---|
| Salmon (Grow-out Sites only) (Republic) | 16 (4.6) | 392 (23.5) | 24.5 |
| Bottom Mussels (Rep.) | 47 (13.6) | 235 (14.1) | 5.0 |
| Rope Mussels (Rep.) | 79 (22.8) | 293 (17.5) | 3.7 |
| Pacific Oysters (Rep.) | 149 (43.1) | 295 (17.7) | 2.0 |
| Native Oysters (Rep.) | 6 (1.7) | 217 (13.0) | 36.1 |
| Clams and Scallops (Rep.) | 15 (4.3) | 55 (3.3) | 3.6 |
| Novel Finfish (Rep.) | 4 (1.2) | 8 (0.5) | 2.0 |
| Northern Ireland (all species) | 30 (8.7) | 174 (10.4) | 5.8 |
| | 346 | 1,669 | 4.8 |

*Note:*

[†]    The data for the Republic of Ireland are based upon 2002 figures; the data for Northern Ireland are based upon 2003 figures.

*Sources*: Republic of Ireland – based upon data contained in Robinson (2003); Northern Ireland – data contained in Lantra (2006)

oyster cultivation. More detail will be provided in the regional variations below, but it needs to be noted here that the cross-border data are for firms in NI and in the RoI that are part of the CBAIT, and does not cover all operations in the two cross-border loughs.

Table 10.2 shows data for firms and employment levels in the aquaculture industry. We do not have complete species-specific data for NI, but given the small production (less than 1,000 tonnes) of finfish there, these data reflect the small-scale nature of Northern Irish aquaculture firms (5.8 FTE for the 30 firms); this pattern compares to the small-scale nature of livestock agriculture in NI (Lantra 2006). The highest FTE (36.1) in the Republic is for the very small native oyster industry, largely based in the West. These data reflect the owner/operators involved in fishing cooperatives rather than employees. The 24.5 FTE for salmon farming operations are actually wage-labourers in the 16 firms that have remained after the corporate consolidation of the late 1990s (Phyne 1999) and early part of this century (see below). The 47 bottom mussel and 79 rope mussel farming operations are mostly small-scale ventures, although some concentration of capital is present in Donegal (bottom mussels), and Cork (rope mussels). The 149 Pacific oyster farms have the lowest FTE (2.0) of the five largest (see Table 10.1) aquaculture producers in the RoI; these firms still most likely represent the small-scale family operations (with heavy dependence on unpaid family labour) noted by O'Connor et al. (1992) in the early 1990s.

*Aquaculture production in the Republic*

Irish salmon farming developed in the 1980s with smaller production volumes than its Scottish and Norwegian competitors. Since then the gap has only increased. By the end of the twentieth century, Norway and Chile (a new arrival) controlled nearly 70 per cent of the world's farmed Atlantic salmon production (Phyne and Mansilla 2003). In light of Norwegian dominance in the EU market, it is not surprising that Irish and Scottish producers are concerned with the 'ability' of Norwegians to be 'price-makers' (Phyne 1999).

The Irish salmon aquaculture industry developed from numerous small firms (24 in the early 1990s [Phyne 1999]) to a position today where there are three large and 13 smaller producers. The distinction here between 'large' and 'small' corresponds broadly to a division between foreign and domestic capital. In the early days Irish entrepreneurs often came from outside the Gaeltacht to coastal areas that were 'best suited' for salmon farming. By the mid-1980s, some native Gaeltacht people were concerned with the absence of local entrepreneurs in local salmon farming. In response to these concerns Údarás na Gaeltachta (The Gaeltacht Authority) encouraged local involvement in the industry by financing the acquisition of cages and smolts. Small producers were paired with larger producers in order to gain access to export markets. Údarás na Gaeltachta helped develop the Irish Salmon Producers Group (ISPG) (Phyne 1999). Originally established to market farmed salmon and other fish products for Gaeltacht producers, ISPG became the largest exporter of farmed salmon in Ireland, even surpassing Ireland's largest fish farming company, Marine Harvest (now Pan Fish), which runs sites in Counties Donegal and Mayo (Irish Salmon Producers Group 2005).[2]

When the Republic of Ireland attempted to disperse salmon aquaculture along the western seaboard in the 1980s, protests over licences and environmental concerns caused most of the production to become centralized in Connemara (within and outside the Gaeltacht). By 1995 nearly two-thirds (n=21) of the 36 licences were located in Connemara alone (Phyne 1999). Licences were also issued to farms in Mayo, Donegal and in southwest Cork. By 2002, ISPG packed salmon for 15 of Ireland's 20 salmon farms. These farms hold 27 licences, 15 of which were located in the Connemara area (Irish Salmon Producers Group 2005).

---

2   Pan Fish started life as Fanad Fisheries in the late 1970s. Fanad Fisheries originally grew rainbow trout but quickly shifted to raising a Norwegian strain of Atlantic salmon. In the 1990s, it changed foreign ownership several times. In late 2006, Marine Harvest (owned by the feed firm Nutreco) was acquired by Pan Fish of Norway.

Irish producers desired to reach 30,000 tonnes by 2000 (Phyne 1999); this target has proved elusive. The volume of processed farmed salmon stood at 16,347 tonnes in 2003 and fell to 12,465 tonnes in 2004. In the summers of 2003 and 2004, salmon mortalities in Donegal (the location of Marine Harvest sites) contributed to the decline in production (Parsons 2005). The SISRG expects production to exceed 30,000 tonnes only by 2015 (BIM 2006).

The Irish experience compares with other countries where salmon farming is both geographically and economically concentrated.[3] Galway (6,300 tonnes) and Donegal (6,300 tonnes) had the most significant clusters of farmed salmon production in the Republic in 2003. These counties accounted for over 77 per cent of all farmed salmon production in 2003 (Parsons et al. 2004). The Irish industry also underwent considerable restructuring in the late 1990s and early part of this century. With the transition from Norsk Hydro to Marine Harvest and now to Pan Fish, Norwegian capital's dominance of the RoI's largest salmon faming company continues. Pan Fish is a 'lead firm' in one of the most geographically concentrated areas of farmed salmon production on the island.

The Republic is also a marginal player where mussel farming is concerned.[4] Spain, the Netherlands, France and Denmark all produce larger volumes of mussels (McLeod 2002). The inability of the Dutch to supply all the needs of the French market in recent years has opened the door further for Irish imports, albeit with significant Dutch involvement (Globefish 2005).

Both rope and bottom mussel production increased between 1990 and 2004, but the value per kilogram of the latter rose much faster than the former.[5] The bottom mussel sector has even come to surpass the farmed salmon sector in volume. The 47 bottom mussel firms located across the Republic in 2003 ranged from the longer-established Wexford and Waterford sites to the more newly established Donegal ones. Spat are collected from the Irish Sea for relaying to the coastal waters of Donegal. These mussels are harvested after two to three years and sold to European markets (O'Carroll 2002). In 2000 the Dutch mussel farming industry, the largest in Europe, invested in the Irish industry to secure more mussels to supply its traditional Belgian and French markets (Evans 2005).

The rope mussel industry emerged after state-financed trials of raft culture in Killary Harbour in the 1970s proved promising. Raft culture was replaced by long-line technology in which mussels are 'socked' and attached to long lines

---

3 An exception is Norway where the social democratic state has nurtured a geographically dispersed and economically decentralized industry. Limits are placed on site location and licence concentrations, thereby resulting in the spread of aquaculture along the Norwegian coast (see Phyne, Hovgaard and Hansen 2006).

4 Due to space restrictions, I will restrict my analysis of the shellfish sector to the larger bottom and rope mussel industries.

5 In 1990, rope mussels were making €0.23 per kilogram as compared to €0.35 per kilogram in 2004. Bottom mussels were in contrast fetching only €0.06 per kilogram in 1990 as compared to €0.33 in 2004 (Parsons 2005).

suspended in the water column (O'Carroll 2002). In contrast to bottom mussels, rope mussels have less 'grit' and, depending on water quality, require less depuration than bottom mussels. Among the drawbacks with rope technology mussels is greater exposure to naturally-occurring toxins (such as 'red tides') and poor water quality stemming from land-based pollutants.[6] Rope mussel operators are mostly small-scale with few employees, though there are some larger concerns such as Bantry Bay Seafoods (4500 tonnes of production), a firm with grow-out sites in southwestern Ireland and a processing plant in Bantry. Like most rope mussel processors, Bantry Bay Seafoods supplies preserved and frozen mussels to the European market, in addition to its sales in the United States (Bantry Bay Seafoods 2005).

Even though over 50 per cent (14,638 tonnes) of bottom mussel production is found in Donegal, the bottom mussel sector has a significant presence in six Irish counties. Over the course of the 1990s, Donegal overtook Wexford as the largest producer of bottom mussels in Ireland. As for rope mussels, 6,100 tonnes (or over 65 per cent of total production) is located in County Cork – the home of Bantry Bay Seafoods (Parsons 2005). Shellfish production consists of smaller firms with a greater degree of geographical dispersion than salmon production; however, the move towards greater economies of scale and geographical concentration is steadily becoming a feature of shellfish aquaculture – especially mussel farming.

*Aquaculture in Northern Ireland and cross-border areas*

Aquaculture production in NI is concentrated in Belfast and Strangford Loughs. Belfast Lough is the centre of bottom mussel aquaculture in NI. The vast majority (6,500 tonnes) of the 7,000 tonnes produced in 2005 were by 25 small-scale firms located in this lough. Strangford Lough has some farmed mussel production, but much recent effort has gone into the promotion of Pacific and Native oyster cultivation (Roberts et al. 2004). Small-scale capital prevails in both sectors albeit with much smaller production levels than in the RoI (see Table 10.1). In the finfish sector, the Fisheries Division of the Department of Agriculture and Rural Development (DARD) reports 750 tonnes of farmed Atlantic salmon (DARD 2006). Most production is most likely held by a small organic salmon farm located in Glenarm (near Belfast Lough). This firm has a marketing arrangement with Young's Bluecrest, a major seafood processor in the United Kingdom (Evans 2006b).

Whereas the Department of Communications, Marine and Natural Resources (DCMNR), BIM and Údarás na Gaeltachta are pivotal for aquaculture in

---

6  Red tides are phytoplankton blooms in which otherwise 'safe' phytoplankton change their chemistry and become toxic to shellfish. This happens in situations where the phytoplankton population increases very rapidly for a given oceanic zone. In 1995 and 2000, water quality issues resulted in the closures of several mussel growing areas in Ireland, resulting in reduced production and losses (see O'Carroll 2002).

the Republic, the same applies to DARD and NIS in NI. DARD provides the regulatory framework (like DCMNR) (DARD 2006). NIS promotes the marketing of products from its 33 members (some of which are engaged in aquaculture), largely to the Northern Irish and wider United Kingdom markets (Northern Ireland Seafood 2004).

The future of aquaculture in Northern Ireland, however, is very likely going to be informed by efforts to promote an 'all island' seafood identity on the part of BIM and NIS (Northern Ireland Seafood 2004). The role of cross-border arrangements is critical here. Kennedy and Magennis (2006) note that cross-border arrangements date back several generations prior to the signing of the NIPA and have always included fisheries matters. Carlingford Lough and Lough Foyle are of importance here. While the vast majority of farmed production in Lough Foyle is in the bottom mussel sector on the RoI's side of the Lough, there is a greater sharing of production for bottom mussels and most notably small-scale Pacific oyster cultivation in Carlingford Lough (Cabinet Office 2003).

A bottom mussel review is being prepared for the island as a whole, and this will have implications for the expansion of this sector in cross-border areas as well. Under the 1964 Voisinage Agreement, NI and RoI have shared access to mussel seed from the Irish Sea for the purposes of relaying into coastal areas (for the production of bottom mussels). A central concern is that the demand for bottom mussels for relaying has outstripped the available supply. In NI, this has resulted in a moratorium on future bottom mussel sites. The bottom mussel review will eventually deal with both the socioeconomic and environmental sustainability of this sector for the island as a whole (DARD 2007a). A controversial matter here is the role played by Dutch mussel dredgers using the Irish Sea in order to deal with the shortfall of Dutch mussels on the main French market (Evans 2005).

In summary, the Republic is characterized by economic and geographical centralization in the salmon sector and economic and geographical decentralization in the shellfish sector, although economies of scale and geographical centralization are moving into the bottom and rope mussel sectors. Northern Ireland and cross-border areas are characterized by small-scale capital in the bottom mussel and Pacific oyster sectors. Despite their differences, each of the Irish regions is subordinate to the European seafood theatre. It is to this issue that I now turn.

## Markets: Prices and traceability

*'Buyers' and the power of pricing*

Busch and Bain (2004) argue that global institutions, such as the WTO and giant food retailers play a pivotal role in the 're-regulation' of food chains. These global players have displaced the dominance once assumed by nation-states and food processors during the Fordist food regime of standardized mass production for mass consumption (Atkins and Bowler 2001). Greater concentration in buying power means buyers can impose pricing demands on food processors and producers further upstream in food chains.

The small Irish industry competes with larger Norwegian and Scottish salmon-producing industries in EU markets. To prevent the Norwegians from dumping 'excess production' in EU markets, a Minimum Import Price (MIP) was applied to Norwegian exports to the EU in 1997. When this MIP expired in 2003 the Irish and Scottish salmon farming associations pressured the EU to impose tariffs on Norwegian production (European Commission 2003). Concerns were also raised over the small but rapidly growing presence of Chilean salmon (Chile has a free trade deal with the EU) in the EU market. The EU held an inquiry that invited representations from the Irish, Scottish and Norwegian producers. In their defence, the Norwegian Seafood Federation presented evidence that, between 2000 and 2001, most of the final price for farmed salmon went to supermarkets and only a fraction of the price to salmon farmers. The implications of this argument are that large buyers, such as EU processors and supermarkets, are in a position to provide all salmon farmers with higher prices. Eventually the EU ruled in favour of the Norwegians and noted that the small Chilean presence was largely in frozen salmon – a lower-priced product that is not sold by EU producers (European Commission 2003). Only in 2006 did the EU decide in favour of a new MIP (Evans 2006a).

For salmon farming nations, debates over price are crucial as end markets are often in other jurisdictions. The seafood industry in the 1990s witnessed an emerging division of labour between producers on the one hand and food processors and retailers on the other hand (Guillotreau 2003). The vertically integrated fish companies in the North Atlantic 'vertically disintegrated' by removing themselves from harvesting and by sourcing seafood from a greater variety of places in the North Atlantic (see Apostle et al. 1998). Leaner seafood companies leveraged themselves to gain access to the 'new power' in the food chain – concentrated retailers.

Smokehouses and retail multiples are the dominant seafood actors in France (McIver Consulting 2001). Smokehouses purchase raw material from salmon farming companies, smoke it to French consumer standards, and then sell it on to retail giants. Small and large salmon-farming nations alike sell semi-processed fish that realizes a greater value elsewhere. In 2001, over 75 per cent of the value of fresh Atlantic salmon fillets accrued to French supermarkets and hypermarkets. The top five retailers controlled 96 per cent of the total retail market in 1999 (McIver Consulting 2001). As producers enter the French seafood chain, they enter a narrowing funnel that is ultimately shaped by the retail giants. Such a commercial context reinforces the importance of processors (such as the ISPG) and marketing institutions (with BIM and NIS) in attaining market access for smaller Irish producers.

France is also the main market for the RoI's mussel exports. A recent increase in exports to the French market has occurred as mussel seed availability has decreased for the much larger Dutch industry. To compensate for the shortfall, some large Dutch mussel farming firms have invested in the Republic. Dutch mussel dredgers are securing seed supply from the Irish Sea and relaying it to bottom mussel sites in the south of the Netherlands. They also have a presence in Lough Swilly in the Republic. The ecological and economic consequences of this practice for the Irish bottom mussel industry remain to be seen.[7]

The fresh mussel market in France absorbs 84 per cent of the Republic's export but this secures the lowest price. The Irish cannot match the volumes and same-day deliveries of their competitors. They deal with six importing companies who in turn sell the product to a retail market dominated by five firms (BIM 2003). This is a factor that the SISRG (BIM 2006) would like to end by promoting greater value-added production and providing for a quality-enhanced 'Seafood Island' product to enable Irish producers to bypass the intermediaries that control distribution channels in the French market. This will be no easy task as French retailers and the EU also drive 'quality'.

*Traceability: The EU, retailers and third-party certification*

In the wake of scares over 'mad cow disease', foot and mouth disease and the risks of Genetically Modified Organisms (GMOs), the EU introduced a mandatory food policy that aims to provide traceability for quality and safety all along the food supply chain (European Commission 2002). The European Food Safety Authority (EFSA) was launched in 2002 as a risk-assessment and communication agency (European Commission 2002). By 1 January 2005 all firms selling food in the EU were required to have traceability arrangements that allowed the EFSA to track food and feed at each point along the supply chain.

Since food supply chains are concentrated downstream, in addition to controlling price, retailers can act as policing agents by ensuring that their suppliers meet certain quality standards (see Busch and Bain 2004). Smaller producers who do not have the means to implement quality control standards (such as HACCP)[8] can potentially be removed from the market. To stay in business, 'process

---

7   A wild oyster development association in Lough Swilly raised concerns in 2005 that Dutch mussel dredgers were transferring bonamia, a disease lethal to native oysters. The fear of oyster farmers is that bonamia may be transferred by contaminated mussel seeds attached to the bottom of dredging vessels (see Evans 2005).

8   HACCP (Hazard Analysis at Critical Control Points) was introduced by the Pilsbury Company in 1964 for the United States space program. Although it is used in the American food industry, its introduction was not without controversy. Since HACCP involves some microbial testing, it was resisted by American beef industry interests which insisted that they only had to adhere to the 'poke and sniff' requirement of the still operational 1906 Meat Inspection Act (see Nestle 2003). However, since the turn of the century all companies exporting food to the United States must adhere to HACCP standards. The Chilean salmon farming industry, for instance, was obliged to upgrade to HACCP in order to maintain access to the American market (Phyne and Mansilla 2003). HACCP is also a part of the EU's food safety criteria (see Phyne, Apostle and Hovgaard 2006).

upgrading' (Humphrey and Schmitz 2001) may therefore be needed by smaller food producers.

Traceability requirements can reconfigure the food chain. In the wake of EC 178/2002 (European Commission 2002), all actors in the food chain had to be compliant by 2005 with the traceability provisions enabling food to be traced from 'farm to fork' (European Commission 2000). Irish and other producers have to engage in 'process upgrading' so as to have their products meet EU standards. Such upgrading does not imply any 'competitive advantage'; it is a necessity for 'market survival' (see Schmitz 2004).

What implications do traceability provisions have for Irish producers in coastal communities? The EU makes use of policies for the classification of shellfish waters (EC 91/492/EEC) and for the control of maximum permissible levels of trace metals such as mercury and cadmium (EC 466/2001/2001) (Parsons et al. 2004). Shellfish waters are classified so that products grown in them can either directly enter the market, require depuration prior to sale or are excluded from the market. In 1995 and again in 2000, the Irish shellfish aquaculture industry was hit by biotoxin problems that cost farmed products market access (O'Carroll 2002). EC 91/492/EEC also deals with the issue of faecal coli form counts; there is provision for shellfish closures in areas where such counts exceed permissible levels (see O'Carroll 2002; Parsons 2005).

EC 96/23 monitors 'substances and residues' in farmed salmon (Parsons et al. 2004). To counter diseases, farmed salmon are often fed antibiotics in their feed. There is a mandatory withdrawal period whereby salmon are starved prior to harvesting in order to remove pharmacological residues from their system. The Marine Institute, on behalf of the DCMNR, monitors this for the Republic.

Besides EU directives, food safety and quality issues arise at the firm level. In light of 'food scares' in France, quality provisions have become more important. McIver Consultants (2001, p. 8) observe how 'the major retailers are increasingly seeking quality systems which are independently audited to comply with their own quality systems'. It is expensive for a firm to attain the Label Rouge mark, but it does guarantee greater market access. In the late 1990s, Scottish farmed salmon met this criterion.

EN45011 was launched by BIM to enable Irish producers meet EU market standards. This provides for the certification of the actual product as well as the quality system in place (McIver Consultants 2001). The Republic's farmed salmon sector was the first food sector to receive EN45011. Irish Quality Salmon (IQS), a label certifying best practice in traceability, food safety and environmental quality, has been achieved by 80 per cent of the salmon-farming sector (BIM 2005). BIM, in conjunction with NIS is also implementing quality assurance schemes for finfish and shellfish producers in NI and in cross-border areas (NIS 2004).

Irish Quality Mussels (IQM), has also been introduced for the mussel-farming sector. Under this, which again endeavours to achieve traceability from harvesting to the packaging process (Parsons et al. 2004), two of the four main mussel-processing companies were certified in 2003. Among IQM measures are protocols that deal with hygiene management, compliance with biotoxin requirements, full traceability and third party monitoring (Parsons et al. 2004).

Third-party certification adds yet another actor to the food chain. International Fish Quality Certification (IFQC) provides a quality assurance certification program that enables firms to upgrade to EU standards, especially the food safety directive (EU 178/2002). The certification process spans the supply chain. By 2006, IFQC had certified 11 production sites and four processing facilities in the RoI's salmon aquaculture industry. Most of these involve Marine Harvest (now Pan Fish) – Ireland's largest producer of farmed salmon (Irish Fish Quality Certification 2006). Given its location in County Louth, IFQC is well positioned to provide quality assurance for cross-border operations in Carlingford Lough.

In addition to industry-wide certification schemes, at the firm level there are measures used to facilitate market access. Since the Irish industry is small-scale, it is well known and still argued (see BIM 2006) that an enhanced high quality and perhaps more expensive product is necessary in order to maintain market access. In the RoI and NI, organic salmon farming is promoted by some firms. Northern Salmon Company in Glenarm in NI has a contract to supply organic salmon to Young's Bluecrest, one of the United Kingdom's largest seafood processors. This firm, in turn, smokes the salmon for sale in a restaurant chain (Evans 2006b). This is a case of supply chain integration from production to consumption. In a minor way, it parallels developments in the British dairy industry after the abandonment of quotas in 1994; dairy farmers entered alliances with dairy processors in order to gain leverage in entering concentrated retail markets with nascent traceability arrangements (Banks and Marsden 1997).

Coordinated marketing by the ISPG does enable smaller producers to meet the challenge of lower prices. However, quality provisions present challenges of a different order. Achieving higher quality in an environment of unstable prices will have consequences at the point of production. This is where the material conditions of labour may change. In order to meet this challenge, product and process upgrading will be necessary (see Phyne, Apostle and Hovgaard 2006). Room to manoeuvre can be expected to be limited in an arena of asymmetric power relations where traceability measures enable retail multiples and EU legislation to 'govern from a distance' (Larner and Le Heron 2004).

## Governing the environment

*Risk, sustainability and coastal environments*

To this point, sustainability has been addressed from the standpoint of the role of capital in coastal communities in the production of farmed fish, and the impacts of prices, quality and traceability upon the delivery of farmed fish to EU markets. The sustainability of capital (foreign and local) and employment (quantity and quality) is based upon the degree to which demands from the EU market can be successfully accommodated at the point of production without leading to the decline of capital and/or the degradation of working conditions. Another dimension of sustainability is the degree to which coastal actors can agree upon what environmental protection is to entail. Here, a number of critical questions arise. To what extent is aquaculture production perceived as having negative impacts upon the environment? Given the legacy of user conflicts, what measures are available for accommodating the interests of aquacultural capital and other coastal actors? What are the implications of EU, national and local-level policies in structuring the accommodation of different coastal interests?

These issues point to the role of risk perception and sustainable development in the development of suitable activities in coastal zones. Risk is a product of social perceptions in an era of heightened environmental awareness (Beck 1992; Beck et al. 1994; Beck 1996; Strydom 2002). Reference to scientific discourse alone is therefore not sufficient to address concerns over environmental risks. Risk anxiety is also connected to a decline in trust levels in risk societies. Environmental activists have questioned the degree to which experts in environmental controversies are able to provide assurances over risks. Of course, environmental organizations provide forms of expertise that are, in their turn, questioned by scientific authorities and their clients in industry and government.

To explore the links between sustainability and environmental regulation, I consider the Coordinated Local Aquaculture Management Systems (CLAMS) in the RoI. After this, I discuss the extension of CLAMS to cross-border areas where there has been a recent move to transfer licensing authority in both cross-border loughs to the FCILC, thereby removing it from the jurisdiction of the DCMNR in the Republic and the DARD in NI. Hence, CLAMS in cross-border areas may eventually be nested inside another agency. Following this, I consider the implications of Integrated Coastal Zone Management (ICZM) at the national and EU levels for the governance of Irish aquaculture.

*The Republic: From single bay management to CLAMS*

In the late 1980s, the expansion of the Irish salmon aquaculture industry coincided with a severe decline in the sea trout population (see Phyne 1996). Salmon farmers agree that sea lice are attracted to salmon farms; they disagree, however, that the sea louse species that affects farmed salmon is the same as the species that affects wild sea trout.[9] To deal with the problem, Irish salmon farmers learned from their Scottish counterparts and engaged in the fallowing and rotation of sites in order to break the life cycle of sea lice. This strategy was also decided on in the aftermath of the controversy provoked by the use of the pesticide Nuvan to deal with sea lice infections. In order for fallowing and site rotation to work, each farm needs three licences: one for year one salmon, a second for year two salmon and a third licence

---

9   For details on early conflicts associated with the sea lice controversy, see Phyne (1999).

for a fallowed site. This arrangement evolved into a wider strategy of single bay management (SBM) in which all salmon farmers in an embayment coordinated fallowing and rotation practices. SBM (a voluntary strategy) is especially important in Kilkieran Bay, which has the largest concentration of salmon farms in Ireland (Phyne 1999). In its review of the Irish seafood sector, the SISRG called for resources to extend access to fallowing sites for the Irish salmon farming industry (BIM 2006).

SBM became the nucleus for CLAMS which developed in the late 1990s under the auspices of BIM. Behind CLAMS is the requirement that all producers (finfish and shellfish) co-ordinate the environmental monitoring of aquaculture in an embayment. Some CLAMS (as in Kilkieran CLAMS) are mainly in the hands of finfish producers; others have greater representation from shellfish farmers (such as Clew Bay CLAMS). The recommendations of a CLAMS group are advisory in nature; legislative authority remains with the DCMNR. The licensing of aquaculture sites is a DCMNR responsibility; the Marine Institute (a DCMNR body) conducts the monitoring of sea lice levels on salmon farms.

CLAMS provides for interested parties to have a consultative role in the drawing up of aquaculture management plans. BIM (n.d.[a]) insists that this participation is merely consultative, and cannot become a basis for launching criticism of aquaculture. Such criticism had been a feature of public hearings in the 1980s, as well as the disputes that developed in the 1990s (see Phyne 1999). In the final analysis, aquaculture producers determine CLAMS plans.

The National Development Plan (NDP) aimed to have a CLAMS group established for all aquaculture producing areas by the end of 2006 (Marine Institute 2000). Each CLAMS group would be linked to a national CLAMS committee. Another objective was to have each CLAMS committee linked to an integrated coastal management plan at the county level.

For CLAMS to be a success, the RoI has introduced a process of environmental quality certification, known as Ecopact, for fish farmers. Ecopact is a BIM initiative that will enable firms to have the certification necessary to adhere to the EU's Eco-Management and Audit Scheme (EMAS). For Ecopact certification, a firm must engage in measures such as the monitoring of environmental impacts, nature conservation, and the management of noise, odours, stock health and waste management on aquaculture sites (BIM n.d.[b]).

Within BIM, Ecopact enables firms to '...reduce risk and to maximize opportunities in a coordinated way' (BIM n.d.[b], p. 7). This certification is difficult to achieve, especially for small shellfish operators. BIM envisages that CLAMS groups will '...provide an implementation and audit of the take-up of Ecopact or accredited EMAS within the individual companies that form their CLAMS groupings' (BIM n.d.[b], p. 30). BIM (in conjunction with the fish farming section of the Irish Farmers Association) guides the implementation and auditing of Ecopact, an environmental equivalent of traceability measures for food sold into EU markets. In contrast to the mandatory requirements of traceability, Ecopact is mainly voluntary. Specific items within Ecopact certification may be mandatory under EU directives and national law, but other items are voluntary. Within the context of CLAMS and ICZM, Ecopact does nonetheless address national and EU objectives. The future may witness both Ecopact and traceability as mandatory requirements for all fish farmers in the EU.

### Northern Ireland and cross-border areas

In the early part of this decade, efforts were made to revitalize aquaculture production in the three loughs entirely in NI (Strangford, Belfast and Larne) and the two cross-border loughs (Foyle and Carlingford). The objective was to assess the carrying capacity of the loughs for the purposes of oyster and mussel cultivation (see Roberts et al. 2004; Institute of Marine Research [IMAR] 2004). More important for our purposes is the transfer of CLAMS to NI and cross-border loughs by the CBAIT (BIM n.d.[c]). CLAMS in the cross-border loughs has the added objective of working with disadvantaged areas and groups on both a '...cross-border and cross-community basis'. The promotion of sustainable aquaculture such as an oyster growers producers' group involving growers from NI and cross-border areas is another objective of the CBAIT.

Recent developments emanating from the NIPA have implications for the operation of CLAMS in the cross-border loughs. FCILC is a product of the North South Ministerial Council (NSMC) and is responsible for the regulation of activities in the two cross-border loughs. In an eight-week period from the end of 2006 to January 2007, a consultative process was used for the purposes of devolving the regulation of inland fishing and aquaculture activities in cross-border areas to the FCILC. The result is a draft bill dedicated to transferring administrative authority from the DCMNR in the RoI and the DARD in NI to the FCILC (DARD 2007a). On the surface, this meets the spirit of cross-border cooperation and the EU's commitment to subsidiarity, but The Draft Foyle and Carlingford Fisheries (Northern Ireland) Order 2007 (which will be accompanied by similar legislation in the RoI) is fraught with controversy.

DARD (2007b) published on-line the submissions to the consultative process; concerns and/or objections were registered by wild fisheries groups, aquacultural interests and state bodies in NI. The raised concerns ranged over questioning the legitimacy of the short consultative period to uncertainty over the scope of jurisdictional authority possessed by the FCILC.[10] Two wild fisheries groups operating in Lough Foyle argue their 'traditional rights' were compromised by a short consultative period promoting aquaculture. One group protests the impact of the introduction of large quantities of mussel seed for bottom mussel aquaculture in Lough Foyle since 1997 – a feature it views as displacing the traditional oyster fishery. Both groups are going to appeal the new legislation under their rights as 'Irish and European citizens'. For their part, aquacultural interests are concerned over the costs of a new layer of administration on small aquacultural

---

10   What follows is just a brief sampling of the more than 30 submissions.

projects. This includes environmental approvals and the fact that Northern Ireland producers are already paying leasing fees to the Crown Estate. A submission from the Crown Estate for Northern Ireland raises concerns over the impact of the FCILC's administration of aquaculture licenses on the proprietary rights of others. Furthermore, the board of the FCILC points to the problem of balancing conservation issues with jurisdictional authority.

In short, CLAMS as a process is going to face more multilayered issues in the cross-border loughs than in the Republic. How the voluntary aspects of CLAMS will mesh with the evolving licensing and regulatory role of the FCILC remains to be seen. What further complicates the promotion of 'sustainability' is the need to incorporate official environmental concerns from the state in NI and the RoI within the context of the EU's move towards ICZM.

## The State, EU and ICZM

Aquaculture in the Republic, NI and cross-border areas is linked to policy discussions over ICZM. Integrated management, like 'sustainable development', has become a policy mantra in many jurisdictions. In the RoI discussion papers on integrated coastal management date back to 1997. Yet the move towards integrated coastal management has not moved much beyond the rhetorical stage. Cummins et al. (2004) note that the management of the Irish coastline is characterized by the predominance of sectoral interests that pre-date coastal management initiatives. The ultimate challenge is to create institutions that provide a horizontal meeting point cutting across sectoral interests. If aquaculture interests are to participate in effective environmental stewardship, they cannot rest upon CLAMS alone. CLAMS, as Cummins et al. (2004) observe, is a basis for but not the endpoint of integrated coastal management. Given this, and the commitment under the NDP to have CLAMS in every embayment by the end of 2006, and the rhetoric of linking CLAMS to integrated coastal management, under what circumstances might these initiatives be realized?

Northern Ireland and the United Kingdom are also engaging in their own ICZM initiatives within the context of the EU's ICZM. In NI, the Department of the Environment's (DOE) biodiversity strategy points to the need to incorporate the integration of aquaculture into biodiversity initiatives in the cross-border loughs (Department of the Environment 2005). Moreover, similar to Cummins et al. (2004), DOE points to the need to move beyond the current sectoral interests governing biodiversity. It also raises concerns over the delay in a coastal development strategy first promoted in NI in 2001 (Department of the Environment 2005).

In the United Kingdom, the proposal for a Marine Bill further points to the difficulty of integrating diverse sectoral interests into a holistic eco-system based framework (Department of the Environment, Food and Rural Affairs 2007). An objective of the Marine Bill is to promote a national Marine Management Organization alongside the devolution of integrated marine management to coastal areas. Activities as varied as aquaculture and maritime shipping would become part of an ICZM scheme that concurs with that of the EU.

Beginning in 2002, the EU engaged in an initiative to promote ICZM (European Topical Centre on Terrestrial Environment 2003). It already had shellfish directives of relevance to aquaculture, a water directive and provision for species protection under measures such as NATURA, but there was an absence of an integrated approach to coastal activities. By the end of 2006, the EU planned to have a directive on integrated coastal management. This has not happened. Under EU law, any directive has to be transposed into national legislation by member states within five years. In a recent update on progress towards the development of national level coastal management plans, the RoI was one of 11 out of 20 EU coastal nations that had not made any significant progress in the way of policy initiatives (European Commission 2005). Given the unevenness of the responses by national governments, it is unlikely that any directive will be in place in the foreseeable future. Most recently, the EU published a Green Paper dedicated to promoting an integrated maritime policy. This combines the usual 'suspects' of competitiveness, economic growth, sustainability and decentralized governance (European Commission 2006).

ICZM involves more than sustainability; it entails a multilayered governance structure stretching from the Irish coast to Brussels. What exists is an elusive goal of 'democratic governance' alongside a more firmly structured process of 'economic governance' emanating from the European seafood chain. We can further contextualize the elusiveness of ICZM by summarizing the institutional structure governing Irish aquaculture (see Table 10.3). The structure presented in Table 10.3 is by no means exhaustive, but it conveys the complexity involved in what is essentially a small industry. The DCMNR, BIM and Údarás na Gaeltactha assume prominent roles in the Republic in matters ranging from developmental assistance to licensing/regulation. The same pattern, albeit with different institutions, is assumed in NI and in cross-border areas. Despite differences, there are a great number of similarities. Moves towards decentralized governance are not equally embraced by all local level actors and institutions. Moreover, the mandatory and voluntary measures mentioned in Table 10.3 mean that the environmental management of Irish aquaculture in conjunction with other coastal interests is an elusive goal.

Existing structures of production are heavily influenced by prices and food quality provisions set by actors further downstream in EU seafood chains. This is a discernible pattern faced by producers in NI as well as in the RoI. Despite the rhetoric of sustainable development and ICZM, no such discernible pattern exists at the local level. There are a plethora of institutions and arrangements (mandatory and voluntary) in the RoI, NI and in cross-border areas that defy categorization into any particular pattern. Producers *must* meet the demands of distant actors in the seafood chains, but often only have to abide to voluntary arrangements at the local level. Traceability is mandatory; CLAMS and Ecopact are voluntary. One has to be incorporated into the bottom line; the other is at worst an inconvenient intrusion made necessary by previous coastal resource conflicts.

**Table 10.3    Some of the Institutional Structures Governing Irish Aquaculture, 2007**

| Institutional Arena | Development and Export Assistance | Licensing Authority | Regulatory Bodies (Mandatory [m] and Advisory [a]) |
|---|---|---|---|
| RoI | BIM; Udaras na Gaeltachta | DCMNR | DCMNR [m]; DELG [m]; BIM (CLAMS [a]); ICZM [a] |
| NI | Northern Ireland Seafood | DARD | DARD [m]; DOE [m]; SMILE [a]; ICZM [a] |
| Cross-Border | CBAIT | NSMC; DCMNR; DARD; FCLIC | NSMC [m]; DCMNR [m]; DARD [m]; FCLIC [m]; SMILE [a]; CLAMS [a]; ICZM [a] |
| EU | EFF | No authority | Shellfish, Habitat and Water Directives [m]; ICZM [a] |

The ultimate irony is that if power is to be reallocated at the 'local level' the 'global', or in this case the 'European', needs to be brought in. Given the recent claims by coastal residents in Lough Foyle to their 'European' as well as to their 'Irish' citizenship in articulating their rights *vis-à-vis* the newly formed legislative authority vested in the FCILC, coupled with moves towards an EU-based ICZM, the European dimension is an institutional direction that will form future discussions over coastal governance. The ultimate issue is the degree to which an EU ICZM directive will be developed and if so how long it will take for RoI, NI and cross-border institutions to incorporate this into local governance initiatives.

### Discussion and conclusions: The future of Irish aquaculture

The integrity of human health and the environment are predominant concerns in food production in the post-Fordist age. Urban institutions and consumers are now 'lead drivers' governing rural production. Over a decade ago I noted (Phyne 1996) that the fledgling Irish aquaculture industry was poised in a delicate balance between environmental integrity and social equity. In the wake of conflicts over the safety of the food supply (including farmed fish) human health has been added to this balance. Yet, some fundamentals remain. Irish producers still face asymmetric relations with larger producers and concentrated retailers in EU markets and tenuous relations with their coastal neighbours.

A number of challenges facing policy makers and social science research can now be identified. Irish public policy needs to forcefully meet the challenges posed by price trends, traceability and environmental stewardship. In a risk society, heightened environmental and food safety awareness means that policy issues are 'driven' by the EU and food retailers. What is needed is an integrated policy approach that deals effectively with supply chain dynamics. There is a need to coordinate the upgrading of food production and processing *via* the environmental stewardship measures being used in Ecopact certification that is part of CLAMS. The Republic, through BIM, is assisting small firms with the latter, though what is being attempted may not be sufficient unless closely co-ordinated with the product traceability issues required by EU law.

An EU ICZM directive may be some years away, but if it comes about then traceability and ICZM rules will both bear heavily on the seafood sector. Moreover, ICZM will mean that land-based polluters will be made more accountable, especially in relation to the health of the more sensitive bivalves in the shellfish sector. An integrated approach to policy coordination today is important as the Irish industry faces more competitors in its attempt to access EU markets. The EU can also be of assistance here: the food safety directive (EC 178/2002) has a provision that recognizes the need of 'developing countries' to be provided with some leeway as they upgrade their food requirements to meet EU standards (European Commission 2002). Although the 'Celtic Tiger' phenomenon has resulted in Ireland's elevation, as a whole, from the Objective 1 criteria of the EU, parts of Ireland, especially the western seaboard counties at the heart of the aquaculture industry, still fall below the 'norm' for wealthy status in the EU. In addition, the fish harvesters and farmers of NI are in a marginal EU region (European Commission 2001). Given this and the plethora of small producers in the aquaculture industry, Irish public and private institutions need to lobby the EU for upgrading assistance for its small producers so that they can meet EU standards. The current small amounts available under the European Fisheries Fund (EFF) may not be sufficient (see BIM 2006).

While the policy prescriptions noted above apply to producers, we need more focussed research on the social composition of the labour deployed in Irish aquaculture, looking for instance at the dynamics of class, gender and country of origin.[11] GCC analysis is rightly criticized for its emphasis on systemic chain dynamics to the exclusion of local-level processes (see Barrett et al. 2002). But what can be said about social relations at the point of production? To date, the literature on the sociology of aquaculture has been notoriously silent on this issue. Many of the concerns over aquaculture have dealt with the user conflicts and environmental disputes at the heart of the 'blue revolution'. Moreover, given the location of salmon aquaculture in a 'buyer-driven' commodity chain, it is not surprising that the research has followed the centres of economic and political power.

---

11   In Nordic countries with low unemployment, migrant workers have been brought in to take the place of local young people unavailable and/or unwilling to work in processing jobs. Tamils are found in processing plants in northern Norway, and workers from several European and African countries are employed on processing lines in the western Norwegian community of Austevoll (Phyne, Hovgaard and Hansen 2006). Even in the Faroes, Thai women have been introduced for processing work on the most southerly island of Suðuroy (Phyne, Apostle and Hovgaard 2006). On Prince Edward Island, Canada's smallest province, which has a history of high unemployment, workers were difficult to find during the spring of 2006. To meet the need for labour, fish plant owners recruited seasonal workers from Russia.

In order to meet traceability and environmental standards in a 'buyer-driven' food chain, producers may be tempted to 'download' costs onto their vulnerable paid and unpaid labour force. We therefore need detailed ethnographic research on the social position of workers (including the dynamics of gender and household relations) in the grow-out sites and processing plants associated with Irish aquaculture. The early research of O'Connor et al. (1992) showed that unpaid labour was more prominent in the shellfish sector. If this is still the case, we may find more involvement by women and children in family-based shellfish enterprises than in their more capital-intensive counterparts in the shellfish and finfish sectors.

Ideally, local-level field research should be used in conjunction with global commodity chain analysis. To combine the two would allow the teasing out of the impacts of class, gender and ethnic relations at the point of production, and how this varies according to the nature of the commodity (shellfish versus finfish), the impact of 'buyer-driven' relations upon the social organization of production, and the degree to which low market prices coupled with the growing reluctance of young people to work in fish production and processing work makes for the recruitment of migrant labour (see Apostle et al. 1998; Phyne, Hovgaard and Hansen 2006). Given the ageing of the fisheries and aquaculture labour force and the difficulty in recruiting young workers, the SISRG (BIM 2006) has pointed to the need for educational upgrading, retention incentives and, if necessary, the use of immigrant labour.

What is clear from our discussion is that the social organization of production, the socio-cultural context of coastal communities and global food chains needs to be integrated by social scientists within the context of local-level risk management initiatives. The economic, environmental and health dimensions of the food chain will increasingly impinge on coastal communities producing for concentrated retail markets. Post-modern tendencies exist, but in the final analysis any consideration of aquaculture needs to deal with the changing dynamics of global capitalism and its regional manifestations in different food chains.

## Acknowledgements

The insights in this chapter would not have been possible without the field research conducted in Ireland (1991, 1993 and 1995), Chile (2000 and 2001) and Norway (2002 and 2003). I thank all of those individuals in each of the settings who gave me, and my fellow researchers (in Norway and Chile), their time for interviews, as well as the several individuals who have provided advice and insights on aquaculture. Finally, appreciation is extended to the editors for inviting me to contribute to this collection and for providing comments on a previous draft of this chapter.

## References

Apostle, R., G. Barrett, P. Holm, S. Jentoft, L. Mazany, B. McCay and K. Mikalsen (1998) *Community, State, and Market on the North Atlantic Rim: Challenges to Modernity in the Fisheries*. Toronto: University of Toronto Press.

Atkins, P. and I. Bowler (2001) *Food in Society: Economy, Culture and Geography*. London: Arnold.

Bair, J. and G. Gereffi (2001) 'Local Clusters in Global Chains: The Causes and Consequences of Export Dynamism in Torreon's Blue Jeans Industry'. *World Development*, 29 (11): 1885–903.

Banks, J. and T. Marsden (1997) 'Reregulating the UK Dairy Industry: The Changing Nature of Competitive Space'. *Sociologia Ruralis*, 37 (3): 383–404.

Bantry Bay Seafoods (2005) *Bantry Bay Seafoods*. http://www.bantrybayseafoods.com. Retrieved 13/04/07.

Barrett, G., M. Caniggia and L. Read (2002) '"There are More Vets Then Doctors in Chiloé": Social and Community Impact of the Globalization of Aquaculture in Chile'. *World Development*, 30 (11): 383–404.

Beck, U. (1992) *Risk Society: Towards a New Modernity*. London: Sage.

Beck, U. (1996) 'Risk Society and the Provident State', in S. Lash, B. Szerszynski and B. Wynne (eds), *Risk Environment and Modernity: Towards a New Ecology*. London: Sage, pp. 28–43.

Beck, U., A. Giddens and S. Lash (1994) *Reflexive Modernization*. Cambridge: Polity.

BIM (n.d.[a]) *Coordinated Local Aquaculture Management Systems (C.L.A.M.S.): Explanatory Handbook*. http://www.bim.ie/. Retrieved 10/06/06.

BIM (n.d.[b]) *Ecopact: Environmental Code of Practice for Irish Aquaculture Companies and Traders*. http://www.bim.ie/uploads/text/_content/docs/Ecopact.pdf. Retrieved 10/06/06.

BIM (n.d.[c]) *Cross-Border Aquaculture Initiative*. http://www.bim.ie/uploads/text/_content.asp?node_id=544. Retrieved 11/03/07.

BIM (2003) *The European Mussel Market*. BIM. http://www.bim2b.com/BIM/markets/report_results.jsp. Retrieved 10/02/06.

BIM (2005) *Irish Farmed Salmon: The Facts*. BIM October 2005. http://www.ie/uploads/text_contents/docs/salmon%20the/%20facts%BIM%20Report%20October%202995.pdf. Retrieved 10/06/06.

BIM (2006) *Steering a New Course: Strategy for a Restructured, Sustainable and Profitable Irish Seafood Industry 2007–2013. Report of the Seafood Industry Review Group*. December 2006.

Busch, L. and C. Bain (2004) 'New! Improved? The Transformation of the Global Agrifood System'. *Rural Sociology*, 69, 3: 321–46.

Cabinet Office (2003) *UK Fisheries Consultation Project – Consultation Paper*. http://www.cabinet office.gov.uk/strategy/downloads/su/fishconsulation/org/cross.pdf/. Retrieved 26/03/07.

Cummins, V, C. O'Mahoney and N. Connolly (2004) *Review of Integrated Coastal Zone Management and Principals of Best Practice.* Prepared for the Heritage Council by the Coastal Marine Resources Centre, Environmental Research Institute, University College Cork, Ireland. www.heritagecouncii. ie/publications/coastal_zone/coastal_zone_reviews.pdf. Retrieved 10/02/06.

DARD (2006) *Fisheries Division: Corporate Plan 2006–09; Business Plan 2006–07.* http://dardni.gov.uk. Retrieved 28/03/07.

DARD (2007a) *Terms of Reference for the Bottom Mussel Review.* http://www. dardni.gov.uk/index/publications/pubs-dard-fisheries-farming-and-food/ publications – fisheries. Retrieved 28/03/07.

DARD (2007b) *The Draft Foyle and Carlingford Fisheries (Northern Ireland) Order 2007.* http://dardni.gov.uk/index/consultations/archived-consultations/ content-foyle-cford-order.htm. Retrieved 11/03/07.

Department for Environment, Food and Rural Affairs (2007) *A Sea Change: A Marine White Paper. Presented to Parliament by the Secretary of State for Environment Food and Rural Affairs By Command of Her Majesty,* March 2007. http://www.defra.gov.uk/corporate/consult/marinebill/index.htm. Retrieved 19/03/07.

Department of the Environment (2005) *Delivery of the Northern Ireland Biodiversity Strategy: The First Report of the Northern Ireland Biodiversity Group.* http://www.doeni.gov.uk. Retrieved 15/03/07.

European Commission (2000) *White Paper on Food Safety.* Brussels: Commission of the European Communities.

European Commission (2001) *Unity, Solidarity and Diversity for Europe, its People and its Territory: Second Report on Economic and Social Cohesion.* Volumes I and 2. Luxembourg: Office for the Official Publications of the European Communities.

European Commission (2002) Regulation (EC) No. 178/2002 of the European Parliament and the Council of 28 January 2002. *Official Journal of the European Communities* L-31/1-24.

European Commission (2003) *Proposal for a Council Regulation terminating the anti-dumping and anti-subsidy proceedings concerning imports of farmed Atlantic salmon originating in Norway and the anti-dumping proceedings concerning Imports of Farmed Atlantic Salmon originating in Chile and the Faeroe Islands.* Brussels. 30.4.2003. COM (2003) 224 Final.

European Commission (2005) *Subject: Progress in Implementation of EU ICZM Recommendation.* Brussels, November 2005.

European Commission (2006) *Green Paper: Towards a Future Maritime Policy for the Union: A European Vision for the Oceans and Seas.* Brussels, 7.6.2006.

European Topic Center on Terrestrial Environment (2003) *Measuring Sustainable Development on the Coast: A Report to the EU ICZM Expert Group on Indicators and Data Under the Lead of the ETC-TE.* http://ec.europa.eu/ environment/iczm/pdf/interim_report.pdf. Retrieved 17/02/06.

Evans, J. (2005) 'Disease threatens Irish oyster beds'. *Intrafish.* http://www. intrafish.no/global/news/article69733.ece?service+print. Retrieved 17/02/06.

Evans, J. (2006a) 'New Irish fisheries minister under pressure to help aquaculture sector'. *Intrafish.* http://www.intrafish.no/global/news/article101096.ece?serv ice=print. Retrieved 17/02/06.

Evans, J. (2006b) 'Young's 'Super-premium' smoked salmon hits market'. *Intrafish.* October 14.

Gereffi, G. (1994) 'The Organization of Buyer-Driven Global Commodity Chains: How U.S. Retailers Shape Overseas Production Networks', in G. Gereffi and M. Korzeniewicz (eds), *Commodity Chains and Global Capitalism.* Westport: Greenwood Press, pp. 95–122.

Gereffi, G. (1999) 'International Trade and Industrial Upgrading in the Apparel Commodity Chain'. *Journal of International Economics,* 48: 37–70.

Gibbon, P. (2001) 'Upgrading Primary Production: A Global Commodity Chain Approach'. *World Development,* 29 (2): 345–63.

Globefish (2005) *Mussels – August 2005. Globefish.* http://www.globefish.org/ index.php?id=2518. Retrieved 17/02/06.

Guillotreau, P. (2003) 'How does the European Seafood Industry Stand after the Revolution of Salmon Farming: an Economic Analysis of Fish Prices'. *Marine Policy,* 28: 227–33.

Humphrey, J. and H. Schmitz (2001) 'Governance in Global Value Chains'. *IDS Bulletin,* 32(3): 19–29.

Institute of Marine Research (2004) *Sustainable Mariculture in Northern Irish Loughs.* http://www.ecowin.org/smile/index.htm. Retrieved 01/25/07.

International Fish Quality Certification (2006) *Fisheries and Aquaculture.* http:// www.ifqc.ie/html/company.htm. Retrieved 02/17/06.

Intrafish News (2005) 'Irish see expanding market for blue mussels'. *Intrafish.* http://www.intrafish.no.global./news/article69339.ece. Retrieved 17/02/06.

Irish Fish Quality Certification (2006) *Fisheries and Aquaculture.* http://www. ifqc.ie/html/company.htm. Retrieved 02/17/06.

Irish Salmon Producers Group (2005) *ISPG Hatcheries, Farms and Packing Stations.* http://www.ispg.ie/about/supplychain_popup.asp. Retrieved 10/02/06.

Kaplinsky, R. and M. Morris (2002) *A Handbook for Value Chain Research.* Sussex: Institute of Development Studies.

Kennedy, M. and E. Magennis (2006) 'North-South Agenda Setting in the 1960s and 1990s: Plus Ca Change?' *Journal of Cross-Border Studies in Ireland,* 2: 34–53.

Lantra (2006) *Northern Ireland Labour Market Intelligence.* Lantra: Lantra House, Coventry, Warwickshire.

Larner, W. and R. Le Heron (2004) 'Global Benchmarking: Participating "at a Distance" in the Globalising Economy', in W. Larner and W. Walters (eds), *Global Governmentality: Governing International Spaces.* London and New York: Routledge, pp. 211–32.

Marine Institute (2000) *CLAMS: Co-ordinated Local Aquaculture Management System: Cuan Chill Chiaráin*. Marine Institute. www.marine.ie/industry+services/aquacultureCLAMS/Kilkierin+bay.htm. Retrieved 25/11/05.

McIver Consulting (2001) *Strategic Directions for Irish Farmed Salmon*. Produced for BIM. http://www. bim2b.com/BIM/MARKETS/report_results. jsp. Retrieved 10/06/06.

McLeod, D. (2002) 'The Life and Times of the "Myti" Mussel'. *Bulletin of the Aquaculture Association of Canada*, December/décembre (102–3): 8–16.

Nestle, M. (2003) *Safe Food: Bacteria, Biotechnology and Bioterrorism*. Berkeley: University of California Press.

Northern Ireland Seafood (2004) *About Us*. http://www.niseafood.co.uk/aboutus/company/background.asp. Retrieved 19/03/07.

O'Carroll, T. (2002) 'The Irish Mussel Industry', *Bulletin of the Aquaculture Association of Canada*. December/décembre (102–3): 25–33.

O'Connor, R., B. J. Whelan, J. A. Crutchfield with A. J. O'Sullivan (1992) *Review of the Irish Aquaculture Industry and Recommendations for its Development*. Dublin: The Economic and Social Research Institute.

Parsons, A. (2005) *Status of Irish Aquaculture 2004*. A Report Prepared by Marine Institute, Bord Iascaigh Mhara and Taighde Mara Teo. Marine Institute. http://www.marine.ie/publications/specialreportsdownloads/index.htm. Retrieved: 16/02/06.

Parsons, A., T. O'Carroll, M. Ó Cinnéide and M. Norman (2004) *Status of Irish Aquaculture 2003*. A Report prepared by Marine Institute, Bord Iascaigh Mhara and Taighde Mara Teo. Marine Institute. http://www.marine.ie/publications/specialreportsdownloads/index.htm. Retrieved: 16/02/06.

Phyne, J. (1996) 'Balancing Social Equity and Environmental Integrity in Ireland's Salmon Farming Industry'. *Society and Natural Resources*, 9: 281–93.

Phyne, J. (1999) *Disputed Waters: Rural Social Change and Conflicts Associated With the Irish Salmon Farming Industry, 1987–1995*. Aldershot: Ashgate.

Phyne, J., R. Apostle and G. Hovgaard (2006) 'Food Safety and Farmed Salmon: Some Implications of the European Union's Food Policy for Coastal Communities', in D. Vanderzwaag and G. Chao (eds), *Canadian Aquaculture Law and Policy: Towards Principled Access and Operations*. Oxon and New York: Routledge, pp. 385–420.

Phyne, J., G. Hovgaard and G. Hansen (2006) 'Norwegian Salmon Goes to Market: The Case of the Austevoll Seafood Cluster'. *Journal of Rural Studies*, 22: 190–204.

Phyne, J. and J. Mansilla (2003) 'Forging Linkages in the Commodity Chain: The Case of the Chilean Salmon Farming Industry'. *Sociologia Ruralis*, 43 (2): 108–27.

Ponte, S. (2001) 'The "Latte Revolution"? Regulation, Markets and Consumption in the Global Coffee Chain'. *World Development*, 30 (7), July: 1099–122.

Roberts, D., Davies, C., Mitchell, A., Moore, H., Picton, B., Portig, A., Preston, J., Service, M., Smyth, D., Strong, D. and S. Vize (2004) *Strangford Lough*

*Ecological Change Investigation (SLECI): Work Package 8. Aquaculture Activity in Strangford Lough*. Department of the Environment. http://www.ecowin.org. Retrieved 26/01/07.

Robinson, G. (2003) *Irish Aquaculture Production 2003*. BIM. http://www.bim.ie/. Retrieved 10/06/06.

Ruddy, M. and T. Varley (1991) 'Sea Farming and Development in North Connemara', in T. Varley, T. A. Boylan and M. P. Cuddy (eds), *Rural Crisis: Perspectives on Irish Rural Development*. Galway: Centre for Development Studies, University College Galway, pp. 77–102.

Schmitz, H. (2004) 'Globalized Localities: Introduction', in *Local Enterprises in the Global Economy*. Cheltenham, UK and Northampton, MA, USA: Edward Elgar, pp. 1–19.

Strydom, P. (2002) *Risk, Environment and Society*. Buckingham and Philadelphia: Open University Press.

Wilson, T. (1984) 'From Clare to the Common Market: Perspectives in Irish Ethnography'. *Anthropological Quarterly*, 57 (1): 1–15.

Winson, A. (1993) *The Intimate Commodity: Food and the Development of the Agro-Industrial Complex in Canada*. Garamond.

# PART III
# Information Technology, Tourism and Sustainability

Chapter 11

# Knowledge-based Competition: Implications for Sustainable Development in Rural Northern Ireland and the Republic of Ireland

Seamus Grimes and Stephen Roper

## Introduction

Promotion of the knowledge-based economy, and more generally, the information society has been prominent in both the Republic of Ireland (RoI) and Northern Ireland (NI) over recent years. In Ireland, the current public spending emphasis on Research and Development (R&D) and innovation through Science Foundation Ireland and other initiatives emphasizes the centrality of 'knowledge' to future development. In Northern Ireland, essentially similar although smaller scale initiatives are underway through investment in R&D Centres of Excellence and the implementation of a regional innovation strategy originally developed in 2003. More broadly, the move from European Union (EU) Framework Programme 6 to Framework Programme 7 marks a substantial increase in EU resources devoted to new knowledge creation, sharing and application.

Past investments in R&D and innovation in the Republic and Northern Ireland, both by indigenous firms and inward investors, have contributed to rapid growth. GNP per capita – a measure of earnings – in the RoI rose from 56 per cent of the Organization for Economic Cooperation and Development (OECD) average in 1989 to 95 per cent in 2003, with Northern Ireland rising from 75 per cent of the OECD average to 86 per cent over the same period. This places the combined economies of the island of Ireland about 14th in the global league of per capita incomes. This is not likely to be the end of the story, however. Indeed, a recent report – Engineering a Knowledge Island (Irish Academy of Engineering/Engineers Ireland [IAoE/EI] 2005) – suggests that, given appropriate policy, future growth in the RoI and NI may be sufficient to put the island economy among the top 5 of the OECD by 2020. As the report suggests: 'when account is taken of the forecast growth of the leading economies, the island economy would need to grow by about 4.5 per cent per annum – slightly less than that achieved over the last decade' (IAoE/EI 2005, p. 1).

The vision outlined in the IAoE/EI (2005) report focuses on developments at national level, or more accurately at the level of the all-island economy. In this sense, it is aspatial and essentially ignores much recent academic thought which has re-emphasized the importance of local or regional dynamics in shaping growth trajectories. In particular, the recent regional development literature emphasizes the central role of knowledge production, at the sub-national or regional scale, in generating sustainable economic development (Lagendijk 2001). Such developments reflect both endogenous and exogenous factors, however. In urban areas, or those where there is a strong concentration of knowledge generating institutions, endogenous processes of knowledge production may drive local competitiveness. In less favoured regions, and particularly those with low levels of urbanization such as the Border, Midland and Western (BMW) Region in the south or more rural areas in the northeast of Northern Ireland, such endogenous processes may be weaker, placing more emphasis on knowledge production resulting from inward investment or policy initiatives. For these regions therefore inward investment attraction is likely to continue to form an important component of regional development strategy (Amin and Tomaney 1995; Border, Midland and Western Regional Assembly 2005).

In this chapter we focus on three areas central to the knowledge economy which illustrate the tensions between endogenously and exogenously generated growth, and between the centrifugal and centripetal forces shaping the location of economic activity in both Northern Ireland and the Republic. In each case the underlying question is the extent to which rural parts of the island of Ireland can be expected to participate in the future growth of the knowledge economy. Firstly we focus on R&D and innovation – both crucial drivers of the knowledge economy – and ask to what extent rural areas are likely to be able to participate in either arena. For example, to what extent are rural areas likely to be able to sustain innovation and endogenous knowledge production? Our view here is not optimistic, with the evidence suggesting that developments in knowledge-intensive industries are tending to be becoming more centralized rather than more dispersed. This emphasizes the potential importance for rural areas of more exogenously-led and policy-led development. One possibility, considered in the second part of the chapter is that new knowledge-based activities may choose to locate – or be encouraged to locate – in more rural areas. Weak local knowledge networks, or infrastructure limitations may, however, limit such activity.

Whether or not developments in the knowledge economy in more rural areas are endogenously, exogenously or policy-led, another key question relates to the ability of smaller firms in these areas to benefit from such economic growth. Adoption of information and communication technologies (ICTs) will be important here, with the potential to enable rural firms to overcome spatial barriers and participate in knowledge economy growth. Again, however, our argument here is that – despite extravagant claims by vested interests in the technology sector – ICTs do not bring about the end of geography; and that the significant effects on rural economies associated with being distant from core markets are not easily overcome by the new technologies.

## R&D, innovation and the rural economy

R&D and innovation are of particular importance in the knowledge economy, providing the basis for sustainable competitive advantage. Recent academic research in this area has emphasized the importance for innovation of dense networks, and the potential advantages of urban locations in generating endogenous knowledge production. Shefer and Frenkel (1998), for example, in their work on Northern Israel, find that an urban location has a strong positive effect on the probability of innovating for high-tech firms in electronics and electrical engineering. Roper and Grimes (2005) also highlight the potential advantages for firms of being located in urban centres which have strong global linkages. Two main sources of endogenous advantage deriving from an urban location are emphasized by these studies: clustering advantages generated by clusters of similar firms (e.g. specialist services) and 'Jacobs' externalities in which firms benefit from the diversity of the urban economy (e.g. Cooke, Davies and Wilson 2001).

Network, clustering and diversity advantages are inevitably strongest in urban locations, with this type of endogenous growth dynamic being more difficult to generate and sustain in rural locations. A key question is, however, how important are such factors in shaping innovation on the island of Ireland? Three recent papers are relevant here. First, Roper and Love (2005) suggest that for firms in both the Republic of Ireland and Northern Ireland network factors – taken as an indication of the importance of endogenous growth processes – are highly significant in firms' innovation success. Links to other firms, links along the supply chain and connections to research institutes all play an important role in shaping firms' innovative capability. This suggests the importance of endogenous growth processes for innovation on the island of Ireland, and perhaps the potential for firms in urban locations to derive particular locational benefits.

Historically, however, other evidence suggests that it may be easy to exaggerate the importance of such endogenous growth processes, at least for manufacturing firms. Roper (2001), for example, suggests that during the 1990s manufacturing firms in the RoI and NI did not benefit from significant locational advantages in terms of innovation. Indeed, based on innovation survey evidence, and controlling for sector etc., this study suggested that rurally-based firms were as effective at innovation as their urban counterparts (see also Davelaar and Nijkamp 1989, 1992; Kleinknecht and Poot 1992; Koschatzky et al. 1998). For industries where activity is highly-R&D or knowledge intensive, on the other hand, endogenous growth processes associated with urban locations may be more important. Shefer et al. (2003), for example, show that the effect of a rural location on the innovativeness of firms depends on their R&D intensity – the more R&D-intensive the activity the greater the negative impact of rurality. Other more anecdotal evidence also suggests the importance of clustering and urban location for firms in knowledge-based activities.

Perhaps the key example here is software development which is strongly concentrated in Dublin and to a lesser extent in Belfast (e.g. O'Malley and

O'Gorman 2001). Here, the evidence from across the RoI and the United Kingdom (UK) suggests significant cluster advantages, with the development of strongly localized entrepreneurial dynamics. In Dublin, this process has been documented in recent work by Crone (2002, 2003), which emphasizes the interdependency of software firms within the Dublin cluster in terms of contracting, recruitment and interlocking ownership. Other location-specific factors have undoubtedly also been important in shaping the concentration of software development in Dublin, such as the availability of venture capital and the strength of the local market from the financial sector (e.g. Roper and Grimes 2005). Evidence for Northern Ireland is less clear here, however, as strong concentration of software development activity and inward investment in the greater Belfast region is also evident.

This evidence is somewhat discomforting in terms of future rural sustainability and participation in the knowledge economy. While historically dispersed manufacturing activity may have been as capable of knowledge production and innovation as its urban counterpart, future knowledge production in more knowledge-intensive activities seems more likely to benefit from stronger endogenous growth processes in urban locations. Development in these more knowledge-intensive sectors is therefore likely to have an urban bias, widening developmental gaps between urban and more rural areas. One possible offset to this general trend is the potential for new urban ventures to develop associated routinized activities such as back-offices and data-processing which can be effectively located in relatively small centres. This has already happened in the RoI context, with a number of small towns benefiting from such investment (Grimes 2003a). Such centres have also benefited from the decentralization of public service activities in the RoI, although the current round of decentralization appears to be meeting with strong resistance. In Northern Ireland, decentralization of public sector activity has been less important, although there have been some notable success in attracting call centre activity to smaller urban centres (e.g. Armagh, Enniskillen, Antrim).

A major weakness of peripheral rural areas is their lack of capacity to exploit opportunities associated with information and knowledge society, and this is reflected in their low levels of participation in EU-funded research projects in these areas. Partly in response to this the EU has sought to bring about greater integration between Framework Programme research activities and Structural Funds which have the objective of promoting social cohesion (Grimes 2003c). A criticism of this approach, however, suggests that policy must seek to move beyond redistributive measures, since the lack of absorptive capacity of these regions is a primary reason for their poor ability to compete for research funding. Among the various factors contributing to the weak absorptive capacity of these regions are the following: low levels of innovation among firms, poor quality services for firms, weak entrepreneurial culture; a primary focus on traditional sectors and on family businesses, little international marketing and a poor quality scientific infrastructure (DeMichelis 1999). Within the Republic, the Institutes of Technology have played an important regional role in diffusing the new technologies into non-urban regions and in facilitating the growth of technology start-ups in these regions. A similar, if less well developed, role is also played by the Further Education Colleges in Northern Ireland.

Because of fears relating to continuity, inward investment strategies continue to be viewed with considerable scepticism and the objective of endogenous development remains an important dimension of regional development strategies. Regional success, however, often depends more on the ability to commercialize technology than actually produce innovative technologies (Grimes and Collins 2003). As a result policy makers are more aware of the need to facilitate the commercial exploitation of technology, regardless of its origins. In light of the island's small size and history of economic peripherality, these issues are particularly resonant. Since it heretofore lacked the capacity for endogenously led development, Ireland instead accesses global sources of knowledge and global demand to spur regional development.

## Inward investment: Private and public

During the early stages of the development of the knowledge economy in the Republic and Northern Ireland, there were widespread and largely unfounded expectations that the new technologies would help to bring about a more even spread of economic development, which would facilitate regional development and would help to integrate peripheral rural areas more effectively into the economy. Twenty years later there is emerging a more sober and realistic view of what can be expected in terms of the link between the knowledge economy, ICTs and the spatial economy (Grimes 2003b). While there is little doubt of the impact of the new technologies on economic development within the RoI and NI generally, the spatial results have been somewhat paradoxical. Whereas during the 1970s inward investment contributed significantly towards the decentralization of manufacturing activity to less-developed regions, the most recent period has been strongly associated with considerable agglomeration of investment in the core urban regions. Employment in multinationals makes this spatial disparity particularly evident, as between 1990 and 2000 the Greater Dublin Area's (GDA) relative share of Multi National Companies (MNCs) employment increased 11.2 per cent from 33.2 per cent in 1990 to 44.4 per cent in 2000. Over the same period, the more disadvantaged BMW Region saw its regional share decline by a combined seven per cent (Forfás 2000).

In terms of private-sector inward investment of commercial activity on the island, the recent trend has therefore been toward centralization rather than dispersal, reflecting similar pressures to those highlighted earlier in terms of R&D. Private sector investment in other arenas underpinning the knowledge economy; however, such as broadband infrastructure development has also favoured urbanized rather than rural areas. Despite the impression given in the media and elsewhere that access to broadband could be provided in all locations relatively easily, the reality is that because of the absence of Points of Presence (PoP) in rural areas such as County Mayo, telecommunications traffic must be routed through urban centres such as Sligo and Galway (Western Development Commission 2002).The significant investment associated with the provision of fibre optic infrastructure suggests that the earlier stages of provision will be directed towards areas of high density population and clusters of development. The continuing high costs of broadband services acts as a disincentive for increasing the necessary level of demand to entice providers to invest in more dispersed infrastructure. Many businesses in rural areas, therefore, continue to depend on the most basic dial-up modem form (PSTN) of access to the internet, with significant limitations in terms of speed and capacity. In other cases internet users who are located within six kilometres of the nearest telephone exchange may have access to ISDN, which is a significant improvement on PSTN, but is now regarded as outdated technology.

Internet users in more urbanized locations, on the other hand, have easy physical access to broadband (DSL), which is 'always on' (as opposed to dial-up) and has access speeds up to 30 times faster than a standard telephone line. DSL services continue to be expensively priced and can only be provided to users within three kilometres of the nearest DSL-equipped exchange. Because of the high costs of broadband, users are reluctant to migrate to the more efficient technologies, and providers argue that there is no commercial case for rolling out broadband to small rural centres (Western Development Commission 2002). This lag in the provision of the most recent technology between urban and rural areas is also the common pattern in more developed regions such as the United States (Malecki 2003). Malecki (2003, p. 212) goes on to remind us that the 'rural penalty' remains: face-to-face 'handshakes' will always be more costly for rural businesses, even if internet transactions and communications impose no additional cost burdens.

Unlike the earlier stages of telecommunications provision which were based on the principle of cross-subsidization, more recently policy at the European level has favoured both privatization and liberalization, leaving the provision of services largely to market forces. The significant general downturn in the telecommunications sector has also worked against the more widespread provision of services such as broadband. Despite this difficult environment the Irish state has promoted a number of public-private partnerships aimed at stimulating the regional rollout of broadband infrastructure, but these have had but limited levels of success. As in many other European regions, the national backbone infrastructure in the RoI continues to be monopolized by Eircom, which previously was the semi-state company responsible for developing this infrastructure. Competing companies must work through this national backbone, making delivery of services for the 'last mile' quite expensive. In 2003, Eircom announced a plan to roll out broadband services to more than 150 towns, some of which have a population of less than 1,500. A minimum of 200 registered users would be the necessary threshold for provision, and potential users who are more than four kilometres from a telephone exchange are unlikely to be connected. To date progress in building up the broadband user population has been quite slow nationally with the RoI significantly behind other European countries in this respect, with 0.19 users per 100 inhabitants in 2003 compared with the EU average of 3.31 (OECD 2003).

In a recent six-country EU study of internet usage by rural firms, it was a little surprising to discover that more than half the firms surveyed were satisfied with technical backup, which might be expected to be problematic in areas with low levels of provision. It is interesting to note that firms who were more likely to express dissatisfaction in Counties Clare and Wexford were those with an ISDN rather than a PSTN dial-up connection (Grimes 2005). One source of frustration was the limited range of choices being offered by service providers. One firm expressed this frustration as follows:

> Eircom are disastrous! They promised an ISDN connection for the past two years and failed to deliver. Our ordinary telephone lines are faulty so internet connection is a nightmare (Grimes 2005).

Unlike the slow rate of progress of broadband rollout in the Republic, particularly to non-urban regions, Northern Ireland is the only UK region which has already achieved parity with London, with 100 per cent of households being covered by broadband by 2005 (Department of Trade and Industry 2005). It is unclear at this point, however, to what extent this degree of broadband coverage has generated significant benefits for rural businesses in the region.

Overall, therefore, we conclude that both private commercial and infrastructural investment is increasingly tending to favour denser urban areas where knowledge production and network density are greatest. In Northern Ireland, infrastructural investments are more uniform but urban areas retain their obvious advantages in terms of knowledge production and networking. Such differences are likely to exacerbate rather than ameliorate urban-rural differences in growth rates, and make it more difficult for firms in rural areas to benefit from knowledge economy growth. In addition to unrealistic expectations in relation to broadband, policymakers also made seriously unfounded assumptions about how electronic business and commerce models might relate to the nature of enterprises in rural areas. Considerable efforts were made to diffuse the necessary IT skills throughout the population at large, and particularly within the small enterprise community, insufficient thought has been given to developing programmes based on a clear understanding of the particular needs of rural enterprise (Southern and Tilley 2000). Policy appeared to be based on the fundamental flaw that new technologies could in some way substitute for the absence of basic entrepreneurial skills (Grimes 2000). Since the commercial exploitation of opportunities offered by the internet can only be one of a number of policy instruments which are required to improve the functioning of rural enterprise, simplistic policy approaches which have suggested higher levels of competitiveness with internet usage have given rise to justified criticism (Southern and Tilley 2000).

## Policy for an information society

Policy in the area of the information society – led by the EU – has long sought to bring about greater levels of social cohesion, particularly in relation to less favoured regions, while at the same time seeking to raise the competitiveness of European business. These objectives have been built into the EU's Framework Programme for research, with thousands of projects being funded to help the population of less favoured regions to exploit the opportunities associated with the new information society. It has been suggested, however, that the operationalization of this policy has proved to be problematic, partly because of the inherent conflict between raising competitiveness, on the one hand, while seeking to promote greater social cohesion between regions (Grimes 2003a). Thus the EU is seeking to bring about greater integration of economic activity, which will see rural areas becoming more exposed to outside competition, and at the same time trying to ameliorate the effects of increasing integration.

More widely, at European Union level, similar aspirations were shared, with considerable expenditure and effort invested in promoting the so-called 'Information Society' (Bangemann 1994). Part of this policy was to seek to ensure that European peripheral regions might benefit from exploiting the new technologies, thus helping to bring about a greater convergence between core and peripheral regions (Cornford et al. 1996; Dabinett 2001; Gibbs 2001). Within this broader regional development framework, the policy also sought to help enterprises in rural areas to acquire the necessary skills to benefit from the opportunities provided by ICTs (Grimes 2005). European Union policy has also been moving away from the older core-periphery models towards a greater emphasis on 'polycentric development'. The thinking here suggests the growing importance of softer and aspatial aspects in our understanding of the changing nature of peripherality. With the rise of ICT usage and the shift toward electronic business, it has been argued that distance from markets no longer has the same impact on remote locations as in previous eras (Copus 2001). Thus while EU Information Society Technology policy has positively shifted away from the more technologically determinist approach of the initial stages towards a greater emphasis on stimulating and facilitating institutional learning and change, a considerable gap remains between the theoretical basis of policy formulation in relation to the promotion of e-commerce and its practical implementation (MacLeod 2000; Gibbs 2001).

The European-wide AsPIRE study of internet use by rural enterprise found that most rural enterprises continue to rely on the most basic dial-up form of internet access, with only limited access being available to broadband. One of the striking findings of this study was that there was only a weak statistical difference between firms in more peripheral and more accessible rural areas in six different European countries including the RoI in terms of their internet usage (Grimes 2005). A more general study looking at issues such as employment growth and other indicators in Northern Ireland also found little significant difference between urban and rural firms (Patterson and Anderson 2003). Care must be taken in the interpretation of such results, and one should not be too surprised that the major differences of firm performance tends to be between countries, reflecting different approaches to policy at the national level. Thus firms operating in rural Ireland were found to be much further ahead in their use of the internet than those in Greece, where access levels remains very low. On the other hand the findings suggest that firms in more peripheral rural locations appear to be using the internet as effectively as their peers in more accessible locations, and perhaps contributing to their ability to compete with these firms by such effective usage. The other main finding, however, suggests relatively low levels of involvement in electronic business models by rural firms in Ireland and elsewhere, with many firms adopting a cautious approach to what the new technologies can offer them.

Despite the absence of statistically significant differences between firms in more or less accessible locations, differences do exist between the profile of business in different regions and also in the approach of development agencies. Rural enterprises in Wexford were found to be more involved in manufacturing activities, and Enterprise Ireland – which was primarily responsible for promoting electronic commerce activity in the region – was rather sceptical about the likely prospects of significant uptake by firms in the short-term. In the Mid-West region on the other hand, the profile of business was more involved in international markets, and Shannon Development has had a long record of integrating ICTs as a significant dimension of their regional development strategies. This was partly reflected in the higher level of interest by firms in requiring support for improving their use of the new technologies; although because of the poor provision of local infrastructure, some firms found it necessary to locate servers in the Dublin region for their requirements. In Northern Ireland such infrastructural issues are less pressing with ICT development therefore balanced by a broader emphasis in recent Government plans on promoting rural diversity and female rural enterprise (e.g. Department of Agriculture and Rural Development [DARD] 2006).

## Conclusions

Recent developments in the knowledge economy across Europe have clearly illustrated the tension between the centralizing tendency which results from agglomeration advantages, and social and political aspirations towards more spatially distributed development. In RoI, these tendencies have perhaps been most evident in the recent concentration of much inward investment in the Dublin area and the rapid, and geographically concentrated, growth of the software sector in Dublin and Belfast. ICT adoption – often suggested as a potential offsetting force in the face of increasing economic centralization – has proved to be limited in its effectiveness in connecting rural businesses in the Republic of Ireland and Northern Ireland to international markets. In part, this may be linked to policy difficulties and the continuing lack of broadband access in some rural areas of the RoI.

Longer-term, if as envisaged by IAoE/EI (2005) rapid growth across the island economy continues, current trends in centralization seem likely to intensify rather than dissipate. In other words, the growth of knowledge-intensive activities in Dublin and Belfast is likely to continue with some displacement of more routinized activities to smaller urban centres.

Potential brakes on this centralization process relate to infrastructure, labour costs and labour availability, particularly related to people's willingness to commute long distances. Potential offsetting factors may include further developments in rural broadband infrastructure alongside policy initiatives designed to encourage ICT usage in rural areas. In general terms, however, the intensification of knowledge-based competition in the knowledge economy seems likely to undermine rather than contribute to sustainable rural development (Grimes and Lyons 1994; Grimes 2003b). Despite the widespread expectations of a greater spread of development associated with an economy which was moving in the direction of exploiting the new technologies, we are only slowly beginning to acknowledge some of the reasons for a continuing and growing pattern of concentrated development.

## References

Amin, A. and J. Tomaney (1995) 'The Regional Dilemma in a Neo-Liberal Europe', *European Urban and Regional Studies*, 2 (2): 171–88.

Bangemann, M. (1994) *Europe and the Global Information Society: Recommendations to the European Council*, Brussels.

Border, Midland and Western Regional Assembly (2005) *New Challenges New Opportunities – Report of the Border, Midland and Western Regional Foresight Exercise 2005–2025*. Ballaghadereen: Border, Midland and Western Regional Assembly.

Cooke, P., C. Davies and R. Wilson (2001) 'Innovation Advantages of Cities: From Knowledge to Equity in Five Basic Steps', *European Planning Studies*, 10 (2): 233–50.

Copus, A. (2001) 'From Core-periphery to Polycentric Development: Concepts of Spatial and Aspatial Peripherality', *European Planning Studies* 9(4): 539–52.

Cornford, J., A. Gillespie and R. Richardson (1996) *Regional Development in the Information Society: A Review and Analysis*. Newcastle: Centre for Urban and Regional Development, University of Newcastle upon Tyne.

Dabinett, D. (2001) 'Mainstreaming of the Information Society in Regional Development Policy', *Regional Studies*, 35 (2): 168–73.

Davelaar, E. and P. Nijkamp (1989) 'Spatial Dispersion of Technological Innovation: A Case Study for the Netherlands By Means of Partial Least Squares', *Journal of Regional Science*, 29 (3): 325–46.

Davelaar, E. and P. Nijkamp (1992) 'Operational Models on Industrial Innovation and Spatial Development: A Case Study for the Netherlands', *Journal of Scientific and Industrial Research*, 51: 253–84.

DeMichelis, N. (1999) 'European Regional Development Policy and Innovation'. Paper presented to the 'Regional Futures, Measuring Success: Models for Sustainable Development in the Information Society', Catholic University of Leuven, Kortrijk, Belgium.

Department of Agriculture and Rural Development (2006) *DARD Rural Strategy 2007–2013*. Belfast: Department of Agriculture and Rural Development.

Department of Trade and Industry (2005) *UK Broadband Status Report*. London: Department of Trade and Industry.

Forfás (2000) *Annual Employment Survey 2000*. Dublin: Forfás.

Gibbs, D. (2001) 'Harnessing the Information Society? European Union Policy and Information and Communication Technologies', *European Urban and Regional Studies*, 8 (1): 73–84.

Grimes, S. (2000) 'Rural Areas in the Information Society: Diminishing Distance or Increasing Learning Capacity?' *Journal of Rural Studies*, 16: 12–21.

Grimes, S. (2003a) 'Ireland's Emerging Information Economy: Recent Trends And Future Prospects', *Regional Studies*, 37 (1): 3–14.

Grimes, S. (2003b) 'The Digital Economy Challenge Facing Peripheral Rural Areas', *Progress in Human Geography*, 27 (2): 174–93.

Grimes, S. (2003c) 'Information Society Technology Programmes, Structural Funds and European Cohesion: The Case of Ireland', in S. Riukulehto (ed.), *New Technologies and Regional Development*, Series A:6. Helsinki: University of Helsinki Seinajokji Institute for Rural Research and Training, pp. 15–32.

Grimes, S. (2005) 'How Well are Europe's Rural Businesses Connected to the Digital Economy?' *European Planning Studies*, 13 (7): 1063–81.

Grimes, S. and P. Collins (2003) 'Building a Knowledge Economy in Ireland through European Research Networks', *European Planning Studies*, 11 (4): 395–413.

Grimes, S. and G. Lyons (1994) 'Information Technology and Rural Development: Unique Opportunity or Potential Threat?' *Entrepreneurship and Regional Development*, 6 (3): 219–37.

Irish Academy of Engineering/Engineers Ireland (2005) 'Engineering a Knowledge Island 2020', Inter*Trade*Ireland, October.

Kleinknecht, A. and T. P. Poot (1992) 'Do Regions Matter for RandD?', *Regional Studies*, 32: 221–32.

Koschatzky, K., A. Frenkel, G. H. Walter and D. Shefer (1998) 'Regional Concentration and Dynamics of Fast Growing Industries in Baden-Wurttemberg and Israel', ISI-Arbeitspapiere Regionalforschung, Nr 14.

Lagendijk, A. (2001) 'Scaling Knowledge Production: How Significant is the Region?', in M. M. Fischer and J. Frohlich (eds), *Knowledge, Complexity and Innovation Systems*. Berlin: Springer, pp. 79–100.

MacLeod, G. (2000) 'The Learning Region in an Age of Austerity: Capitalizing on Knowledge, Entrepreneurialism, and Reflexive Capitalism', *Geoforum*, 31: 219–36.

Malecki, E. J. (2003) 'Digital Development in Rural Areas: Potentials and Pitfalls', *Journal of Rural Studies*, 19: 201–14.

OECD (2003) 'Broadband Access in OECD Countries per 100 Inhabitants', December (www.oecd.org/sti/telecom).

O'Malley, E. and C. O'Gorman (2001) 'Competitive Advantage in the Irish Indigenous Software Industry and the Role of Inward Foreign Direct Investment', *European Planning Studies*, 9 (3): 303–21.

Patterson, Z. and D. Anderson (2003) 'What is Really Different about Rural and Urban Firms? Some Evidence from Northern Ireland', *Journal of Rural Studies*, 19 (4): 477–90.

Roper, S. (2001) 'Innovation, Networks and Plant Location: Evidence for Ireland', *Regional Studies*, 35 (3): 215–28.

Roper, S. and S. Grimes (2005) 'Wireless Valley, Silicon Wadi and Digital Island – Helsinki, Tel Aviv and Dublin and the ICT Global Production Network', *Geoforum*, 36 (3): 297–313.

Roper, S. and J. H. Love (2005) 'Innovation Success and Business Performance – An All-Island Analysis', *All Island Business Model Research Report*, July.

Shefer, D. and A. Frenkel (1998) 'Local Milieu and Innovations: Some Empirical Results', *Annals of Regional Science*, 32: 185–200.

Shefer, D., A. Frenkel and S. Roper (2003) 'Public Policy, Locational Choice and the Innovation Capability of High Tech Firms: A Comparison Between Israel and Ireland', *Papers in Regional Science*, 82 (2): 203–21.

Southern, A. and F. Tilley (2000) 'Small Firms and Information and Communication Technologies (ICTs): Toward a Typology of ICTs Usage', *New Technology, Work and Employment*, 15 (2): 138–54.

Western Development Commission (2002) *Update on Telecommunications in the Western Region.* Ballaghadereen: Western Development Commission.

Chapter 12

# Conflict to Consensus: Contested Notions of Sustainable Rural Tourism on the Island of Ireland

Ruth McAreavey, John McDonagh and Maria Heneghan

'(T)he sustainable tourism debate is patchy, disjointed and often flawed with false assumptions and arguments' (Liu 2003, p. 459).

## Introduction

It is perhaps an under-statement to suggest that sustainability has become the buzzword of the last two decades. Despite its nebulous characteristics, this concept has nonetheless become the key phrase in both policy and political arenas and a term that must be engaged with at various levels and across myriad interests – public and private. Indeed, in talking about sustainability in the context of tourism, Weaver (2004) interestingly concedes that it is because 'of the oxymoronic nature of the term 'sustainable tourism' and its amenability to appropriation by supporters of various ideologies ... (that) ... it can be used to represent and support just about any model of development' (p. 518).

In the literature on sustainable tourism there is however some consensus, namely the currency that this is tourism that is 'economically viable, but does not destroy the resources on which the future of tourism will depend, notably the physical environment, and the social fabric of the host community' (Swarbrooke 1999, p. 13). This normative representation focuses on the inter-relationship between the human and physical environment with its competing aspects and interests, priorities and negotiations. Indeed sustainable rural tourism has often been depicted as a key area in fulfilling expectations in terms of rural development (Sharpley 2000; Garrod, Wornell and Youell 2006; Saxena and Ilbery 2008).

This is nowhere more evident than on the island of Ireland where rural tourism has for a number of years attempted to benefit from its perceived market advantage in terms of its relative 'clean and green' countryside. The changing nature of rural areas however has led to new demands, conflicts and priorities for rural communities. The restructuring of agriculture and its decline from a former dominant position in the economy allied to changing social, economic and settlement patterns has brought in to being a 'new' countryside where traditional practices of production are being replaced by consumption practices. While this change is not new, what is new is how the small-scale, almost passive tourist activity of the past, has been transformed by globalization and the emergence of the 'post mass tourist' into a more demanding and central part of the rural economy. Between 1950 and 2007 international tourism arrivals grew by 6.5 per cent annually, culminating in 898 million visitors in 2007 (UNWTO 2008). That figure is set to grow to 1.6 billion by 2020. Just over half of international tourist arrivals in 2006 were for recreation, leisure and holidays. UNWTO estimates that worldwide receipts from international tourism reached US$733 billion (€584 billon) in 2006.

In 2005 European states recorded in excess of 440 million visitor arrivals accounting for 10 per cent of European GDP and 20 million jobs (Actions for More Sustainable European Tourism 2007). Not surprisingly then, tourism is one of the world's largest and fastest growing industries (Wallace and Russell 2004; Saarinen 2006) and this is likely to be the case for the foreseeable future. Consequently the tourism sector must respond to the pressures placed on it directly, from increased visitor numbers, and indirectly, from negative impacts on the environment and on destination communities, as a 'business-as usual approach will not provide a more sustainable tourism industry' (Gössling, Hall, Lane and Weaver 2008, p. 123). It is therefore fitting in the context of this book to explore rural tourism through the lens of sustainability as this arena is perhaps an ideal vehicle in which to tease out what exactly this concept means and to whom or what we are referring when we enlist its use.

This chapter then will endeavour to critique the different ways in which rural tourism and sustainability is engaged with on the island of Ireland by considering the understandings and challenges that many rural communities face and how rural tourism is often employed to address such challenges. To do this, case studies are drawn on to analyse the extent to which a sustainable development approach can usefully be applied to rural tourism. The chapter concluded by considering some pressing implications for stakeholders in rural tourism.

**Sustainability and tourism**

Although the notion of sustainable tourism is a relatively modern concept, it has been accused of repackaging old ideas under a new format, thus provoking questions over whether current discussions actually offer anything new (Hunter 1997; Butler 1999). It remains unclear whether the new labelling is a form of self-preservation for the sector given its reputation as the 'big bad wolf of the modern era' (Wallace and Russell 2004, p. 236). Indeed sustainable tourism could be described as an ambiguous and malleable term with multiple interpretations (Weaver 2004). Despite this vagueness what we also see in practice and policy is that sustainable tourism is both a desirable and widely embraced principle.

Consequently, just as actual tourism has rapidly expanded, so the volume of literature on sustainable tourism has grown. While past research provided wish lists that never explored the meaning of sustainable tourism, more recent research has analysed the dynamics of sustainable tourism focusing on issues of power, knowledge, development, growth, equity and discourse (Hall 1997, 2003; Hunter 1997; Sharpley 2000; Saarinen 2006; Wallace and Russell 2008). Indeed some commentators would argue that this has gone far enough and the debate needs to move on now to consider the implementation of sustainable tourism (Garrod and Fyall 1998).

'Development' and the notion of 'carrying capacity' consumed tourism studies during the 1960s through to the early 1980s. The mass tourism that epitomized this era was accompanied by visible negative impacts such as the degradation of the Spanish coast. In response to this and as post-Fordist economies enjoyed more flexible forms of production and consumption, the idea of sustainable tourism was moved from the margins to assume centre stage in tourism debates. Ecology, conservation and economic development played a role in this process (Bramwell and Lane 1993) all of which resonate with the ubiquitous Brundtland Report, otherwise known as the World Commission on Environment and Development (WCED 1987), with its emphasis on development, inter-and intra-generational equity and environmental responsibility.

Part of the problem with the notion of sustainable tourism however is that it imposes the contested concept of sustainable development (Redclift 1987; 2005) onto a specific sector, i.e. tourism, often in an unquestioning way. It also superimposes external ideas onto a local context. A collection of difficulties arises as a result of tensions between the principles of tourism and that of sustainable development. Born out of a network of conservationists, environmentalists and international conference delegates, the whole notion of sustainable development can be considered as representing an elitist perspective that does not necessarily take account of local conditions, values and specificities. Frequently it serves to provide a framework for government to manage and monitor the implementation of a particular set of policies. This is exemplified in the *UK Government Sustainable Development Strategy* (DEFRA 2005) which set out plans for central Government departments and executive agencies to produce sustainable development actions plans with resulting principles and checklists 'trivializing' critical issues (Garrod and Fyall 1998, p. 202).

When sustainable development is applied to the tourism sector a less than straightforward amalgamation occurs. Tourism is a notoriously fragmented, multi-sector activity which is predominantly privately owned, and so motivated by short term gains typically in the form of profit (McKercher 1993; Sharpley 2001). Tourists are not homogenous. They exhibit differing levels of 'greenness' according to the situation (McKercher 1993; Turner, Pearce and Bateman 1994) and they seek an array of experiences (Sharpley 2001). Consequently the aspiration of the intra- and inter-generational equity of the sustainable development paradigm quickly becomes a challenge to sustainable tourism and rather than being the starting point for discussion sustainable tourism is often presented as the end point (Hunter 1997).

Sustainable tourism, as a socially constructed and idealized set of aspirations, is thus dynamic in the sense of constantly being constructed and reconstructed by different stakeholders. It is a political process that depends on value systems and ethical judgments which are related to knowledge and power (Hall 1997; Hunter 1997; National Research Council [NRC] 1999; Saarinen 2006; Bramwell and Lane 2008). What we see at its core are issues of economic efficiency, equity and environmental protection and indeed it could be argued that there is a special relationship between tourism and the concept of sustainability as it is very rooted in environment and society. Indeed the Tourism Sustainability Group (TSG) (2007) suggests that because of this relationship manifest in quality environments, cultural distinctiveness and such like, 'tourism can be a destroyer of these special qualities which are so central to sustainable development … (or) … can be a driving force for their conservation and promotion' (p. 2). Consequently tourism and its integration into the rural product can be very much part of developing employment opportunities; increasing local prosperity; raising awareness of the importance of environment and its conservation and maintenance and generally ensuring a greater spread in terms of who can benefit (economically and socially) from a well thought out, planned and managed tourism sector. Tension however can emerge between different interest groups representing the different facets of sustainable tourism, in particular between those that emphasize a development approach and those who highlight the ecological perspective (McKercher 1993; Caffyn 2000). Sustainability itself may become a commodified product by the tourist sector (Hughes 1996). The failure to consider the wider aspects of sustainable tourism results in an approach that is 'overly tourism-centric and parochial', grasping only fragments of the total, namely visible processes and impacts relating to the industry (Hunter 1997; Gössling 2000).

## Tourism on the island of Ireland

Despite the growth in tourism at a global level, European tourism has dropped from 70 per cent in 1979 to 58 per cent in 2000 of its world share (Irish Rural Tourism Federation 2007). Nonetheless a growing fascination with destinations that are more exotic or different suggests that some rural areas at least have the resources to meet an emergent trend. Certain rural places can strive to provide a special appeal to tourists because of the mystique associated with their distinct culture, history, ethnic and geographic characteristics (Heneghan 2002).

Since 1995 the tourist industry in the Republic of Ireland (ROI) has been regarded as one of the country's economic success stories. Six million overseas visitors spent €4 billion in the country in 2002 (Tourism Policy Review Group 2003). On a national scale RoI did reach its targets on volume and expenditure in 2006 but this growth has not been distributed to all areas of the country; and some key areas or destinations are experiencing demand which is threatening their carrying capacity and giving a huge spatial imbalance. Some 5.3 million tourists stayed in Dublin generating €1.67 billion while the North West, with its breathtaking scenery had 1.35 million tourists generating revenue of €301.4million (Failte Ireland 2006). Taking figures for three of the western counties of Ireland, the imbalance is also obvious with numbers of overseas tourists for 2006 in Roscommon being 58,000 (yielding revenue of €58m); for Mayo, 308,000 (yielding revenue of €92m) and Galway, 1,1798,000 (yielding revenue of €356m) (Failte Ireland 2007).

The *New Horizons for Irish Tourism 2003–2013* report (Tourism Policy Review Group 2003) states that after a very successful decade Irish tourism faces a number of challenges and that new strategies are now required. Even though policies that measure success by crude visitor number statistics are currently seen as outdated (Dunne and Leslie 2002), this 'New Horizons' document rather worryingly states that one of the challenges for the future is to increase visitor numbers to 10million, almost three times the size of the population of the country, by 2012 with an associated spend of €6 billion while also affirming that respect for the natural and built environment and support for their conservation and enhancement must not disappear.

In a similar vein in Northern Ireland, the Tourism Minister describes how Government has set 'ambitious targets for tourism over the next three years when we aim to increase visitor numbers by 25 per cent and the amount they spend by 40 per cent' (Tourism Minister, NITB website, 21 February 2008, accessed 25 April 2008). But the background in NI is somewhat different to that in the south. Since the signing of the historic Northern Ireland Peace Agreement in 1998, NI society has experienced considerable change. One of these changes has been the so-called 'peace dividend' which has resulted in a growth in tourism. In 2006 tourism continued to grow with visitor and domestic revenue exceeding £0.5 billion (NITB – Tourism Facts 2006), and this upward trend is set to continue (Tourism Barometer 2007, Wave 3 December – NITB). Supporting almost 30,000 full-time equivalent jobs (Tourism Minister ibid.), it is now on a par with the agricultural sector which was traditionally seen as the mainstay of the rural economy.

Contextualized by these past tourist numbers and desires to increase visitor numbers in the future the next section explores how this aspiration fits with the notion of sustainability. Through the lens of the adaptive approach whereby consideration of economic, environment, social and cultural issues are central, the use of case studies yields a number of insights, none more relevant than the need for further supported actions that go beyond measuring success in terms of volume of visitors and expenditure. What these debates also illustrate is that rural tourism must be cognizant of other significant issues if it is to play its part in local and national economies both North and South.

## Sustainable tourism: The adaptive approach

Sustainable tourism derives from a consensus that it is 'economically viable, but does not destroy the resources on which the future of tourism will depend' (Swarbrooke

1999, p. 13). The fluidity of this resource, namely the inter-relationship between the human and physical environment and the competing aspects and priorities, present a flexibility and adaptability according to various situations (Kernel 2005). In this way a paradigm is offered which allows options to be explored through the identification of priorities and the selection of favoured choices. Recognition is given to the fact that one person's balance may be another's imbalance (Hunter 2002). Drawing from the International Guidelines for Sustainable Tourism (Convention on Biological Diversity, www.retour.net accessed June 2008) the notion of adaptive management seems a useful concept in addressing tourism, as it provides an arena of 'uncertainty, complexity and potential for conflict' (Reed 1999). The key to adaptive management is that it embraces uncertainty in that where a policy is successful, the approach is validated but when there are problems or a policy is seen to fail, then the adaptive approach 'is designed so that learning occurs, adjustments can be made, and future initiatives can be based on the new understanding' (Lee 1993 cited in Reed 1999, p. 335).

In such an entity the whole notion of collaboration between different interest groups and the identification of shared desires is central. Somewhat reflective of the way in which there are myriad opportunities for rural areas to benefit from their natural resources, or indeed where rural areas compete for visitors largely because they are drawing from the same resource, it can be argued that with changing demands on rural areas, the move from agriculturally based activities to more tourist and recreational ones, the need for collaboration and some form of unified thinking becomes more crucial. Consequently adopting what has been called an 'adaptive paradigm' (Hunter 1997, p. 864) facilitates application to very different situations and so supports the articulation of different goals in terms of the use of natural resources as ultimately determined by the circumstances and needs of the destination. Priorities will change in line with particular situations, but development will be sensitive to economic, environmental, social and cultural impacts. This adaptive approach pays attention to the fact that different groups can have different values and needs as, for example; ecological conservation objectives may be incompatible with the desires of local communities (Stocking and Perkin 1992).

Using case studies, the remainder of this chapter considers how notions of sustainable tourism are played out in practice and how different groups or stakeholders might interact in a sustainable tourism paradigm.

### The Wicklow Uplands, Mullaghmore and the proposed Mourne National Park

The way in which the adaptive approach is played out in similar yet different spaces provides an interesting insight to the contested notion that is sustainable rural tourism. Focusing on the proposed Mourne National Park, the Wicklow uplands and the Mullaghmore controversy, what becomes apparent is that in areas where there are multi and diverse stakeholders, issues of land ownership, conflicts both within tourism/recreation interests (mountain bikers and hill walkers) and with other interests (environmental or farming lobby), the need for a successful collaboration between stakeholders (an adaptive approach) is both necessary and achievable. Among the commonalities of the areas is the obvious recognition of the scenic qualities of the areas and their consequent ability to generate economic dividend through tourism development.

The Wicklow uplands are described by Phillips (1999) as representing a 'microcosm of Ireland's landscapes of outstanding quality' (p. 88), containing not only scenic landscapes of mountains, valleys, woodlands and the most afforested county in Ireland, but also being rich in architectural heritage 'ranging from the vernacular to great houses and estates of the eighteenth and nineteenth century' (ibid.). Further it is an area with a strong agricultural tradition with farming being a central source of local employment within the county. What is also of significance is the threat in recent years from the continued expansion of Dublin and the implications that increased urbanization and greater demand for access to the countryside is having on this region.

The second area is that of Mullaghmore, situated in the unique limestone plateau of the Burren region on the west coast of Ireland. This glaciated karst area with its rare plants of Alpine, Artic and Mediterranean species forms part of the Burren National Park established in 1991. It is classified as a World Conservation Union (IUCN) Category II protected area (state owned and managed lands). Mullaghmore is a very striking geological feature in this landscape with its easily visible folded strata formed in a downward symmetrical shape. Mullaghmore, like many other parts which have remained untouched by technological and scientific progress, is often presented as a space which symbolizes difference and authenticity (McGrath 1996 cited in Healy and McDonagh in press). However, as Healy and McDonagh (in press) suggest, 'the states proposal to commodify Mullaghmore Mountain, to transform it from a place of nature into a space for mass tourism ... expose how nature has now become a tourist product, a quality of a certain destination that can be commoditized and exploited for commercial purposes'.

The final case study, the proposed Mourne National Park, is designated an Area of Outstanding Natural Beauty (AONB) and covers around 570km². A focal point for visitors ever since the Victorian era, the Mourne area is the second most popular attraction in NI today, surpassed only by the Giant's Causeway and Causeway Coast in Antrim. Evans (1967) noted its special qualities in highlighting the areas many historical and cultural customs along with the rich land use traditions. These characteristics continue to be valued in the 21st century and it is one of five signature projects selected by the NI Tourist Board (NITB). Its landscape comprises a spectacular coastline, twelve significant peaks, a high granite wall, walking trails, state forests and interesting topography. Its archaeological landscape reflects an area rich with heritage and tradition and one where 'unifying geological, natural and cultural factors ... have shaped the living landscape we see today' (Alison Farmer Associates and Julie Martin Associates 2005, p. 28).

What emerges in all these cases is the need for a collaborative approach to meet the growing demand for access to the countryside by a wide range of rural and non-rural dwellers. What is also apparent in these cases is that this demand is very often permeated by conflict (in terms of who should have access) and concern (in terms of conservation for example) within and between public and private interests. The role of land ownership is also very central to the debate. We see from these examples that different levels of ownership exist in the North and South of Ireland. In relation to the proposed Mourne National Park local communities are not excluded from living within its boundaries, whereas a different approach exists in the South where State owned National Parks very much exclude the community in terms of for example farming activities being allowed to continue. The complexity of sustainability is thus easily mirrored in the many themes that emerge, from issues of access to the countryside to the role of farming and farmers, to exploitation versus conservation, and to the instances of conflict between/within different interests and the debates over landownership.

In terms of this latter issue, in the case of the Wicklow uplands van Rensburg, Doherty and Murray (2006) demonstrate how partnerships were major assets in addressing recreational conflicts involving issues of land-ownership and determining who should have access to the uplands of this region. While it was recognized by all that the scenic lands of the Wicklow region offered great possibilities in terms of attracting tourists to the area, the issue of landownership and benefits accruing were central to the conflict between those wishing to gain access (hill walkers for example), the farmers who owned the lands and their concerns regarding liability and the direct benefit received by allowing such access. What develops during the course of the van Rensburg et al. (2006) study, is an emergent partnership between the diverse stakeholders and a 're-positioning of incentives' that avoided changes in land ownership and property rights, bringing about an 'inclusive approach of encouraging all to participate ... demonstrat(ing) the potential for multi-stakeholder cooperation to address conflicts between landowners and recreationists and ... enhance the latter's compatibility with landowner interest' (p. 32).

This outcome contrasts with the experience of the proposed development at Mullaghmore in Co. Clare in the early 1990s. Here, as Healy and McDonagh (in press) note, the Irish Government's desire to expand a largely unfilled tourism sector through the development of Interpretative Centres, saw a much more exclusive approach whereby this project was presented to communities as *fait accompli* rather than with any desire for a more inclusive partnership approach. The ten year Mullaghmore controversy typified the top-down nature of planning and development and how commercial imperatives were the key drivers. Strategies such as collaborative planning where stakeholders are 'included in the planning, design and operation of visitor centres and where the consultation process is negotiated with communities and visitors' (ibid., p. 5) were largely ignored. What the Mullaghmore controversy did instigate however was 'a significant constitutive force of change in the governmental power structure of Ireland' (Peace 2005, p. 508, cited in Healy and McDonagh in press). The debates and legal proceedings over Mullaghmore had 'significant consequences for the exercise of power at a national level' (ibid.) and expanded the authority and responsibility of Irish environmental authorities particularly as it related to institutional cultures, planning practices and especially in negotiations with those communities (Healy and McDonagh in press).

In the case of the proposed Mourne National Park more questions emerge. The objective here very clearly meets the over-arching criteria for sustainable tourism with attention paid to physical attributes as well as to socio-economic characteristics. Its official sponsors assert that the 'designation should reflect the national importance of the Mourne landscape. It must also have the support of the people who live and work in the Mourne area, as well as those who visit the area for recreation and in doing so, support the local economy' (EHS 2004). In this way the whole notion of sustainable tourism being achieved through the national park seems promising but yet what we see emerging are a number of very different contestations with no overall agreement on how the specific sustainable tourism paradigm should transpire within the Mourne area, in other words, the degree to which environmental, social, economic and cultural aspects are to be addressed remains nebulous.

In contrast to Mullaghmore or the Wicklow Uplands, much of the uplands and High Mournes are in large holdings with ownership residing with Mourne Trustees, Water Service, Forest Service and the National Trust. Some 53 per cent of the land is actively farmed and is in small holdings (average farm size is 15 hectares) with approximately 1500 landowners (Haydon 2007). Consequently the process of consultation was central to the National Parks development. In 2002 a report commissioned by the Environment and Heritage Service identified the Mourne area as being the most suited for National Park status. Following this study, the former Minister of the Environment, Dermot Nesbitt, announced that he would be working towards creating a National Park in Mourne, while a Department of the Environment (DoE) (2004) report stated that its intention was to 'take forward proposals for the designation of a national park in the Mournes' (2004, p. 3).

In conjunction with this political backing, the Mourne National Park Working Party (MNPWP) was established in 2004 following public consultation. While very much in contrast to the process followed at Mullaghmore where the Government took a 'DAD' approach (decide, announce and defend), disregarding the views of the local communities (and other interest groups) and ultimately leading to the formation of the Burren Action Group (BAG) in opposition to the development, the MNPWP had all of the outward appearances of a consultative process. There were however, questions remaining concerning the extent of the consultation. The remit of the MNPWP was to consult on proposals regarding boundaries and a management structure for the park and to make recommendations to Government. It ceased operating in 2007 following the closure of the consultation process. Despite stated aspirations it was puzzling that the MNPWP was not charged with consulting on whether or not the public wished to see a national park and, not surprisingly, there was a perception that the Working Party was a 'smoke screen' for a 'done deal' (Meeting 2, 28 October [2004] item 3f).

Again reinforcing the need for an adaptive approach and somewhat similar to the conflict that was evident in Van Rensburg et al. (2006) study of the Wicklow Uplands, frustration and discord was also evident among Working Party members associated with the Mourne project, but went largely unaddressed. Indeed the group felt that there were avoidable delays that held back the development of a legislative framework for establishing a National Park in NI and that 'the gap (was) widening between the consultative and legislative processes' (Meeting 19, 19 October [2006] item 5). The disharmony went further with some members resigning from the MNPWP as they disagreed with the final recommendations that were made.

This consultation process provides a useful lens to examine sustainable tourism, the diverse interest groups involved and the political nature of the sustainable tourism paradigm. However if we look at the confusion and vagueness that cloud the meanings conferred on sustainable tourism other insights emerge. The DoE highlight the fact that 'suggested aims for Northern Ireland's national parks draw on recognition of several well-developed sustainable tourism and rural socio-economic development initiatives in areas of special landscape significance in Northern Ireland' (DoE 2004, p. 14). In this way environmental and ecological aspects are set aside from socio-economic development. Different values were held by people with different interests. The following excerpt from the minutes of the MNPWP illustrate that many from the farming sector were resistant to the idea of a national park. 'At the open meeting held in Newry in early December farmers gave a resounding 'no' to the proposal for a national park in the Mournes although at the moment they did not know what they were saying 'no' to' (Meeting 4, 16 December [2004] item 6).

The tension between the different objectives of sustainable tourism is evident in the recommendations made by the Working Party. Support for 'the retention and appropriate development of existing and new industries and a diverse economy' (Mourne National Park Working Party 2007, p. 14) is specified in the first recommendation. Meanwhile recommendation 3 draws attention to environmental protection while supporting existing farming and other businesses and also diversifying the economy. Achieving all of these objectives concurrently would appear to be extremely challenging. On the one hand how can environmental features be protected while also allowing economic diversification and development? The concept of development was contentious and the Working Party was clear that the consultation provided a divided response on planning issues: 'Some wanted to see a proposal that would stop inappropriate developments being built in the area, while a significant number were concerned that there would be increased planning restrictions' (ibid., p. 29).

Indeed fears over planning restrictions were not limited to possible future developments, but like the Wicklow Uplands they extended to the issue of public access. In particular the nature of the landowners' liability raised concerns among the farming community. The consultation indicated a belief among landowners that they were liable for any injury experienced by anyone entering their land and it revealed a strong fear of litigation among this group of stakeholders. Meanwhile the Government did 'not consider that the current provisions for occupiers' liability are a barrier to access ... (and) ...(t)here is no known reported case of adult trespassers successfully suing a landowner because of an injury caused due to natural features arising in the countryside' (DoE DFP Information Leaflet ND, pp. 1, 7). Nonetheless only months after making this statement, the Government announced that it was allocating £500,000stg for access management in the Mournes, to include helping landowners deal with their access problems (press release NI executive, online accessed 24 April 2008).

Finally, one of the interests that appeared to be missing from the consultation process was that comprising people living outside the designated area. Due to lack of resources the Mobile Information Unit did not travel outside the area and this caused concern for some members of the Working Party (Minutes, 19 October 2006; 18 April 2007). Somewhat similar to the token consultation gestures that took place during the Mullaghmore debate and what Healy and McDonagh (in press) described as the overt but again largely tokenistic references to public consultation in the more recent Cliffs of Moher Visitor Centre project, we see that engagement with public consultation represents a fundamental flaw in the processes followed and highlights a limited application of an adaptive approach and more broadly the sustainable tourism paradigm.

## Conclusion

It is apparent, even within this short discussion, that there are many different aspects to sustainable rural tourism and consequently many different issues and interests to negotiate between. As Butler (1998) emphasized 'in the context of tourism and recreation, competition is often as fierce between interests as it is between tourism/recreation and other interest' (p. 227). The potential conflict between hill walkers and ski slope developers, or fishermen and boating enthusiasts or those looking for quietness and wilderness as opposed to theme parks or interpretative centres often provide much contested arenas. Van Rensburg et al. (2006) referred to such conflicts in their discussion on the role of partnerships in minimizing the conflict between landowners and hill walkers in some of the upland areas of Ireland. Consequently one could argue that there is a need for the right measures and supports at policy level which can seek to achieve compatibility between the needs and resources of the local community, its residents and the tourists.

In terms of what these measures or supports could entail, it would seem that a requirement for a multi-faceted approach, involving partnership at a number of levels, international, national and local is necessary. The Comhairle na Tuaithe strategy established in 2004 is perhaps one of the ways in which some of the issues dealt with in this chapter could potentially be addressed. This strategy brought together various interest groups, farming organizations, state bodies and recreational users of the countryside in efforts to develop a National Countryside Recreation Strategy in the South of Ireland. What largely emerged from this body was the need to recognize the value of countryside recreation; the need for protection of the natural and cultural landscape and the need to be cognizant of the legal rights and concerns of landowners (both private and public) (Department of Rural Community and Gaeltacht Affairs no date).

The case studies considered in this chapter have revealed how policy must be cognizant of the political nature of achieving consensus within sustainable tourism; conflict is inevitable as particular groups strive to achieve dominance. Lees (2003) has shown how sustainability is used to justify gentrification in an urban context; it is imperative that sustainability is not used to rationalize ulterior and unreasonable motives.

In the 2007 document entitled 'Actions for More Sustainable European Tourism' the EU reinforced some of this thinking in stressing the need for governments and their agencies to take a more holistic and balanced approach to planning and development, giving due consideration to the needs of future generations. In fact, whether it is at global, national or local level, the development of sustainable rural tourism however defined suggests at minimum the need for synergy of purpose within and between communities, vested interests, individuals, state bodies and other stakeholders. What needs to be realized is that while a structured group approach may be the way to develop and promote rural tourism, creating inter-community cooperation and collaboration will be a complex and difficult process (Heneghan 2002).

If we accept that different groups value different aspects of sustainable tourism, we must also accept that there will be conflict between these various groups. Therefore how to negotiate between them and having a political framework in place to aid this process will inevitably determine how sustainable the 'consensus' reached will be. Consequently, as Crouch (2006, p. 355) suggests, while it is familiar to 'point to the increasing significance of tourism in the rural economy, and tourism's agencies as producer, generator and power for change', it is also fair to suggest that despite its growing importance there is still a dearth of specific rural tourism policies or appropriate political frameworks in place. While the UK has been among the more active in terms of countryside planning, the RoI has been far less dynamic. In past decades the absence of such policies was less significant in that rural areas were largely dominated by agriculture and production. In a contemporary sense this is no longer the case with rural areas being recognized for their complexity, multifunctional capacities and as places with a multiplicity of interest groups representing farmers, environmentalists, new rural dwellers, tourists, etc. all claiming their rights to, and use of, different rural spaces (see Butler 1998).

This changing configuration of the rural brings into focus the myriad debate that surrounds the concept of sustainability and how it is engaged with and interpreted in different ways, at different levels and by different groups. It is perhaps only in more recent decades that tourism and the particular challenges that this sector poses for rural areas has been linked with the notion of sustainability. What is perhaps ironic in this association is that while the traditional occupation of rural areas in terms of agriculture has been pilloried in past decades due to its perceived unsustainable and environmentally damaging impacts, we increasingly place emphasis on the economic potential of tourism in rural areas, almost ignoring that tourism is also an extractive industrial activity (Pechlaner et al. 2002) with wide ranging impacts on environmental, social, human, heritage and cultural resources. Indeed the focus on the economic gains of rural tourism very often sees discussion fixed on how to attract tourists to rural areas (for example what marketing tools to use; what range of activities need to be provided) rather than showing concern with their likely impact on arrival. A further paradox in this debate is the seeming desire to replace one vulnerable activity (agriculture) with another (tourism). The wisdom of this choice remains to be seen.

What this chapter has attempted to highlight then is that in terms of a future rural tourism strategy for the island of Ireland, the process of identifying a sustainable tourist initiative is no easy task. What is clearly evident is that there are multiple stakeholders and a wide array of interest groups that can all make legitimate claims on the concept of sustainability. Understanding this and realizing the interrelationship between – tourism, the environment and local communities is of crucial importance. It leads us to conclude that adaptation, collaboration, consultation with stakeholders, however complex these activities may be – are critical to any long term perspective of what could be termed a successful sustainable rural tourism approach.

## References

Alison Farmer Associates and Julie Martin Associates (2005) *Mourne National Park Boundary Recommendations*. Belfast: Environment and Heritage Service.

Bramwell, B. and B. Lane (1993) 'Sustainable Tourism: An Evolving Global Approach', *Journal of Sustainable Tourism*, 1: 1–5.

Bramwell, B. and B. Lane (2008) 'Priorities in Sustainable Tourism Research', *Journal of Sustainable Tourism*, 16, 1: 1–4.

Butler, R. (1993) 'Tourism: An Evolutionary Perspective', in R.W. Butler, J. W. Nelson and G. Wall (eds), *Tourism and Sustainable Development: Monitoring, Planning and Managing*. Ontario: Department of Geography Publication Series, University of Waterloo, pp. 27–43.

Butler, R. (1998) 'Rural Recreation and Tourism', in B. Ilbery (ed.), *The Geography of Rural Change*. Harlow: Longman, pp. 211–32.

Butler, R. (1999) 'Sustainable Tourism: A-State-of-the-Art Review', *Tourism Geographies* 1: 7–25.

Caffyn, A. (2000) 'Developing Sustainable Tourism in the Trossachs, Scotland', in G. Richards (ed.), *Tourism and Sustainable Community Development*. London: Routledge, pp. 83–100.

Crouch, D. (2006) 'Tourism, Consumption and Rurality', in P. Cloke, T. Marsden and P.H. Mooney (eds), *Handbook of Rural Studies*. London: Sage, pp. 255–64.

DEFRA (2005) *The UK Government Sustainable Development Strategy*. London: The Stationery Office.

Department of Community, Rural and Gaeltacht Affairs (no date) *Comhairle na Tuaithe – National Countryside Recreation Strategy*. Dublin: Department of Community, Rural and Gaeltacht Affairs.

Department of the Environment for Northern Ireland (2004) *National Parks and Other Protected Landscape Areas: A Discussion of Options for Establishing National Parks and Managing Other Outstanding Landscapes in Northern Ireland*. Belfast: Department of the Environment.

Department of the Environment (NI) and Department of Finance and Personnel (NI) (no date) *Occupiers' Liability Law in the Context of Access to the Countryside in Northern Ireland: Information Leaflet*. Belfast: Environment and Heritage Service.

Dunne, M. and G. Leslie (2002) 'An Overview on Carrying Capacity Indicators and Visitor Management Techniques'. *Tourism and the Environment – Sustainability in Tourism Development*. Dublin: Dublin Institute of Technology.

Environment and Heritage Service (NI) (2004) *A Mourne National Park*. Belfast: EHS Corporate Communications.

Evans, E. E. (1967) *Mourne Country: Landscape and Life in South Down*. Dundalk: Dundalgan Press, Second Edition.

Failte Ireland (2007) *A Future for Rural Tourism in Ireland*. Report by Irish Rural Tourism Federation. Ireland: Dublin.

Garrod, B. and A. Fyall (1998) 'Beyond the Rhetoric of Sustainable Tourism?' *Tourism Management*, 19, 3: 199–212.

Garrod, B., R. Wornell and R. Youell (2006) 'Re-conceptualising Rural Resources as Countryside Capital: The Case of Rural Tourism', *Journal of Rural Studies*, 22: 117–28.

Gössling, S. (2000) 'Sustainable Tourism Development in Developing Countries: Some Aspects of Energy Use', *Journal of Sustainable Tourism*, 8: 410–25.

Gössling, S., C. M. Hall, B. Lane and D. Weaver (2008) 'The Helsingborg Statement on Sustainable Tourism', *Journal of Sustainable Tourism*, 16, 1: 122–4.

Hall, C. M. (1997) 'The Politics of Heritage Tourism: Place, Power and the Representation of Values in the Urban Context', in P. Murphy (ed.), *Quality Management in Urban Tourism*. New York: John Wiley and Sons, pp. 91–102.

Hall, C. M. (2003) 'Politics and Place: An Analysis of Power in Tourism Communities', in S. Singh, J. T. Dallen and R. K. Dowling (eds), *Tourism in Destination Communities*. Wallingford: CABI, pp. 99–114.

Healy, N. and J. McDonagh (in press) 'Commodification and Conflict: What can the Irish Approach to Protected Area Management Tell Us?', *Society and Natural Resources*, 22: 381–91.

Heneghan, M. (2002) *Structures and Processes in Rural Tourism. Signpost to Rural Change*. Conference proceedings, Teagasc Agriculture and Food Development Authority, Rural Economy Research Centre.

Henry, I. P. and G. A. M. Jackson (1996) 'Sustainability of Management Processes and Tourism Products and Contexts', *Journal of Sustainable Tourism*, 4: 17–28.

Hughes, G. (1996) 'Tourism and the Environment: A Sustainable Partnership', *Scottish Geographical Magazine*, 112, 2: 107–13.

Hunter, C. (1995) 'On the Need to Re-conceptualize Sustainable Tourism Development', *Journal of Sustainable Tourism*, 3: 155–65.

Hunter, C. (1997) 'Sustainable Tourism as an Adaptive Paradigm', *Annals of Tourism Research*, 24, 4: 850–67.

Hunter, C. (2002) 'Aspects of the Sustainable Tourism Debate From a Natural Resources Perspective', in R. Harris, T. Griffin and P. Williams (eds), *Sustainable Tourism: A Global Perspective*. Butterworth-Heinemann, pp. 3–23.

Irish Rural Tourism Federation (2007) *A Future for Rural Tourism*. Irish Rural Tourism Federation.

Kernel, P. (2005) 'Creating and Implementing a Model for Sustainable Development in Tourist Enterprises', *Journal of Cleaner Production*, 13: 151–64.

Laws, E., B. Faulkner and G. Moscardo (1998) 'Embracing and Managing Change In Tourism', in E. Laws, B. Faulkner and G. Moscardo (eds), *Embracing and Managing Change in Tourism: International Case Studies*. New York: Routledge, pp. 1–10.

Lees, L. (2003) 'Visions of 'Urban Renaissance': The Urban Task Force and the Urban White Paper', in R. Imrie and M. Raco (eds), *Urban Renaissance? New Labour, Community and Urban Policy*. Bristol: Policy Press, pp. 61–82.

Liu, Z. (2003) 'Sustainable Tourism Development: A Critique', *Journal of Sustainable Tourism*, 11, 6: 459–75.

McKercher, B. (1993) 'Some Fundamental Truths about Tourism: Understanding Tourism's Social and Environmental Impacts', *Journal of Sustainable Tourism*, 1: 6–16.

Mourne National Park Working Party (2007) *Report to the Minister*. Belfast: Environment and Heritage Service.

National Research Council (NRC) (1999) *Our Common Journey: A Transition to Sustainability*. National Research Council.

Phillips, A. (1999) 'From Conflict towards Partnership – Testing Landscape Management Systems in County Wicklow', in *Policies and Priorities for Ireland's Landscape*, Conference Papers, Tullamore, Co. Offaly, Ireland, pp. 87–102.

Pretty, J. and M. Pimbert (1995) 'Beyond Conservation Ideology and the Wilderness Myth', *Natural Resources Forum*, 19: 5–14.

Redclift, M. R. (1987) *Sustainable Development: Exploring the Contradictions*. Routledge: London.

Redclift, M. R. (2005) 'Sustainable Development (1987–2005): An Oxymoron Comes of Age', *Sustainable Development*, 13, 4: 212–27.

Reed, M. G. (1999) 'Collaborative Tourism Planning as adaptive Experiments in Emergent Tourism Settings', *Journal of Sustainable Tourism*, 7, 3 & 4: 331–55.

Saarinen, J. (2006) 'Traditions of Sustainability in Tourism Studies', *Annals of Tourism Research*, 33, 4: 1121–40.

Saxena, G. and B. Ilbery (2008) 'Integrated Rural Tourism: A Border Case Study', *Annals of Tourism Research*, 35, 1: 233–54.

Sharpley, R. (2000) 'Tourism and Sustainable Development: Exploring the Theoretical Divide', *Journal of Sustainable Tourism*, 8: 1–19.

Sharpley, R. (2001) 'Sustainable Rural Tourism Development: Ideal or Idyll?', in L. Roberts (ed.), *Rural Tourism and Recreation: Principles and Practice*. Cambridge, MA: CABI Publishing, pp. 57–8.

Stocking, M. and S. Perkin (1992) 'Conservation-with-Development: An Application of the Concept in the Usambara Mountains, Tanzania', *Transactions of the Institute of British Geographers*, 17: 337–49.

Swarbrooke, J. (1999) *Sustainable Tourism Management*. Oxon: CAB International.

Tourism Policy Review Group (2003) *New Horizons for Irish Tourism – An Agenda for Action 2003–2013*. Dublin: Department of Arts, Sports & Tourism.

Tourism Sustainability Group (2007) 'Action for more Sustainable European Tourism'. http://ec.europa.eu/enterprise/services/tourism/doc/tsg/TSG_Final_Report.pdf, accessed 6th March 2007.

Turner, R.K., D. Pearce and I. Bateman (1994) *Environmental Economics: An Elementary Introduction*. Hemel Hempstead: Harvester Wheatsheaf.

UNWTO (2008) *World Tourism Barometer*, 6, No.1.

Van Rensburg, T., E. Doherty and C. Murray (2006) *Governing Recreational Activities in Ireland: A Partnership Approach to Sustainable Tourism*. Working Paper No. 113, Department of Economics, National University of Ireland, Galway.

Wallace, G. and A. Russell (2004) 'Eco-cultural Tourism as a Means for the Sustainable Development of Culturally Marginal and Environmentally Sensitive Regions', *Tourist Studies*, 4, 3: 235–54.

Weaver, D.B. (2004) 'Tourism and the Elusive Paradigm of Sustainable Development', in A. A. Lew, C. M. Hall and A. M. Williams (eds), *A Companion to Tourism*. Oxford: Blackwell Publishing, pp. 510–24.

World Commission on Environment and Development (1987) *Our Common Future*. Oxford: Oxford University Press.

www.retour.net/Resourcecenter/WebDocuments/documents/CBDdocs/Internatio nal%20Guidelines%20CBD%20-%20tourism.htm, accessed June 2008.

# PART IV
# Social Differentiation
# and Sustainability

## Chapter 13

# Demography of Rural Decline and Expansion

### Trutz Haase

## Introduction

This chapter considers the place of demographic factors in an understanding of rural decline and expansion. Until recently, the overriding historical pattern of population change in Ireland has been one of sustained emigration. In rural areas, this pattern was frequently associated with higher rates of economically-dependent population groups, gender imbalances, a loss in ability to create new employment opportunities and, as a result, overall weakened communities.

However, the general ideal of demographically-balanced, self-sustaining and economically-viable communities may be more a product of ideology than of actual historical reality. That reality, stable as it may appear in the aggregate, is made up of enormous diversity when a detailed look is taken. On either a larger world scale or a smaller local scale these imbalances are part of a dynamic process in which it is recognized that there is no necessity for any particular area, community or country to be continually in demographic balance, or indeed in a situation of decline.

The central question we must therefore ask in this chapter is to what extent can poverty in rural Ireland explain weak demography or to what extent is poverty in rural Ireland the outcome of weak demography? This question is rendered even more pertinent if we consider whether Ireland's rural areas have received a fair share in the country's improved economic fortunes over the past decade, and whether the improved economic conditions for the country as a whole have resulted in an improvement in the demographic, social and economic sustainability of rural communities.

The chapter is divided into three sections, the first looks at the appropriate conceptualization of deprivation, particularly when considered in a rural context. The second examines empirical data for Ireland tracing the main changes in population and settlement patterns and the resulting variations in the geographical distribution of affluence and deprivation over the 1991 to 2002 period. The final section draws conclusions about the changes that are necessary in order to make progress towards sustaining rural communities in Ireland.

## Conceptualizing deprivation

Any attempt to describe the extent and distribution of disadvantage in Ireland, and in rural Ireland in particular, encounters the problem of appropriately defining poverty and deprivation. Within the extensive literature that has been produced on this subject over the past 20 years, the dominant approaches have built on Townsend's definition of poverty, which highlights the relative character of the concept by comparing how people experience their lives relative to the community they are living in. People are 'relatively deprived if they cannot obtain, at all or sufficiently, the conditions of life – that is, the diets, amenities, standards and services – which allow them to play the roles, participate in the relationships and follow the customary behaviour which is expected of them by virtue of their membership of society' (Townsend 1993, p. 36).

Townsend places considerable emphasis on lack of income, and income poverty is undoubtedly an essential element of deprivation. However, exclusive reliance on income poverty as a measure of deprivation is problematic for a number of reasons: firstly, it assumes that the only unit of analysis is the individual; secondly, it assumes that deprivation should be measured solely in terms of outcomes as opposed to risks, conditions or opportunities; and thirdly, it does not consider broader aspects of the quality of life, such as, for example, health, education, environment, or access to transport and services. A definition of deprivation which is overly-reliant on individual measures of income poverty unduly narrows the focus of policy and may deflect attention away from those areas where the most effective interventions towards building sustainable communities can be made. Given the emphasis on building sustainable rural communities, it is worth noting that successive deprivation indices in the UK, all of which have closely followed Townsend's emphasis on the lack of income experienced by the individual, have subsequently been shown to exhibit substantial urban bias, essentially as they have failed to adequately conceptualize and measure rural deprivation.

This critical view is incorporated into the broader definition proposed by Coombes et al. (1995) who state that '(t)he fundamental implication of the term deprivation is of an absence – of essential or desirable attributes, possessions and opportunities which are considered no more than the minimum by that society' (p. 5). This author believes this to be a preferable definition of deprivation and the following paragraphs briefly discuss some of the issues involved.

### Focus on the individual

At least in the European and Anglo-Saxon context, the debate on poverty and social exclusion has, over the past two decades, been characterized by an increasing focus on the individual. This is not only the case in relation to the development of transfer mechanisms within the tax and social welfare systems that aim to alleviate poverty, but extends to the growing emphasis on counting the number of individuals targeted under various area-based initiatives and on 'counting the poor' in the construction of spatial deprivation indices. The argument is that one should precisely estimate the number of people suffering deprivation in a given area, before directing resources to people residing in these areas in order to minimize deprivation.

However, questions in relation to the value of 'counting the poor' have begun to emerge. Most commentators agree that the majority of poor people are unlikely to live in designated disadvantaged areas, at least not if designated areas are defined narrowly enough to make any sense for the targeting of scarce resources. Once this fact is acknowledged, it follows that the principal policy instrument for targeting the poor (as individuals or households) must be the tax and social welfare system, with area-based initiatives functioning as a complementary intervention that is particularly suited to enhancing the infrastructure and services available to particular communities.

*Who exactly is deprived?*

The second issue is the question of whom or what exactly is deprived. The question as to whether deprivation is suffered by individuals, households or communities is a difficult one. The dominant view amongst commentators has been that it is the individual who is deprived and, as such, the individual is the appropriate building-block for all definitions of deprivation.

There are, however, a number of caveats associated with this assumption, particularly given that the individual's experience is also shaped by household (e.g. race and class) and neighbourhood factors (e.g. environment and social conditions). There is now a substantial body of international research that shows the influence of neighbourhood characteristics on individuals; i.e. that characteristics which are shared by groups of individuals (e.g. in schools, neighbourhoods, communities, etc.) have an impact on the individual's well-being over and above what could be predicted from his or her socio-economic characteristics alone.

The size of these neighbourhood effects is generally small when compared to individual-level effects; nonetheless, they are both statistically and substantively significant, and shed considerable light on the question of why after years of tackling social exclusion and deprivation primarily through individually-targeted responses certain communities remain unable to escape from a vicious cycle of deprivation.

*Actually or potentially deprived*

The third question, and one that is closely linked with the previous one, relates to whether our definitions should be confined to those who are 'actually deprived', or whether they should include considerations of the 'risk' of deprivation. Most commentators emphasize outcomes; that is, the actual experience of deprivation of individuals or households. However, as Coombes et al. (1995) note, in practice, this distinction may not be sustainable:

The notion of a 'cycle of deprivation' illustrates the problem: individuals who are poor are also more likely to live in unsatisfactory housing conditions and to suffer health problems, thereby endangering their employment status and thus reinforcing their poverty. In this way, each outcome is also a condition which makes the sufferer more vulnerable to other aspects of deprivation...[and] the tendency for individuals to thus experience more than one form of deprivation has been simplified in the term multiple deprivation (ibid., p. 7).

Summing up the points made above, definitions of deprivation must clearly go beyond considerations of income poverty conceptualized at the individual level, to relate the experience of individuals, groups and communities to the prevailing social context. Our definitions must further reflect the fact that the socio-economic context has an impact on people's quality of life and that neighbourhood effects play an important role in this context. Finally, as it becomes increasingly clear that deprivation indices are inappropriate tools for targeting poor individuals, but derive their *raison d'être* from their ability to inform initiatives aimed at the level of communities, they cannot be reduced to poverty outcomes alone, but must also include measures of the risk of poverty.

## What is different about rural deprivation?

Having highlighted some of the conceptual issues that underpin the study of poverty and deprivation in general, this section looks at the specificities of rural deprivation. Rural deprivation distinguishes itself from its urban counterpart in terms of its underlying causes, the forms through which it can be conceptualized and measured, and in the policy responses that it demands. Haase and Pratschke (2005, p. 7) describe the distinctiveness of rural deprivation in the following terms:

> Unlike their manifestation as unemployment black-spots in urban areas, long-term adverse labour market conditions in rural areas tend to manifest themselves either in agricultural underemployment or in emigration. The former occurs due to the strong social incentives that encourage farmers to maintain small landholdings, even where these do not provide a full income. Moreover, individuals who are unable to find paid employment in disadvantaged rural areas may withdraw from the labour market in order to assist a relative engaged in farming. Where agricultural employment is scarce, long-term adverse labour market conditions generally lead to emigration. Emigration is also, and increasingly, the result of mismatches between education and skill levels, on the one hand, and available job opportunities, on the other. In both cases, the (rural) unemployment rate is likely to vastly understate the real extent of labour market disadvantage.

In general, little attention has been paid to the identification and examination of the distinctive features of rural deprivation, poverty and exclusion, and even less to their measurement. Commins (2004) notes that research has tended to focus on identifying 'poor areas', 'poor communities' and 'poor farmers', without tackling the following key questions: 'What, if anything, is distinctively different about poverty in rural areas? Who within rural communities face disproportionate risks of poverty and what factors affect their life chances, and how is poverty generated and reproduced?' (p. 60).

In common with Haase and Pratschke (2005), Commins identifies some of the principal characteristics of rural life and rural poverty, including the invisibility of rural poverty, its dispersed nature, perceptions that rural life is 'problem-free', the outmigration of younger people, high proportions of elderly people, the high level of property ownership which, although it accords status, often masks the existence of low farming incomes.

Shaw (1979, cited in Asthana et al. 2002) identified the three most important features of rural disadvantage as (i) resource deprivation, that is, low income and lack of adequate housing; (ii) opportunity deprivation, arising from lack of availability of health, education and recreational services, and (iii) mobility deprivation, that is, lack of transport and the inaccessibility of jobs and services. The first of these is clearly common to both rural and urban disadvantage, whilst the second and third are specific to rural disadvantage.

Evidence of the failure of existing indicators of disadvantage to identify rural deprivation can be found in the work of Frawley et al. (2000) who, studying the incidence and features of low-income farm households, observe that 'low-income farm households as a single group were found to be indistinguishable from all farm households on a basic lifestyle deprivation index' (quoted from Commins 2004, p. 65). This study also reports that low-income farming households have lower levels of deprivation than all low-income households in Ireland. Frawley et al. (2000) suggests that possible reasons for a lower level of deprivation among low-income farming households, compared to low-income households in general, might be linked to the types of deprivation indicators used, which include possession of strong footwear and a waterproof coat. Possession of these items on a farm would be an absolute essential and their presence as such could not be considered a suitable indicator for the absence of deprivation. Another typical example of essential items in a rural area might be a car. While a car in an urban area might be considered a non-essential or even a luxury item, it becomes an item of absolute necessity in the absence of public transport in a rural area.

There are clearly differences in the types of deprivation experienced in rural and in urban areas. This in turn raises questions about the indicators that should be used to measure deprivation and, secondly, whether different indicators should be used to measure deprivation in rural and urban areas. Noble and Wright (2000, cited in Commins 2004) acknowledge that it might be necessary to treat rural deprivation separately from urban deprivation. They also, however, argue that some comparability between urban and rural areas is required, particularly in relation to the targeting of area-based 'regeneration' funds.

*Measuring deprivation*

The final issue in the discussion of conceptual issues is the actual measurement of deprivation. The section starts with a discussion of the dominant approaches to the measurement of income poverty and other related measures of deprivation. Following this, it looks at the constraints of individual measures of deprivation in the context of spatial analysis, particularly at higher levels of spatial disaggregation.

The Irish National Action Plan for Social Inclusion (NAP/inclusion) distinguishes between two types of poverty, relative poverty and consistent poverty. Whether someone is living in relative poverty is determined by comparing their income to a particular income threshold; if they fall below this threshold, they are deemed to be experiencing poverty. Generally, this threshold is set at 50 per cent, 60 per cent or 70 per cent of median income. The 'standard threshold adopted by the European Union is below 60 per cent of median income. Median income is the middle point of the income distribution, i.e. the middle point if all incomes were lined up, from the lowest income to the highest income' (Combat Poverty Agency Strategic Plan 2005–2007, p. 38).

'A person is said to be in consistent poverty when he or she has both a low income and lacks at least one of a number of specified basic necessities such as warm clothes, adequate food and heating' (ibid., p. 34). As of 2007, the list of deprivation items has been extended and a person is deemed to be consistently poor if at-risk-of poverty and lacking two or more of a list of eleven items.

Deprivation or social exclusion is further defined as 'the process whereby certain groups are shut out from society and prevented from participating fully by virtue of their poverty, discrimination, inadequate education or lifeskills. This distances them from job, income and education opportunities as well as social and community networks and they have little access to power and decision-making bodies' (ibid., p. 39).

There has been a recent trend towards using similar definitions of poverty, articulated at the individual level, in the analysis of the geographical distribution of deprivation, notably the current *Indices of Multiple Deprivation* (IMD) for England, Scotland, Wales and Northern Ireland (Noble et al. 2000). However, this approach is not without its critics, as it falls short in relation to almost all of the conceptual aspects outlined in the previous sections. Above all, it assumes that the level of deprivation in an area is simply the sum of the poor individuals within it, that the risk of poverty is largely irrelevant and that there is no conceptual difference between urban and rural deprivation. A more detailed discussion of these indices is provided in Haase and Pratschke (2005).

Whatever the merits of approaching spatial analysis in terms of the number of people living in (income) poverty within a given area, this approach is practically unfeasible when implemented at the level of small areas. The relative income poverty indicators utilized by the EU and by the Irish Government rely on in-depth household surveys and, because of the limited sample size involved, are first and foremost geared towards providing reliable national indicators. At best, they may be able to provide reliable comparison at the regional (NUTS II and NUTS III) level. County-level indicators cannot be reliably obtained from such an approach. Similarly, the IMD approach in the UK is heavily dependent upon administrative data records, many of which can only be obtained at ward level. Below this level, small area analysis, as for example at Enumerative District (ED) level, must continue to rely on the analysis of data available at that level, notably the Census of Population (see European Commission 2003).

## Demographic patterns of change and settlement

Following the arguments of Commins (2004) and Haase and Pratschke (2005) outlined above, the census indicators most relevant to rural deprivation are those related to emigration; i.e. population loss and increased age and economic dependency rates. Indeed, population decline has been the major issue in Irish demography since the Famine. The effects of emigration, which accounted for a decline from a population of 8 million in 1841 to the present four million, have characterized Ireland's demographic experience over the course of the last century at least, and set it apart from other European countries, with the exception of the past decade. Late marriage, high marital fertility and a low rate of illegitimacy in conjunction with emigration sustained a specific demographic pattern in Ireland long after a low fertility pattern had developed elsewhere in Europe in response to population pressures. The major losses throughout the period in question were experienced by the rural Western counties, a drain of population which led to the demographic distortion and demoralization captured in books such as Brody's *Inishkillane* (Brody 1973).

Considerable attention was paid to the question of poor demography in the *Report of the Commission on Emigration and other Population Problems* (Government of Ireland 1955). This report was especially significant in establishing the rural West as a region of special demographic disadvantage, with very low ratios of females to males and relatively few persons of working age supporting the elderly and young. Indeed, the rural West was also associated with the additional disadvantages of remoteness and a physical topography which effectively cuts it off from the main centres and transport routes. As McCleery (1991, p. 146) writes in relation to the Highlands and Islands of Scotland:

> Constrained by populations that are both numerically low and highly dispersed, remote rural areas invariably also demonstrate unfavourable population structures. In that it at once results from and contributes to marginality, the population condition of such an area is not only especially sensitive as an indicator of socio-economic health but it is also particularly complex to interpret.

*Recent population trends: Republic of Ireland*

Despite the historic turnaround in prevailing emigration patterns since the early 1990s, population decline continues to affect many of Ireland's remote rural areas. Between 1986 and 1991, for instance, population decline continued in 23 out of 34 Local Authority Areas. Between 1991 and 1996, however, migration patterns had started to change, with just over half of the Local Authority areas now experiencing a growth in population. Between 1996 and 2006, every Local Authority area throughout the State experienced population growth.

However, studying emigration and settlement patterns using local authority level statistics may, in fact, conceal more than it reveals. The reason for this lies with the particular type of population movements that have taken place: the thinning out of populations in both rural and inner city areas and the development of new settlements in outer urban belts within commuting distance of the larger cities and towns. This development holds not only for the five major cities – Dublin, Galway, Limerick, Cork and Waterford – but also for effectively every town throughout the country. As planning regulations and the rezoning of land favour the expansion of urban commuter belts, each of the growing towns have come to be surrounded by a rural hinterland which continues to experience population decline.

Table 13.1 shows the average annual population changes for each of the past four inter-census periods at the level of aggregate town and rural areas, and Table 13.2 shows the resulting population distribution across the different categories of the settlement hierarchy. The Irish Central Statistics Office (CSO) defines an Aggregate Town Area as those persons living in population clusters of 1,500 or more inhabitants. The population residing in all areas outside clusters of 1,500 or more inhabitants is classified as belonging to the Aggregate Rural Area.

The tables clearly demonstrate that, despite the turnaround in migration patterns and the overall experience of population growth in Ireland, urban and rural areas have undergone a distinctly different growth experience over the past decades.

As built-up urban areas are constrained in their population growth, the Greater Dublin Area and the other four cities have, until the turn of the century, grown at rates similar to the national averages for each of the inter-censal periods. By far the greatest growth occurred in larger towns (those with over 10,000 inhabitants), partly reflecting the settlement preferences of planning authorities and successive national development plans. Medium-sized towns (above 1,500 but under 10,000 inhabitants), as well as smaller rural towns (below 1,500 inhabitants) have again grown at rates similar to those prevailing at national level. Over the past four years, this trend has further shifted towards smaller towns (those with less than 10,000 population) becoming the fastest growing entities.

The only areas that have consistently failed to share in the national growth experience are rural areas, where population growth has been below the national average for each of the five inter-censal periods considered. In 1981, 36 per cent of the country's population resided in the open countryside. Within the space of only twenty-five years, this share dropped by four percentage points to less than one

**Table 13.1    Average Annual Population Change**

| Year | 1981–1986 | 1986–1991 | 1991–1996 | 1996–2002 | 2002–2006 |
|------|-----------|-----------|-----------|-----------|-----------|
| Greater Dublin Area | 0.2 % | -0.1 % | 0.5 % | 0.9 % | 1.0 % |
| Other Cities | 0.6 % | 0.2 % | 1.0 % | 1.1 % | 1.1 % |
| Towns > 10,000 | 1.4 % | 0.6 % | 1.6 % | 2.7 % | 2.8 % |
| Towns 5,000-10,000 | 0.9 % | 0.1 % | 1.2 % | 2.5 % | 4.5 % |
| Towns 3,000-5,000 | 0.8 % | -0.1 % | 1.3 % | 3.2 % | 5.0 % |
| Towns 1,500-3,000 | 0.6 % | -0.2 % | 0.7 % | 2.0 % | 4.7 % |
| **Aggregate Town Area** | 0.6 % | 0.1 % | 0.9 % | 1.6 % | **2.1 %** |
| Towns 1,000-1,500 | 0.2 % | -0.3 % | 0.4 % | 1.7 % | 4.0 % |
| Towns 500-1,000 | 0.9 % | -0.2 % | 0.2 % | 1.3 % | 4.3 % |
| Towns 50-500 | 1.0 % | -0.3 % | 0.2 % | 2.8 % | 4.8 % |
| Open Countryside | 0.5 % | -0.3 % | 0.1 % | 0.6 % | 1.3 % |
| **Aggregate Rural** | 0.6 % | -0.3 % | 0.1 % | 0.9 % | **1.8 %** |
| **State** | 0.6 % | -0.1 % | 0.6 % | 1.3 % | **2.0 %** |

*Source*: CSO Population Classified by Area, Volume 1, various years

**Table 13.2    Population Shares for Settlement Hierarchy**

| Year | 1981 | 1986 | 1991 | 1996 | 2002 | 2006 |
|------|------|------|------|------|------|------|
| Greater Dublin Area | 27 % | 26 % | 26 % | 26 % | 26 % | 25 % |
| Other Cities | 10 % | 10 % | 10 % | 10 % | 10 % | 10 % |
| Towns > 10,000 | 8 % | 9 % | 10 % | 12 % | 13 % | 15 % |
| Towns 5,000-10,000 | 7 % | 6 % | 6 % | 6 % | 6 % | 6 % |
| Towns 3,000-5,000 | 3 % | 3 % | 3 % | 2 % | 3 % | 3 % |
| Towns 1,500-3,000 | 3 % | 3 % | 3 % | 3 % | 3 % | 3 % |
| Aggregate Town Area | **56 %** | **57 %** | **57 %** | **59 %** | **60 %** | **61 %** |
| Towns 1,000-1,500 | 2 % | 2 % | 2 % | 2 % | 2 % | 2 % |
| Towns 500-1,000 | 2 % | 3 % | 3 % | 3 % | 3 % | 3 % |
| Towns 50-500 | 3 % | 3 % | 3 % | 2 % | 3 % | 2 % |
| Open Countryside | 36 % | 36 % | 35 % | 35 % | 33 % | 32 % |
| Aggregate Rural | **44 %** | **43 %** | **43 %** | **41 %** | **40 %** | **39 %** |
| State | **100 %** | **100 %** | **100 %** | **100 %** | **100 %** | **100 %** |

*Source*: CSO Population Classified by Area, Volume 1, various years

third of Ireland's population, highlighting the continued urbanization of Irish life. Most interestingly, whilst the depopulation of the Irish countryside was historically linked to the poor performance of the economy as a whole, the decline in its share of population accelerated most rapidly during the period of the Celtic Tiger. This raises questions as to whether the poor demographic experience of rural areas can be explained by poor labour market conditions alone, or whether it increasingly results

**Table 13.3    Average Annual Population Change and Shares, Northern Ireland**

| Year | 1971–1981 | 1981–1991 | 1971 | 1981 | 1991 |
|---|---|---|---|---|---|
| Belfast Urban Area (a) | -1.6 % | -0.6 % | 39 % | 33 % | 30 % |
| District Towns (b) | 1.5 % | 0.4 % | 26 % | 31 % | 31 % |
| Aggregate Town Area | **-0.3 %** | **-0.1 %** | **66 %** | **64 %** | **61 %** |
| Rural Towns (c) | 2.4 % | 2.7 % | 7 % | 8 % | 11 % |
| Small Settlements (d) | 0.0 % | -0.3 % | 6 % | 6 % | 5 % |
| Countryside (e) | -0.1 % | 0.6 % | 22 % | 22 % | 23 % |
| Aggregate Rural | **0.5 %** | **1.0 %** | **34 %** | **36 %** | **39 %** |
| NI | **0.0 %** | **0.3 %** | **100 %** | **100 %** | **100 %** |

*Source*: NI Census of Population 1971, 1981 (as subsequently adjusted) and 1991. Ward based data used for (a) and (b), grid square data used for (c) and (d), derived from *A Planning Strategy for Rural Northern Ireland*, Appendix 8 (1993). Unfortunately the Table cannot be extended to 2001, as the area definitions underlying it cannot easily be applied to the latest Census.

from a growing disparity in life-style expectations, where many people feel that they can no longer satisfy their aims within the rural communities in which they grew up.

*Recent population trends: Northern Ireland*

Whilst we have thus far discussed the differential population growth with respect to the Republic of Ireland, broadly similar observations can be made with regard to Northern Ireland. Table 13.3, provides consistent data for the 1971 to 1991 period and shows the considerable decline of population in the Belfast Urban Area, the significant growth of the District and Rural Towns, whilst the combined share of Small Settlements and the Countryside has remained unchanged.

Due to the unreliability of the 1981 Northern Ireland Census, it is not possible to include small level area data in the first map (Figure 13.1) showing the 1981 to 1991 population change. The second map (Figure 13.2) is based on estimates of the 1991 and 2001 data in line with the 1984 ward boundary definitions. When interpreting the 1991–2001/2 population change (the Figure 13.2) it appears, that the population decline in the remote rural areas over the past decade has been more pronounced in the Republic of Ireland than in Northern Ireland, though this observation may also be influenced by the larger size of wards than EDs and the resulting modifiable areal unit problem.

**Explaining rural depopulation**

There are no recent Irish studies dealing with life choices in relation to where people choose to set up home which might allow us to understand contemporary settlement

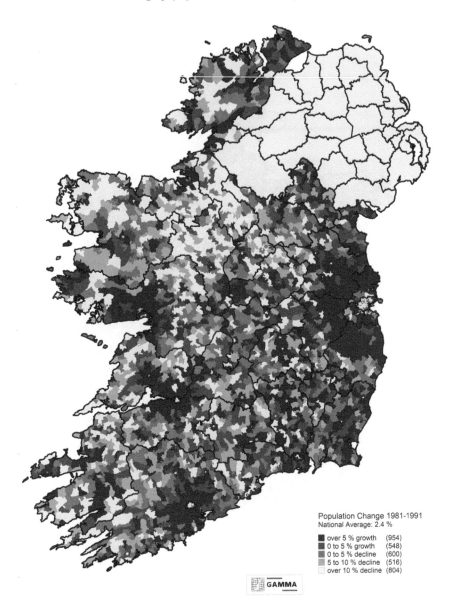

**Figure 13.1   Population Change, 1981–1991**
*Source*: Boundary data by permission of the Ordnance Survey of Ireland

patterns. Historical studies, as quoted above, largely deal with times of economic hardship and their social consequences over previous decades and may not easily be extended to the contemporary period. To understand the continuing urbanization

Population Change 1991-2001
Average: RoI +11.1% , NI +6.8%

- over 15 % growth
- 5 to 15
- 0 to 5
- -5 to 0
- over 5 % decline

GAMMA

**Figure 13.2  Population Change, 1991–2001/2**
*Source*: Boundary data by permission of the Ordnance Survey of Ireland

of Irish life, we must therefore look at the prevailing push and pull factors which might account for the migratory patterns that are currently observable. Some of these will be discussed in greater detail by other contributors to this book.

## Continued decline in agricultural employment

The first aspect is the long-term decline in the number of people who are able to derive a direct income from working on the land. In 1912, 647,000 men were mainly involved in farming. Fifty years later, in 1964, the number had roughly halved, to 344,000, and by 1979 it had further shrunk to 212,000. Recent censuses provide more accurate estimates, that combine both male and female farmers, as well as taking into account whether they derive their income from farming in a full-time, part-time or seasonal capacity. By measuring farming inputs in full-time equivalents, the numbers engaged in farming are thus seen to have fallen from 81,000 in 1991 to 78,000 in 1996 and to 59,000 in 2002. This not only confirmed the long-term decline in the number of people engaged in farming, but also its staggering acceleration during the period of the Celtic Tiger.

## Decline of manufacturing industry and shift to services

The second push factor that currently applies in rural areas is the decline in manufacturing industry and the shift of employment towards services. The accession of Ireland to the European Union in 1973 initially brought about some relief for rural employment, as a significant amount of inward investment was successfully directed towards rural areas. So, for example, the share of manufacturing employment in the aggregate rural area (with town sizes below 1,500 inhabitants) rose within one decade from 17.0 per cent in 1973 to 21.6 per cent in 1982 (Boylan, cited in Curtin, Haase and Tovey 1996, p. 184). However, over the past decade (and partly as a result of the success of the Irish economy as a whole) manufacturing industry has tended to relocate to other low labour-cost countries, for much the same reasons as they originally came to the RoI. Fruit of the Loom, a company located in Co. Donegal, is one of the prime examples of how foreign manufacturing companies have provided alternative employment opportunities for a whole rural region for close on two decades, and are now finally pulling out to relocate elsewhere.

Whilst rural areas were reasonably successful in attracting manufacturing companies during the 1970s and early 1980s, they have found it much more difficult to provide an attractive location for services in the subsequent period. For one thing, many rural areas do not have the necessary skills mix within their workforce. Furthermore, as service firms are particularly mobile, not only nationally but also internationally, most governments, including the Irish one, have dropped their regional employment strategies with a view to maximizing the attraction of firms towards the national economy as a whole.

## Growing gap between aspirations and local employment opportunities

The third factor relates to the choices which individuals make, based on their educational attainments and resulting aspirations, as well as the opportunities that exist within the local labour market. We may start by looking at a discussion generated by the most recent higher education survey *Who Went to College in 2004*, in which Garret FitzGerald observes how:

> [The study] confirms the geographical pattern of entry to higher education that has emerged from earlier studies of this subject. All but one of the six poorest counties in our State are still to be found among those counties which have an entry rate to higher education that is at least 10 percentage points above the national average… (*Irish Times* 18/03/06).

As educational achievement is generally highly correlated to social class and income levels, FitzGerald goes on to explain the apparent conundrum:

> I would feel that the absence of good employment opportunities within a county might also be a factor encouraging school-leavers to enter higher education, with a view to qualifying them eventually for employment outside their area. By contrast, ready local access to post-school employment may discourage entry to higher education…In western counties, high rates of emigration in the past may have left a residue of greater parental concern for children to 'better' themselves, and also, perhaps, a greater willingness on the part of children to leave home for this purpose (ibid.).

### Enhanced mobility

The fourth factor affecting rural areas is enhanced mobility. This relates mainly to the vastly increased rates of private car ownership since the advent of the Celtic Tiger, but also to improved road networks and, to a lesser extent, enhanced commuter rail networks. Enhanced mobility need not necessarily work to the detriment of rural areas. On the contrary, this provides a possibility for people to access more distant job vacancies and thus to continue to live a rural lifestyle. However, it can also work in the opposite direction, as many young people leave their rural homes to move to the larger cities, whilst continuing to commute back to their hometown for the weekend, a pattern familiar throughout Ireland.

This pattern is encouraged by the fact that Ireland is a highly centralized economy, with about one third of all jobs being located within the Greater Dublin Region, rendering daily commuting from the more peripheral counties all but impossible. Moreover, the ready availability of living space in the cities and towns further exacerbates this trend, which brings us to our final consideration.

### Prevailing planning doctrines

The prevailing planning doctrines strongly favour the construction of new homes on the periphery of existing cities, towns and villages. Building new homes in the open countryside has grown increasingly difficult over the past two decades, as planners have voiced concerns about the long-term feasibility of one-off housing. Although recent changes have made the pursuit of planning applications for such houses somewhat easier, the vast majority of new homes form part of substantial developments on rezoned land on the outskirts of existing cities and towns, thus making these the most accessible and affordable locations for young people setting up home.

## Demographic change and deprivation

Having defined deprivation and its spatial articulation, and looked at the issues of population change and settlement over the past four census periods, one can now return to the key question: are these two phenomena linked and if so, in what way?

A useful starting point for this consideration is provided by the current Irish *Index of Relative Affluence and Deprivation* (Haase and Pratschke 2005). In contrast to the authors of all recent British deprivation indices, Haase and Pratschke argue that the use of indicators from different domains should not lead us to neglect the different dimensions of deprivation, most notably its rural form. Based on a review of a large number of deprivation indices throughout OECD countries (Haase 1998), they conclude that overall deprivation can adequately be described by three underlying dimensions: social class disadvantage, acute labour market deprivation and demographic decline. While the social class dimension differentiates affluent and poor areas in both urban and rural locations, acute labour market deprivation, as measured by the prevailing unemployment rate, is a predominantly urban phenomenon. Rural areas which experience prolonged labour market difficulties, by contrast, seldom exhibit high unemployment rates. Instead, people from deprived rural areas tend to emigrate and this effectively reduces the measured unemployment rate. However, as emigration is socially selective, in as much as it is highest amongst the relatively well-educated core working-age cohorts, it is possible to measure its effects in terms of higher age dependency rates and lower educational achievements amongst the remaining adult population.

Econometric analysis and geographical analysis both provide strong support for the close correlation between population decline and resulting population characteristics such as higher age and economic dependency rates and lower educational attainments within the workforce. The latter has not only an effect in terms of current employment levels – and thus income-generating potential – but also in terms of the capacity of local areas to successfully attract new firms and to provide sustainable employment.

Unfortunately, it is not yet possible to fully capture the latter aspect of deprivation and well-being, as the only reliable data for constructing a deprivation index in Ireland derive from the Census of Population. As highlighted by Coombes et al.'s (1995) definition of deprivation, the fundamental implication of the term deprivation is of an absence of essential or desirable *attributes*, *possessions and opportunities* which are considered no more than the minimum by a given society.

The Census of Population can provide insights into population characteristics (e.g. demographic attributes, education and social class) and possessions (as measured through social class, employment and the quality of housing). However, rural areas face a particular risk of falling behind urban areas in relation to the opportunities that they offer for their inhabitants. If it was possible to accurately measure the degree of opportunity which each locality offers, it might help to explain the persistence, and even accentuation, of emigration during times of relative economic affluence, as well as to assist in developing appropriate policies which enhance the opportunities presented to Ireland's rural dwellers.

## References

Asthana, S., J. Halliday, P. Brigham and A. Gibson (2002) *Rural Deprivation and Service Need: A Review of the Literature and an Assessment of Indicators for Rural Service Planning*. Bristol: South West Public Health Observatory.

Brody, H. (1973) *Inishkillane: Change and Decline in the West of Ireland*. London: Allen Lane.

Callan, T., R. Layte, B. Nolan, D. Watson, C. T. Whelan, J. Williams and B. Maitre (1999) *Monitoring Poverty Trends: Data from the 1997 Living in Ireland Survey*. Dublin: Stationery Office and Combat Poverty Agency.

Combat Poverty Agency (2004), *Strategic Plan 2005–2007*. Dublin: Combat Poverty Agency.

Commins, P. (2004) 'Poverty and Social Exclusion in Rural Areas: Characteristics, Processes and Research Issues', *Sociologia Ruralis*, 44, 1: 60–75.

Cook, S., M. A. Poole, D. G. Pringle and A. J. Moore (2000) *Comparative Spatial Deprivation in Ireland: A Cross Border Analysis*. Dublin: Oak Tree Press.

Coombes, M., S. Raybould, C. Wong and S. Openshaw (1995) *Towards an Index of Deprivation: A Review of Alternative Approaches. Part 1*. London: Department of the Environment.

Curtin, C., T. Haase and H. Tovey (eds) (1996) *Poverty in Rural Ireland – A Political Economy Perspective*. Dublin: Combat Poverty Agency & Oak Tree Press.

Department of the Environment (1983) *1981 Deprivation Index*. London: HMSO.

Department of the Environment (1993) *A Planning Strategy for Rural Northern Ireland*. Belfast: Department of the Environment.

Department of the Environment (1995) *1991 Deprivation Index: A Review of Approaches and a Matrix of Results*. London: HMSO.

Department of the Environment and Local Government (2002) *The National Spatial Strategy: People, Places and Potential*. Dublin: Department of the Environment and Local Government.

Department of the Environment (2005) *Statistical Classification and Delineation of Settlements*. Belfast: Department of the Environment.

European Commission (2003) 'Regional Indicators to Reflect Social Exclusion and Poverty' (VT/2003/43), http://europa.eu.int/comm/employment_social/social_inclusion/docs/regionalindicators_en.pdf.

Frawley, J., M. O'Meara and J. Whirisky (2005) *County Galway Rural Resource Study*. Galway: Teagasc, Galway Rural Development Company.

Gleeson, J., R. Kitchin, B. Bartley, J. Driscoll, R. Foley, S. Fotheringham and C. Lloyd (2008) *The Atlas of the Island of Ireland – Mapping Social and Economic Change*. Armagh: International Centre for Local and Regional Development and the All-Island Research Observatory.

Government of Ireland (1955) *Report of the Commission on Emigration and other Population Problems*. Dublin: Stationery Office.

Government of Ireland (1997) *National Action Plan against Poverty and Social Exclusion 2003–2005*. http://www.socialinclusion.ie/publications/napincl_plan0305.pdf.

Government of Ireland (1997) *Sharing in Progress – National Anti-Poverty Strategy*. Dublin: Stationery Office.

Haase, T. (1998) 'The Role of Data in Policies for Distressed Areas', in *Integrating Distressed Urban Areas*. Paris: OECD.

Haase, T. and K. McKeown (2003) *Developing Disadvantaged Areas through Area-based Initiatives – Reflections on over a Decade of Local Development Strategies*. Dublin: Pobal.

Haase, T. and J. Pratschke (2005) *Deprivation and its Spatial Articulation in the Republic of Ireland – New Measures of Deprivation based on the Census of Population, 1991, 1996 and 2002*. Dublin: Pobal. The full report is available at: http://www.pobal.ie/media/Deprivationanditsspatialarticulation.pdf.

McCleery, A. (1991) 'Population and Social Conditions in Remote Areas: The Changing Character of the Scottish Highlands and Islands', in T. Champion and C. Watkins (eds), *People in the Countryside – Studies of Social Change in Rural Britain*. London: Sage Publications, pp. 144–59.

Noble, B. (2000) *Measuring Multiple Deprivation at the Local Level: The Indices of Deprivation 2000*. London: Department of the Environment, Transport and the Regions.

Noble, B., Smith, G., Wright, G. Dibben, C. and M. Lloyd (2001) *Measures of Deprivation in Northern Ireland*. Belfast: Northern Ireland Statistics and Research Agency.

Nolan, B., B. Gannon, R. Layte, D. Watson, C. Whelan and J. Williams (2002) *Monitoring Poverty Trends in Ireland: Results from the 2000 Living in Ireland Survey*. Dublin: Economic and Social Research Institute.

Nolan, B., C. T. Whelan and J. Williams (1998) *Where are Poor Households? The Spatial Distribution of Poverty and Deprivation in Ireland*. Dublin: Oak Tree Press in association with Combat Poverty Agency.

Robson, B., M. Bradford and I. Deas (1994) *Relative Deprivation in Northern Ireland*. Manchester: Centre for Urban Policy Studies, Manchester University.

Robson, B., M. Bradford and R. Tye (1995) *A Matrix of Deprivation in English Authorities, 1991. Part 2.* London: Department of the Environment.

Townsend, P. (1993) *The International Analysis of Poverty.* London: Harvester Wheatsheaf.

Chapter 14

# 'A Growing Concern': Youth, Sustainable Lifestyle and Livelihood in Rural Ireland

Brian McGrath

## Introduction

Whatever the idealized appeal of Irish children and youth as the hope for the future, in reality growing up in Ireland has always been a notably unequal experience when account is taken of social class, gender, ethnicity, religious affiliation and location. Typically, young people have occupied a subordinate position and have had little say in policy decisions impacting on their lives (see Lynch 1998). While the transitions that revolve around school, work, migration, housing, sexuality and lifestyle have always been complex for young people, modern society arguably raises even more contingencies in how young people live their lives. In an increasingly 'individualized' world (Beck 2001), people must grapple with the uncertainties of increased mobility, non-traditional family formation, new patterns of consumption as well as new political economies of work and inequality (see Bauman 2005). Young people may encounter an expanding array of liberating choices, but they also face new challenges that reflect the riskier nature of modern times (see Beck 2001; Cieslik and Pollock 2002). In particular, new forms of vulnerability and social exclusion await those whose capital (human, economic, social) resources fall seriously short of what's required to find security and inclusion.

All this helps to contextualize how 'sustainability' is to be understood for young people. We must, of course, also consider what sustainability might mean from the perspective of young people themselves. Drawing on new survey data, I will attempt to show that despite the mainly positive views young people express about rural social relations, there are distinct constraints associated with rural living which, when added to young people's own family resources or capital, make the prospect of achieving a sustainable livelihood and lifestyle complicated and problematic.

## Rural youth and 'sustainability'

Perhaps a useful starting point when contemplating sustainability from the perspective of the individual is to focus on vulnerability. For Furlong et al. (2000, p. 9) vulnerability, born out of a combination of subjective and objective conditions, is shorthand for 'severely restricted opportunities for secure employment, social and economic advancement and personal fulfilment'. Those most vulnerable are those with the least capacity to 'confront, adapt or cope' with the different challenges they routinely face (ibid.). In contrast, sustainability evokes the possibility of substantive opportunities for secure employment, social and economic advancement and personal fulfilment. Secure and meaningful employment provides the main ingredient of a sustainable *livelihood* while the possession of social capital is necessary for achieving a sustainable *lifestyle*.[1]

In rural society, there are distinct processes over which young people, albeit 'active' subjects in principle have little control. Prominent among such processes are the dynamics of local labour markets and the distribution of resources, information and services in society. These processes frame much of the experience of rural youth and have major implications for their vulnerability/sustainability prospects. Such prospects are also influenced by the degree to which young people are isolated from locations where resources, information and services are mostly likely to exist. Rural areas vary in their level of remoteness, which in turn shapes the range of livelihood and lifestyle options available to young people. A young person's vulnerability can further be viewed as either exacerbated or relieved by kin and non-kin social supports, community functioning (schools, neighbourhood supports) and wider welfare policies (Jack 2000; Jack and Jordan 2001). How well young people do in the face of the challenges they face is very often contingent on the strength and nature of these social relationships, networks and interactions.

Psychosocial elements – subjectivity, identity – are also relevant to 'sustainability'. What Giddens (1991) describes as 'self-actualization' may apply pressure to acquire new experiences beyond one's locality. In an increasingly mobile society, rural life may thus be viewed by youth in ambivalent ways (Wiborg 2004; Haugen and Villa 2005). For many, rurality can be particularly constraining in the kinds of interactions and forms of stimulation on offer. On the other hand, the availability of social capital in rural communities can be particularly appealing for many (including youth) (see Jamieson 2000; Ní Laoire 2001; Wiborg 2004). Of course the realities here can be often shifting, with those in the early adolescent years and those entering early adulthood having different priorities from the younger age groups.

So, can growing up in the countryside, villages and small towns of Ireland provide sustainable lifestyles – meaningful and valued social relationships and forms of interaction – for young people? More specifically, what is young people's sense of living in rural communities and what is their perception of lifestyle advantages and drawbacks?

---

1   'Social capital', a much debated concept in the social sciences (see Shortall 2004; Leonard 2005), is taken here to refer to the nature and quality of the social networks available to individuals.

*Rurality, social capital and lifestyle*

The extent to which growing up in the countryside provides the basis of an idyllic lifestyle – by virtue of its naturalism, sense of community and tranquillity – has been the subject of some recent examination (e.g. Matthews et al. 2000; Wiborg 2004). What this research reveals is how much people's (dis)connection to rural space reflects the gendered and class-based divisions inhering in rural communities. While living in rural society can generate multiple understandings (Wiborg 2004), whether one views the rural as 'idyllic' or 'dull' (by no means mutually exclusive characterizations) depends in large part on one's stock of economic and cultural capital resources, gender, education and incomer/native status (Rye 2006). In relation to rural Ireland, two recent major surveys, one based in the Republic (the Health Behaviour in School-aged Children (HBSC) Survey 2002 and 2006)[2] and the other in Northern Ireland (Young Life and Times Survey 2003),[3] show distinct significant differences in how young males and females typify the areas in which they live (see Tables 14.1 and 14.2).

The HBSC survey findings illustrate the gendered nature and urban/rural division in young people's understanding of where they live. In Table 14.1, a variety of social capital measures are provided which reveal differences in attitudes about the nature of trust, safety, friendliness and support. Apparently rural youth – boys especially – have more positive views about the nature of social interactions with others in their communities. Boys also feel a lot safer than their urban counterparts. Other social capital type measures, such as being able to ask for help from neighbours and being able to trust others not to take advantage, also score higher among rural youth. There is a notable difference among rural and urban boys in terms of being able to ask help from their neighbours.

While all social capital measures are stronger among rural youth in Northern Ireland, one particular aspect – the experience of trust relations – appears again to be especially stronger among rural boys. There are statistically significant differences among girls in terms of how safe they felt in their communities, with almost twenty per cent more rural girls suggesting that they felt safe in their areas during the daytime, though in relation to night-time safety this difference drops dramatically for all girls.

---

2    The HBSC in 2002 and 2006, which surveyed 8,316 and 10,334 pupils respectively, in the Republic of Ireland is part of an international research programme since the 1980s involving 41 countries (see Currie et al. 2004). The Irish database is currently being analysed by the present author and the principal Irish investigator, Dr Saoirse Nic Gabhainn, Department of Health Promotion, NUI Galway, in terms of rural/urban patterns. Comparable data for Northern Ireland from the HBSC was unavailable for 2002.

3    The Young Life and Times Survey is undertaken every year in Northern Ireland by ARK, a division of Queen's University Belfast. The 2003 survey was chosen for analysis purpose in this chapter as it provides more data on 'social capital' measures than subsequent surveys. The sample size is 902 respondents. For more on the survey, see: http://www.ark.ac.uk/ylt.

**Table 14.1     Perceptions of Life in Urban and Rural Areas[†] among Boys and Girls Aged 10–18 Years in the Republic of Ireland, 2002 and 2006**

| Statement | Urban boys % | Rural boys % | Urban girls % | Rural girls % |
|---|---|---|---|---|
| I feel safe in local area ('always') 2006 | 42.2 | 66.7*** | 42.2 | 55.0*** |
| Local area is good place to live ('really good') 2006 | 35.5 | 54.7*** | 36.5 | 49.8*** |
| Safe for children to play outside in the day[a] 2006 | 80.2 | 88.9*** | 76.1 | 85.5*** |
| Good places to spend free time[a] | 57.0 | 36.8*** | 52.9 | 28.5*** |
| Can ask for help from neighbours[a] 2006 | 73.8 | 81.3*** | 76.6 | 81.5*** |
| People say 'hello' and often stop to talk[a] 2002 | 71.3 | 81.5*** | 75.7 | 83.0*** |
| Most people would take advantage if they had a chance[a] 2002 | 56.9 | 63.1*** | 61.2 | 67.5*** |

*Notes*: [†]'Urban' is derived from respondents' description of where they live as 'city or town' while 'rural' indicates 'village or country'.
[a] response of 'strongly agree or agree; **significant difference, $p<0.01$; ***significant difference, $p<0.001$ (Chi squared test)

*Source: Health Behaviour in School-Aged Children Survey (HBSC)*

**Table 14. 2    Perceptions of Life in Urban and Rural Areas[†] among Boys and Girls Aged 16 Years in Northern Ireland**

| Statement | Urban boys % | Rural boys % | Urban girls % | Rural girls % |
|---|---|---|---|---|
| Feel very safe in local area during the day | 72.4 | 80.3 | 56.3 | 75.1*** |
| Feel very safe in local area after dark | 35.2 | 46.5** | 13.8 | 17.1 |
| Local area is a close tight-knit community[a] | 34.8 | 57.0*** | 34.8 | 57.9*** |
| Local area is a friendly place to live[a] | 65.7 | 83.7*** | 66.0 | 83.3*** |
| Local people look after each other[a] | 37.2 | 65.5*** | 41.6 | 65.1*** |
| Most people in area trust one another[a] | 34.8 | 62.2*** | 34.0 | 51.1*** |
| Trust most or many people in local area[a] | 46.9 | 64.5*** | 40.8 | 59.5*** |

*Notes*: 'Urban' in this survey is based on respondents' description of where they live as 'big city, suburbs or small city or town' and 'rural' as 'home in village, country or farm'.
[a] response of 'strongly agree or agree; **significant difference, $p<0.01$; ***significant difference, $p<0.001$ (Chi squared test)

*Source: The Northern Ireland 2003 Young Life and Times Survey (NLTS)*

One obvious contextual feature that distinguishes between growing up in the north and the south of Ireland is the impact of ethno-sectarian division in Northern Ireland and in the Border region (Maguire and Shirlow 2004). Maguire and Shirlow's research among rural children shows that while children and parents' typical concerns tend to revolve around safety and traffic, ethno-sectarian concerns continue to be transmitted in the post-Troubles environment, especially in areas where conflict has long been the experience of parents. For many children, being spatially confined by parents to their own Catholic or Protestant enclaves tended to perpetuate a legacy of suspicion regarding those from the other community (see also Youth Action Northern Ireland 2002).

While rural life seems in general to be viewed in positive terms, this has to be qualified when we consider the spaces that young people can occupy and the activities they can pursue (Panelli et al. 2002; Auclair and Vanoni 2004). This is especially so for rural girls in the HBSC with 28.5 per cent agreeing that there were good places to go (e.g. parks, shops, leisure centres) in rural areas compared with over half the urban female sample. The lack of youth provision appears as an almost universal theme in young people's accounts of rural life. Most of those in Geraghty et al. (1997) study of over 500 rural Northern Ireland youth felt that, apart from sport activities, there were no places for young people to meet. Consequently, most young people spent their time in activities in the home, such as listening to music (92 per cent) or watching television (83 per cent) (p. 44).

In the absence of suitable outlets for the development of young people's lifestyle, the prominence of pub culture appears in several accounts of rural life (e.g. Campbell 2001). Socializing in pubs can occur at a relatively young age, with the pub effectively being the de facto 'youth club' for many young people. In rural areas, it has been argued that pubs tend to be a male domain and a site in which masculinity becomes expressed (Campbell 2001; Ní Laoire 2001). What this again shows is how lifestyle issues are gendered – what appears as 'sustainable' will look quite differently to boys and girls. Thus, young women in Geraghty et al.'s (1997) study were least satisfied with the leisure opportunities available to them, pointing to the dominance of a 'pub and football' culture and its creation of a male dominated public sphere.

Similarly, sport outlets, such as hurling or football, are seen as the preserve of young men, particularly in Catholic communities in the North, while the local marching band represents the equivalent in many Protestant communities (ibid., p. 64). Whatever the religious orientation, such activities are viewed as heavily gendered zones of inclusion and exclusion.

These various practices can begin the process of disaffection with rural life quite early on and can be especially problematic for young women (Haugen and Villa 2005). Almost two thirds of young girls in Geraghty et al.'s study felt it was harder to be a young woman than man in a rural area. On the other hand, some young males may be particularly vulnerable in the context of wider societal and economic changes. In Ní Laoire's (2001) analysis of male suicide rates in Irish rural society, several factors peculiar to rural communities that might explain the recent

**Table 14.3    Time Use (Three Evenings or More per Week Spent with Friends) among Urban, Rural and Farm Youth**

| Urban boys % | Rural boys % | Urban girls % | Rural girls % | Farm boys % | Farm girls % |
|---|---|---|---|---|---|
| 75.6 | 46.7*** | 60.1 | 38.2*** | 36.5*** [a] | 26.1*** [b] |

*Notes*: ***significant difference, p<0.001; [a] significant between farm and non-farm boys; [b] significant between farm and non-farm girls (Chi squared test)

*Source: HBSC survey, 2002*

increase in suicide among men (and young men are especially prone to suicide) are discussed. Against the backdrop of unprecedented rural restructuring, and many challenges to traditional forms of masculinity, 'young men are struggling to find coherent identity formation in the face of isolation, spatial confinement, lack of support networks, declining self-esteem' (2001, p. 233).

The prominence of pub culture in rural areas may not, however, imply a greater problem of underage drinking in rural communities. Hannaford's (2005) analysis of drink and drug taking behaviour among teenagers in Northern Ireland shows distinct differences among rural and urban youth. The Rural Development Council (2005) has expressed concern about underage drinking and links it to a lack of alternative activities across rural locations.

While there are several organizations and services catering for the developmental needs of young people throughout Ireland, coverage can be quite uneven in terms of provision. By the time many teenagers reach 16, the attractions of youth clubs have declined and young people generally feel they have little input into what youth activities are provided and how they are run (Geraghty et al. 1997, pp. 79–80). When young people become parents, rural places can become even more isolating as activities centre upon their children (McGrath and Canavan 2001).

What the lack of public recreational space for young people also means is that behaviour in rural areas appears highly visible and scrutinized, especially by adults (Tucker and Matthews 2001). Analysis of the 2002 HBSC shows that there are also distinct differences in the amount of evening and after school time that rural and urban youth spend with their friends (Table 14.3). Three quarters of urban boys indicate that they spent three evenings or more per week with friends, compared with 47 per cent of rural boys. There is a marked decline in this time use among the rural girls. Some 38 per cent of them spend three evenings or more with friends compared with 60 per cent of urban girls. For girls from farming backgrounds the proportion meeting friends further declines to just over a quarter.

Geographical isolation constitutes a strong barrier for young people's ability to mix socially in rural communities (Youth Action Northern Ireland 2002). The findings presented in Table 14.3 indicate that school itself forms an important social milieu for rural youth, especially young females. This is borne out in Geraghty et al.'s (1997) study of rural youth in Northern Ireland which shows that it is less acceptable for young women to 'hang out' on the streets (see also Tucker and Matthews 2001), and that they have more domestic responsibilities and less time available for themselves (Geraghty et al. 1997, p. 74).

Geraghty et al. make it clear that young people are not necessarily as free as might be assumed, and in particular that their leisure time tends to be absorbed by study, employment, childcare, elderly care, helping in the family business or farm and domestic responsibilities (ibid., p. 47). Despite the importance of school, evidence from Northern Ireland shows that while primary schools tend to be catering reasonably well for rural young people in numerical terms, the same cannot be said for secondary schools. In the 2005–6 enrolment year, primary schools classified as 'rural' accounted for 57 per cent of schools, and for one third of all pupil enrolments as against over one fifth of secondary schools, mostly non-grammar, catering for just 14 per cent of school goers (Department of Education 2006). As many young people spend considerable time commuting to and from school, this not only curtails the time available to them but, as recent high profile road accidents involving rural secondary school pupils in the Irish Republic suggest, leaves them prone to an inferior and risk-laden service.

In summary, even though young rural dwellers have a generally more positive view of their communities than their urban counterparts, many youth encounter obstacles in meeting their needs for social participation. Limited recreation and opportunities for social engagement mean that lifestyle tends to be a heavily problematic feature of growing up in rural Ireland.

*Rurality and sustainable livelihood*

Recent accounts of broader social change suggest a possible increased 'individualization' of experience (Beck 1992, 2001; Giddens 1990; Cieslik and Pollock 2002). Although there are disagreements as to its precise nature (see Shucksmith 2004b; McGrath 2005), it has been argued that people face new sets of institutional dependencies in labour markets, welfare provisions and in their consumption possibilities (in education, for instance). The operation of these institutional elements in young people's lives helps inform our understanding of how sustainable livelihoods are achievable or otherwise.

*Education*

In recent times, education has come to assume a hugely powerful role in shaping life chances. Historically, the nature of rural society has had particular implications for patterns of educational participation. Within a predominantly agrarian society, where both patrilineal farm succession and the gender-based division of farm labour have been defining features, education has assumed particular importance for young women. Curtin and Varley's (1984) review of rural ethnographic evidence, north and south of the border, shows that girls often faced 'harsh treatment…in the male-dominated farm households' (pp. 40–41). Yet educated young women seemed more self-assured and 'their understanding that they have no future on the land' (p. 41) led to higher aspirations and emigration. This pattern has persisted. Girls from farming backgrounds have out-performed their male contemporaries in educational attainment and, compared with other social groups, they have a distinctly higher representation in third level education (O'Hara 1997).

Participation rates in full-time education for 16–24 years olds in the RoI rose steadily over the 1982 – 1998 period (Canny 2001, p. 138; Breen et al. 1995) and it is evident that educational credentials act as a form of protection against the threat of unemployment in risk society. Nowhere is this more evident than among contemporary farming families. The percentage of farm children completing the Leaving Certificate in the Republic is close to 90 per cent, which is almost as high as those from the higher professional class (Gorby et al. 2005). Similarly, the rate of new entrants to third level institutions, particularly the Institutes of Technology and Colleges of Education, has been high among farm youth relative to their share of the population (O'Connell et al. 2006, pp. 48–50).[4] Recent evidence from Gorby et al. (2005, p. 45) also shows that the children of farmers were less likely to work while in school, compared with unemployed groups. Comparisons between rural and urban areas of Northern Ireland show that, in general, the rural wards have a higher proportion of school leavers with at least five GCSEs (Rural Development Council 2002, p. 30).

Like many middle class families, parents are seen as taking what might be seen as an active approach in forging their children's life chances in the individualization stakes (see Gillies 2005). Evidence of 'active individualization' is particularly apparent in O'Hara's (1997) study, which distinctly shows how mothers in farm families concentrate particular efforts on their daughters to gain educational qualifications. Their rationale is that education can remove them from the burden of farming and allow them to pursue financial independence irrespective of whether or not they marry a farmer. O'Hara sees this sphere of influence in education as a form of resistance and power which 'represents an attempt to ensure that the cycle of dependent farm wife is fractured and that the next generation have better choices' (p. 153). This commitment was particularly noted among respondents in the western region. On smallholdings where mothers were disenchanted with farming, all children are encouraged to concentrate on education as a means of 'resistance'. For such families, further education becomes the means to achieving a sustainable livelihood, which typically means being 'educated out' of rural communities.

Moving away from the rural locality, typically around 18 years of age, means disengagement from the parental home, the community and the young person's social support networks (see also Jentsch and Shucksmith 2004). For many

---

4   Looker and Dwyer (1998) show how rural youth in Canada and Australia tend not to stay on at school for as long as their urban counterparts. Those that did were more inclined to take up non-university programmes, with the higher status university route being more a preference of urban youth.

youth in Northern Ireland, pursuing further and higher education has often meant moving outside the country. In their relocation decisions, young people can face considerable accommodation costs. Leaving home may pose difficulties, but it can also provide claims to 'adult status' and independence at an earlier stage than for urban youth. Rural youth are also more likely to have encountered what they considered to be the symbolic markers of 'adult status', namely marriage and/ or parenthood, at an earlier age (Looker and Dwyer 1998). Such patterns have implications for the housing needs of young people, a further obstacle for many in securing a sustainable standard of living in the countryside (see McGrath 2001).

*Work*

Apart from farming and employment based on natural resources, our precise knowledge is limited about the kinds of jobs that young people hold in rural areas, their average earnings, duration of employment, and so forth. Building a livelihood in agriculture in recent years has been defined largely by its unevenness, with sustainability linked to farm system, farm size and region (Commins 1996; Rural Development Council 2005).[5] Among young people in the 15 to 19 year age group, the attrition rate from agricultural and related activity continues apace, especially in the Republic. Census figures reveal that while 6.9 per cent of this youth cohort held occupations in farming, fishing and forestry in 1991, this dropped to 5.4 per cent in 1996 and more than halved again to 2.0 per cent in 2002. In Northern Ireland, the fallout among 16 to 19 year olds in agricultural occupations has been less dramatic, from a figure of 2.9 per cent in 1991 to 1.6 per cent in 2001.

It is clear that contemporary patterns of urbanization and population change in rural Ireland are creating new occupational opportunities for some (Tovey 1999), and that this contributes to young people's calculations about livelihood. As Tovey points out, the main winners from contemporary rural change include those with land for sale, those in the construction industry and local businesses, such as shopkeepers. The Republic's 2002 Census figures show that of those living in rural areas, 10.2 per cent had jobs in the construction industry, compared with 6.1 per cent for those in more urban areas.[6] Many young people in rural Ireland are no doubt seizing the opportunities provided by an up to recently booming construction industry and indeed for those from families with property, access to housing becomes far more attainable.[7] In Ireland north and south, 14 per cent of youth aged 15–19 years (16–19 years in Northern Ireland) were working in

---

5   Evidence from Northern Ireland (Rural Development Council 2003) shows that self-employment is far more important in rural wards than urban ones, with farming accounting for much of employment and self-employment in the west.

6   Here 'rural' is defined as referring to the open countryside and to towns and villages with a population of less than one thousand people.

7   Notwithstanding the problems that occur in many rural locations in terms of planning regulations, such as the Gaeltacht or in areas of high conservation value.

construction at their respective census periods (2001 and 2002), compared with 7.9 per cent in 1991 in the north and 7.7 per cent in the south. Within construction we need to distinguish between those with skilled and unskilled jobs, since the economic returns as well as future viability of these can be quite uneven. While construction jobs can provide many youth with adult status, the problem persists that, without a creditable credential and skill base, these youth become entrapped by poor labour market conditions over the longer term (McGrath 2005).

Many studies attest to the dominance of 'secondary' labour markets in rural locations (Cartmel and Furlong 2000; McGrath 2001; Hodge et al. 2002; Shucksmith 2004a and 2004b). Here firms tend to be smaller with employment typically being low paid, insecure and unrewarding. The jobs on offer provide fewer prospects with low demand for educationally qualified workers. In addition, problems relating to transport and childcare are more acutely felt, especially in remote locations where bus services tend to be more infrequent and crèches/ childminding services less available.

Findings from O'Shea and Williams (2001) that early school leavers tend to rely on informal job-search strategies assumes particular resonance with conditions in many rural areas (Shucksmith 2004; Lindsay et al. 2005). Of course, for those with poor reputations finding work can prove to be highly problematic in small communities where 'fitting in' is of particular importance (Pavis et al. 2001). A university student in McGrath and Canavan (2001, p. 93) reflected on this aspect in the following terms:

> Well if there was anything against your family it would never be forgotten. They would go into every aspect of your life. You wouldn't be able to find any place to live or get a job. If your reputation precedes you at all, that's it.

The gender dimension to labour market opportunities is again highly relevant, with young women who remain in rural areas, especially those with children to rear, quite often falling into the category of 'detached stayers' (Jamieson 2000). Access to childcare (and assumptions about gender roles) can pose major difficulties for many, especially those without family or friendship support networks to draw upon.

**Conclusion**

This chapter has focused on what sustainability might mean in the context of providing access to lifestyle and livelihood possibilities for young people. The picture of rural life looks remarkably similar for young people both north and south of the border, as it does for rural youth elsewhere in Europe. Whether rural living can sustain a reasonable lifestyle and decent livelihood standards for youth depends on the nature of resources that are unique to young people themselves and the conditions provided by their rural communities and the wider, increasingly globalized, society.

Research findings strongly suggest that rural youth experience more social capital in their communities but face definite obstacles in trying to achieve a sustainable lifestyle. In terms of opportunities for recreation and social engagement, young women can find themselves especially disadvantaged. Rural areas can be adult-centric as well as providing social opportunities that are often ill suited to the needs of children and youth. Poor experience of having lifestyle needs met within one's community, particularly in the formative years, should be an important public policy concern.

Proximity to employment opportunities, transport, access to childcare, educational credentials, housing opportunities, family and friendship networks are all factors that help to define whether sustainable livelihoods are available to young persons. Restricted opportunities in secondary labour markets often characterize rural employment patterns. Social networks, reputation, marketable skills and, of course, a strong local economy, are key to making one's way in local labour markets. Education tends to be the strategy that provides access to opportunities beyond rural communities, especially for girls. Those with the least options are those whose resources fail to equip them in either local or non-local labour market opportunities. In the final analysis sustainability for youth will depend on the confluence of subjective and objective conditions. That this is so can give rise to an often complex configuration encompassing young people's own strengths, their family and community networks, the character and dynamics of the local labour market and the wider public policy environment.

## Acknowledgements

My thanks to Dr Saoirse Nic Gabhainn, Department of Health Promotion, NUI Galway, for permission to use the HBSC data and for her assistance with the analysis of the Young Life and Times Survey. I wish to thank ARK, Queen's University, Belfast, for permission to use the Young Life and Times dataset, which is available online; and Caroline Breakey for direction to Rural Development Council publications. Thanks also go to Francis McCann of the Central Statistics Office, Dublin, and to the Northern Ireland Statistical and Research Agency (NISRA), Belfast for the provision of census data.

## References

Auclair, E. and D. Vanoni (2004) 'The Attractiveness of Rural Areas for Young People', in B. Jentsch. and M. Shucksmith (eds), *Young People in Rural Areas of Europe*. Aldershot: Ashgate, pp. 74–104.

Bauman, Z. (2005) *Work, Consumerism and the New Poor*. 2nd edition. Berkshire: Open University Press.

Beck, U. (1992) *Risk Society: Towards a New Modernity*. London: Sage.

Beck, U. (2001) 'Living Your Own Life in a Runaway World: Individualisation, Globalisation and Politics', in W. Hutton and A. Giddens (eds), *On the Edge: Living with Global Capitalism*. London: Vintage, pp. 164–74.

Breen, R., D. F. Hannan and R. O'Leary (1995) 'Returns to Education: Taking Account of Employers' Perceptions and Use of Educational Credentials', *European Sociological Review* 11, 1: 59–73.

Campbell, H. (2001) 'The Glass Plallus: Pub(lic) Masculinity and Drinking in Rural New Zealand', *Rural Sociology* 64, 4: 551–603.

Canny, A. (2001) 'The Transition from School to Work: An Irish and English Comparison', *Journal of Youth Studies* 4, 2: 133–54.

Cartmel, F. and A. Furlong (2000) *Youth Unemployment in Rural Areas*. York: Joseph Rowntree Foundation.

Cieslik, M. and G. Pollock (eds) (2002) *Young People in Risk Society: The Restructuring of Youth Identities and Transitions in Late Modernity*. Aldershot: Ashgate.

Commins, P. (1996) 'Agricultural Production and the Future of Small-Scale Farming', in C. Curtin, T. Haase and H. Tovey (eds), *Poverty in Rural Ireland. A Political Economy Perspective*. Dublin: Oak Tree Press, pp. 87–125.

Currie, C., C. Roberts, A. Morgan, R. Smith, W. Settertobulte, O. Samdal and V. B. Rasmussen (2004) *Young People's Health in Context: Health Behaviour in School-aged Children (HBSC) Study: International Report from the 2001/2002 Survey*. Copenhagen: World Health Organisation.

Curtin, C and T. Varley (1984) 'Children and Childhood in Rural Ireland: A Consideration of the Ethnographic Literature', in C. Curtin, M. Kelly and L. O'Dowd (eds), *Culture and Ideology in Ireland, Studies in Irish Society II*. Galway: Galway University Press, pp. 30–45.

Department of Education (2006) *Schools Located in Urban and Rural Areas 2005/06*. Available online: http://www.deni.gov.uk.

Furlong, A. (1999) 'Lifelong Learning and the Reproduction of Inequalities: A Pessimistic View', in A. Walther and B. Stauber (eds), *Lifelong Learning in Europe; Differences and Divisions, Strategies of Social Integration and Individual Learning Biographies. Volume 2*. Tubingen: Neuling Verlag.

Furlong, A., B. Stalder, and A. Azzopardi (2000) *Vulnerable Youth: Perspectives on Vulnerability in Education, Employment and Leisure in Europe, International Expert Report*. Strasbourg: Council of Europe Publishing.

Geraghty, T., C. Breakey, and T. Keane (1997) *A Sense of Belonging. Young People in Rural Areas of Northern Ireland Speak about their Needs, Hopes and Aspirations*. Belfast: Youth Action Northern Ireland.

Giddens, A. (1990) *The Consequences of Modernity*. Cambridge: Polity Press.

Giddens, A. (1991) *Modernity and Self-Identity: Self and Society in the Late Modern Age*. Cambridge: Polity Press/Blackwell.

Gillies, V. (2005) 'Raising the "Meritocracy": Parenting and the Individualization of Social Class', *Sociology* 39, 5: 835–53.

Gorby, S., McCoy, S. and D. Watson (2005) *2004 Annual School Leavers' Survey of 2002/2003 Leavers*. Dublin: Economic and Social Research Institute.

Hannaford, S. (2005) *Drinking, Smoking, Drugs and Sexual Intercourse – Education and Influences for Young People in Northern Ireland*. Research Update 37. Belfast: ARK Social and Political Archive. Available online at: http://www.ark.ac.uk/ylt/.

Haugen, M. and M. Villa (2005) 'Rural Idylls or Boring Places?' in B. B. Bock and S. Shortall (eds), *Rural Gender Relations: Issues and Case Studies*. Oxfordshire: CABI Publishing, pp. 181–95.

Hodge, I., J. Dunn, S. Monk and M. Fitzgerald (2002) 'Barriers to Participation in Residual Rural Labour Markets', *Work, Employment and Society* 16, 3: 457–76.

Jack, G. (2000) 'Ecological Influences on Parenting and Child Development', *British Journal of Social Work* 30: 703–20.

Jack, G. and B. Jordan (2001) 'Social Capital and Child Welfare', *Children & Society* 13: 242–56.

Jamieson, L. (2000) 'Migration, Place and Class: Youth in a Rural Area', *The Sociological Review* 48: 203–23.

Jentsch, B. and M. Shucksmith (eds) (2004) *Young People in Rural Areas of Europe*. Aldershot: Ashgate.

Leonard, M. (2005) 'Children, Childhood and Social Capital: Exploring the Links', *Sociology* 39, 4: 605–22.

Lindsay, C., M. Greig and R. W. McQuaid (2005) 'Alternative Job Search Strategies in Remote Rural and Peri-urban Labour Markets: The Role of Social Networks', *Sociologia Ruralis* 45, 1/2: 53–70.

Looker, E. D. and P. Dwyer (1998) 'Education and Negotiated Reality: Complexities Facing Rural Youth in the 1990s', *Journal of Youth Studies* 1, 1: 5–22.

Lynch, K. (1998) 'The Status of Children and Young Persons: Educational and Related Issues', in S. Healy and B. Reynolds (eds), *Social Policy in Ireland: Principles, Practice and Problems*. Dublin: Oak Tree Press, pp. 351–4.

Maguire, S. and D. Shirlow (2004) 'Shaping Childhood Risk in Post-conflict Rural Northern Ireland', *Children's Geographies* 2, 1: 69–82.

Matthews, H., M. Taylor, K. Sherwood, F. Tucker and M. Limb (2000) 'Growing Up in the Countryside: Children and the Rural Idyll', *Journal of Rural Studies* 16, 2: 141–53.

McGrath, B. (2001) 'A Problem of Resources: Defining Rural Youth Encounters in Education, Work and Housing', *Journal of Rural Studies* 17, 4: 481–95.

McGrath, B. (2005) 'The Structuration of "Choice" Biography: Exploring the Biographical-institutional Interface of Educational (Non-)Engagement'. Unpublished Ph.D dissertation. Department of Land Economy: University of Aberdeen.

McGrath, B. and J. Canavan (2001) *Researching Rural Youth in the West of Ireland: A Mixed-Methods Approach*. Galway: Social Sciences Research Centre, NUI Galway.

Ní Laoire, C. (2001) 'A Matter of Life and Death? Men, Masculinities and Staying "Behind" in Rural Ireland', *Sociologia Ruralis* 41, 2: 220–36.

O'Connell, P. J., D. Clancy and S. McCoy (2006) *Who Went to College in 2004? A National Survey of New Entrants to Higher Education.* Dublin: Higher Education Authority.

O'Hara, P. (1997) 'Interfering Women – Farm Mothers and the Reproduction of Family Farming', *The Economic and Social Review* 28, 2: 135–56.

O'Shea, C. and J. Williams (2001) *Issues in the Employment of Early School Leavers.* Dublin: The Economic and Social Research Institute.

Panelli, R., K. Nairn and J. McCormack (2002) '"We Make our own Fun": Reading the Politics of Youth with (in) Community', *Sociologia Ruralis* 42, 2: 106–30.

Pavis, S., G. Hubbard and S. Platt (2001) 'Young People in Rural Areas: Socially Excluded or Not?', *Work, Employment & Society* 15, 2: 291–309.

Rural Development Council (2002) *A Picture of Rural Change 2002.* Cookstown: Rural Development Council.

Rural Development Council (2003) *A Picture of Rural Change 2003.* Cookstown: Rural Development Council.

Rural Development Council (2005) *Rural Policy Perspectives: Assessing Rural Sustainability and Change.* Cookstown: Rural Development Council.

Rye, J. F. (2006) 'Rural Youths' Images of the Rural', *Journal of Rural Studies* 22, 4: 409–21.

Shortall, S. (2004) 'Social or Economic Goals, Civic Inclusion or Exclusion: An Analysis of Rural Development Theory and Practice', *Sociologia Ruralis* 44, 1: 109–23.

Shucksmith, M. (2004a) 'Conceptual Framework and Literature Review', in B. Jentsch and M. Shucksmith (eds), *Young People in Rural Areas of Europe.* Aldershot: Ashgate, pp. 8–25.

Shucksmith, M. (2004b) 'Young People and Social Exclusion in Rural Areas', *Sociologia Ruralis* 44, 1: 43–59.

Tovey, H. (1999) 'Rural Poverty: A Political Economy Perspective', in D. G. Pringle, J. Walsh and M. Hennessy (eds), *Poor People, Poor Places: A Geography of Poverty and Deprivation in Ireland.* Dublin: Oak Tree Press, pp. 97–122.

Tucker, F. and H. Matthews (2001) '"They Don't Like Girls Hanging Around There": Conflicts over Recreational Space in Rural Northamptonshire', *Area* 33, 2: 161–8.

Wiborg, A. (2004) 'Place, Nature and Migration: Students' Attachment to their Rural Home Places', *Sociologia Ruralis* 44, 4: 416–32.

Youth Action Northern Ireland (2002) *Distant Voices. An Action Research Project on Young People and Rural Isolation in the Southern Education and Library Board Area.* Belfast: Youth Action Northern Ireland.

# Chapter 15

# Rural Ageing and Public Policy in Ireland

Eamon O'Shea

## Introduction

The issue of rural ageing has gained increased attention in the last few decades, perhaps due to the fact that over half of the world's older people live in rural areas (Wenger 2001). As populations age, the need to provide services to the increasing number of older people will become more acute, even if the share of older people living in rural areas is likely to decline in the future. The reciprocal relationship between space and the individual becomes important with age, particularly where there is physical incapacity. Older people need access to services and facilities at a local level, but this is not always possible or feasible due to supply-side inadequacies arising from concerns about the cost of provision. Older people living in rural areas may, therefore, become disadvantaged simply because of where they live.

There is a tendency for cumulative cycles of decline to occur in rural areas: poor employment opportunities lead to out-migration which in turn leads to a reduction in population, unbalanced age structures and falling demand, which reinforces the poor employment potential of the area. This is quickly followed by a reduction in social services provision. Thus may begin what Lynch (2001) described, in a Northern Ireland (NI) context, as a cycle of deprivation incorporating: mobility, isolation, income, health, opportunity and accessibility. It is now largely taken for granted that social services provision in rural areas cannot be provided to the same level as in urban areas due to economies of scale arguments. Optimality with respect of provision is usually measured in efficiency terms only, without reference to the distributional consequences of different policies, particularly for older people living in rural areas. It is particularly challenging, therefore, to develop policies and programmes to support older people living in rural areas where small, dispersed populations mean that urban models of care and support may not be feasible or appropriate.

One of the difficulties of undertaking research on rural ageing is the absence of a standard definition of what constitutes either a rural or urban area. While rural can be understood as a spatial concept, most simply defined as all that is not urban, it is increasingly difficult to draw a complete dichotomy when delineating urban and rural areas (Marcellini et al. 2006). For example, the Organization for Economic Co-operation and Development (OECD) (1993) classify rural areas according to their degree of accessibility to urban centres, thereby leaving room for considerable ambiguity. In that respect, rurality is an increasingly fluid concept, particularly in Ireland given the spread of cities and towns into the countryside in recent years. Moreover, rurality and population dispersion take on different meanings across countries making it difficult to get agreement or shared understandings of rural space for international comparison. A different approach evokes a rural identity and associates rurality with having a distinct culture of strong community ties, long history, and ethnic and cultural connections (Ponzetti 2003). Even within stable rural communities, however, there may be age cohort differences in philosophical interpretations of rurality. For instance, older rural people may identify more with the traditional values than younger rural people.

There may be as many different kinds of older people living within rural areas as there are differences between rural and urban older populations. For example, Eales, Keating, Rozanaova (2006) identified four distinct groups of rural older adults in Canada: community active, stoic, marginalized, and frail, each with different levels of participation and needs. While recognizing that older people in rural areas are likely to be as different from one another as they are from their counterparts in urban areas, there are important elements of rural life that all older people share as a result of where they live, thus requiring a different policy approach from what is appropriate in urban areas (Kim 1980). Low population density, spatial inequality and poverty may give rise to a number of problems including remoteness and isolation, poor housing condition and the lack of good transport systems to access local communities and services (Ansello 1980). Using this approach, rurality can be explored in terms of differential need and provision.

This chapter will seek to identify some of the important issues for older people living in rural areas. As there is a scarcity of relevant Irish data on rural ageing, both cross-sectional and longitudinal, all we can do is document existing literature and knowledge, signal some of the most important problems and highlight themes for further investigation. The focus is on highlighting some of the most pressing issues facing older people living on the island of Ireland. We begin with a brief discussion of rural older populations, followed by some consideration of their broad health needs. The social and physical infrastructure of rural areas is then explored covering: health and social care provision, transport, housing and technology. The potential of social entrepreneurship in addressing social need is then examined. The chapter concludes with a discussion of policy and practice that might underpin future reform.

## Rural older populations

One of the difficulties in estimating how many older people are living in rural areas is to agree on the definition of urban and rural for the purposes of classification in the Republic of Ireland (RoI). Based on Central Statistics Office (CSO) definitions of what constitutes rural and urban/town areas (population clusters of 1,500 or more inhabitants), there are more older people living in urban/town areas than in rural areas in. the RoI. According to the 2006 Census, 203,971 older people live in rural areas and 263,955 live in urban/town areas. The national ratio of rural elderly to urban elderly is therefore 0.77, down from 0.83 in 2002; the ratio rises in linear fashion from 0.73 for the age category 65–69 to 0.84 for the age group 85+. At present, older people have an above-average share of the population in the Border, Mid-West, South-East, South-West and West regions in the Republic of Ireland. In NI, the older population also shows differences by geographical area. The North East, Coleraine, Moyle, Ballymoney, Ballymena and Larne areas have a marked older population (Rural Development Council 2003). Similarly, there is a marked older population within Fermanagh and parts of Dungannon, Belfast, Ards, North Down and East Lisburn.

Age dependency ratios provide crude but useful summary measures of the age structure of the population at a particular point in time. Because of outward migration by the working age population, the dependency ratio in rural areas is generally higher than in urban areas; the rural old age dependency ratio is lower than the urban figure in only five counties in the Republic of Ireland. The overall dependency ratio in the RoI is 16 per cent, with Leitrim having the highest overall old age dependency ratio at 26 per cent and Kildare the lowest at 10 per cent. While there is a general East-West divide with respect to old age dependency ratios, the highest absolute numbers of older people are still found in the Eastern counties of Dublin, Kildare and Meath.

*Health needs of rural older people*

There is no comprehensive information available on the health and social care needs within older populations living in either rural or urban areas. We are only now beginning to think about cross-sectional information on disability and the first longitudinal study in RoI is, at the time of writing, not yet at its pilot stage. Consequently, it is difficult to say anything with certainty about the health needs of older people living in rural settlements.

Self reported health is a common method of measuring perceived health status among older people. It has been found to predict a range of health outcomes, including health care utilization, morbidity, recovery from illness, decline in functional ability and mortality (Benyamini and Leventhal 2003). The evidence from self-reported health surveys for the Republic of Ireland suggests that the majority of older people are in reasonable health. When asked to rate their own health status, 90 per cent of older people considered their health to be 'fair' to 'good', with only 3 per cent saying it was 'very bad' (Fahey et al. 2007). Data from the *One Island-Two Systems* study (McGee, O'Hanlon et al. 2005) suggest that there is no significant difference between urban and rural areas across the island of Ireland in respect of self-reported health, with the majority of respondents in both areas again perceiving their own health as 'good/excellent'. The proportion of older people reporting their health as 'poor' or 'very poor' is the same in urban and rural areas at 8 per cent (see Table 15.1).

**Table 15.1    Perceived Health Status**

| Ratings of health | Urban (%) | Rural (%) | Total (%) |
|---|---|---|---|
| Good/Excellent | 67 | 62 | 66 |
| Fair | 25 | 30 | 26 |
| Poor/Very Poor | 8 | 8 | 7 |

*Source*: McGee et al. (2005)

Similarly, the Health and Social Services for Older People (HeSSOP) report (McGee et al. 2001) reported that there was no difference in anxiety levels between urban and rural older people, but that the rural older people were more likely to be depressed. While the latter carries worrying implications, given that depression can impair quality of life among older people and have huge consequences for morbidity and mortality (Blazer 2003), it would be incorrect to paint a picture of significant mental health problems among older people living in rural areas. Social isolation is often associated with loneliness leading to potential mental health problems for people living in remote areas. While this may be true for some people, the evidence suggests that social contact among older people is high for older people living in both rural and urban areas. Two thirds of older people living in rural areas meet friends and relatives most days, while a further 29 per cent meet friends one to two times a week (McGee et al. 2005).

Functional ability is another important predictor of need amongst older populations and evidence from McGee et al. (2005) suggest that older people in rural areas may experience slightly more difficulties in this respect. There are a greater proportion of rural people who have difficulties with activities of daily living (ADLs) such as personal care and washing compared to urban older people (13 per cent rural versus 8 per cent in urban). Differences also exist with respect to more complex activities such as shopping (18 per cent rural versus 13 per cent in urban). There is some suggestion, therefore, that for some ADLs people in rural areas may be more dependent than urban older people. But we need a lot more data before we can say anything with certainty about differences between rural and urban older people, particularly in relation to trends.

## Health and social care provision in rural areas

Studies of service provision in rural areas have commented on the decline in provision rates brought about by an ongoing process of rationalization of public and private sector services (Higgs and White 1997). The problem is that services tend to follow population, leaving many de-populated rural areas bereft of important public and social amenities. Provision of adequate services has major consequences for older rural people. As people age or experience greater disabilities they have a greater reliance on the provision of local health and social services. Older people are, therefore, one of the social groups most likely to be affected by mobility deprivation and by changes to service provision.

Social services provision in rural areas is subject to market failure, as the cost involved often does not make it possible for local service providers to operate efficiently. Service uptake in rural areas may not meet the 'usage' criterion that is often adopted to ensure that provision occurs. For economic reasons, therefore, social services have tended to concentrate in urban areas, where the costs of social services provision can be satisfied by economies of scale. These differences in social services provision mean that those living in marginalized rural areas can find themselves deprived of a satisfactory level of provision. Not surprisingly, arguments for social services provision in rural areas have focused on the equity principles of fairness and justice. Equity arguments have featured more prominently in the last few decades in Government and EU policy, where there has been a much stronger focus on decentralization and rural regeneration.

Health and social care services are undoubtedly the most important element in public social services for older people, not only because older people are more likely to suffer from illness and disability but also because even those who are fit and well tend to worry about the availability, quality and cost of health and social care services in the future. The availability of local informal care support structures is also important given that the majority of older people have a greater preference for living at home rather than moving to institutional care. Keeping people at home is an important policy target for dependent older people living in RoI. Achieving this target is only possible, however, if older people have access to care within their families and in their local communities.

Community care services for older people are very poorly developed in the Republic of Ireland (National Economic and Social Forum [NESF] 2005). Some places are worse than others in this regard. For example, when asked to rate social services for older people in their area, 40 per cent of respondents in the O'Shea, Keane and O'Connor (1998) survey of Gaeltacht rural areas said that care services were either 'poor' or 'very poor', while a 'middling' verdict was returned by 32 per cent of respondents. Only 28 per cent of people felt services were 'good' or 'very good'. The gaps in social care provision were also evident in the priority which respondents felt should be given to social services provision in their areas. When asked about the need for additional public expenditure in their area, over two fifths of respondents prioritized the need for additional health and social care services, particularly for older people. Similarly, the HeSSOP study (McGee at al. 2001) compared service utilization between urban ERHA and rural WHB and found that usage of chiropody, dental, dietician, social work services and meals on wheels was higher in the urban ERHA. However, as reported use of these services was low in general, these rural/urban differences should be treated with caution. Poor access to health and social care was also mentioned as a key disadvantage for older people living in rural areas in Northern Ireland by 39 per cent of voluntary and community groups responding to a survey on ageing and rural poverty in that jurisdiction (Rural Community Network 2003).

The availability of informal care services from families and friends is one of the most important elements of community support, especially for older people in rural areas who are deprived of easy access to local health and social services. It is generally accepted that the availability and use of informal support is a key element in preventing or delaying institutionalization (Palmore 1983). There is no evidence of significant urban-rural differences in the proportion of people receiving support from their spouses or relatives. However, changes in the age structure of rural populations in the future are likely to impact on caretaker potential as the proportion of women aged 40–59 decline relative to people aged 65 years and over. This may change the balance of care in rural communities and place much more of a burden on the statutory care system, which currently is poorly equipped to respond.

Based on the limited research available, there appear to be no major differences in respect of primary care between urban and rural populations (McGee et al. 2001). General practitioner (GP) use between the ERHA (urban) and WHB (rural) areas were similar, although there was more of a focus on preventative health service delivery in the WHB (rural area). McGee et al. (2005) found high levels of satisfaction with GP services generally, with 95 per cent of rural respondents saying that they were satisfied with the availability of the GP in their areas. Barriers to GP use, where they existed, related more to psychological issues than to spatial patterns of provision. Some older people were reluctant to attend their GP under any circumstances. A culture of independence and self-sufficiency among older people in rural areas may reduce uptake of all services, not just primary care. There may be a 'stigma' associated with the use of some support services which can lead to a lower uptake for some categories of older people, especially rural men.

A similar picture is evident with respect to hospital use, which shows no major divide between urban and rural areas. When respondents were asked if there were any factors that prevented them from availing of hospital services, no significant barriers were reported in either urban or rural areas. The biggest difference between people living in urban and rural areas was in relation to outpatient appointments, where only 15 per cent of rural residents availed of these services compared to 22 per cent in urban areas. While lower usage of outpatient appointments may lead to a greater reliance on Accident and Emergency (A&E) admissions, there was no evidence of this in the available data.

## Transport

Transport is a particular problem in rural areas for people who do not own a car. Due to demand deficiencies, for-hire private transport is unlikely to flourish in rural areas. Rail transport is now mainly confined to linking major cities and towns. Public bus services in rural areas have been rationalized, especially those serving unprofitable and remote rural areas. While cutbacks in public provision of bus services causes little or no problem for people with cars, considerable hardship is experienced for those who continue to depend solely on public transport for getting around. Transport for older people is particularly important in terms of ensuring access to local services and facilities and engaging in social activities (Banister and Bowling 2004). Transport is also required for access to other resources that form the basis of social inclusion. In this sense, lack of access to transport can affect the physical, mental and emotional well-being of older people, thus impacting on their overall quality of life.

Data from the Rural Community Network (2003) in Northern Ireland suggests that existing rural transport is largely ineffective mainly due the limited geographical coverage, which in turn is related to the financial viability of bus routes. Over 70 per cent of voluntary and community groups cited access to public transport as a key disadvantage for older people living in rural areas in Northern Ireland. The absence of suitable public transport is also a key issue for older people, living in rural areas in the Republic (NESF 2005) – particularly for people living a long distance away from centralized facilities. For example, people living in the three highest older age dependency District Electoral Divisions (DEDs) in Gaeltacht areas have to travel an average of 33 miles to the nearest hospital facility (O'Shea et al. 1998). The distances people have to travel to visit their GP, or to attend day care, while less, are still significant at 8 miles and 7 miles respectively. More than one third of people do not own a car; while one fifth do not have access to one. It is little surprise, therefore, that when asked what changes would improve their quality of life, almost one quarter of respondents highlighted better public transport as the number one priority.

McGee et al. (2005) associate problems with transport in rural areas with a lack of available convenient and regular routes. People living in rural areas are clearly more reliant on their own methods of transportation, with 41 per cent of older people availing of lifts from friends/neighbours and 25 per cent driven by family/relatives. Research by Gilhooly et al. (2003) in the United Kingdom (UK) found that asking for a lift from friends or neighbours was socially problematic for some older people. Unless a reciprocal relationship was involved, older people are often reluctant to ask for help. Research from Northern Ireland suggests that older people often see themselves as a burden on their family and friends in relations to transport needs (Rural Community Network 2003). Having to rely on family and friends for transport ultimately compromises their independence, particularly in the absence of opportunities for reciprocal behaviour.

Free travel schemes for older people have existed in the Republic of Ireland and in Northern Ireland for a number of years and have been lauded as an important innovation in relation to meeting the mobility needs of older people. Since September 2006 restrictions on travelling times under the scheme, whereby those eligible could not travel at peak times in the mornings and evenings, have been removed in the Republic. The All Ireland Free Travel Scheme was initiated in 2007 with older people now able to avail of free travel on public transport throughout the island of Ireland. While these schemes have enabled older people to live a more active life than would otherwise be possible, older people in rural areas do not always reap the benefits given the lack of suitable rural public transport. If the public transport system is weak or non-existent, then it matters little that entitlement is universal.

In recognition of this reality, a specific programme to launch a Rural Transport Initiative (RTI) (now the Rural Transport programme (RTP)) was developed as part of the Republic's National Development Plan (NDP) 2000–2006 (see Rau and Hennessy below). The aim of this programme was to:

> encourage innovative community-based initiatives to provide transport services in rural areas, with a view to addressing the issue of social exclusion in rural Ireland, which is caused by lack of access to transport (www.adm.ie).

The RTI was particularly aimed at people in rural areas who are excluded or may become excluded because transport is not available, accessible or affordable to them locally. A report by Fitzpatrick Associates (2004) on the progress of the RTI up to that time showed that the group availing most of the RTI assisted services were older women, with two thirds of users aged 66 and over. The higher service use among women reflects the fact that more women than men live in rural areas, although it may be that older men are reluctant to use public transport for reasons of stigma. With regard to the purpose of the trips made by users, shopping (61 per cent), leisure activities (34 per cent), pension collection (24 per cent) and health appointments (18 per cent) were amongst the most popular uses. When users were asked about the impact of the service on their lives, most respondents cited 'independence' and 'social contact' as the most important benefits of the RTI. The latter is clearly not just a transport service but is also a social service for rural communities.

The benefits of this scheme are also tangible in terms of health care usage, given that 18 per cent of trips are for the purposes of accessing medical services. The availability of transport services allows older people to access services, thereby preventing possible unnecessary admissions into hospital or long-stay care facilities and allowing older people to continue living independently in their own homes. The advocacy organization Age Action Ireland (2007) has recently called for extension of these services to rural areas not currently covered by the scheme and for a voucher scheme so that older people can avail of private transport where no public transport is available. An additional provision of €90 million was made available for this scheme in the 2007–2013 National Development Plan. In Northern Ireland, community groups felt that Statutory Agencies and Local Health Trusts should have a key role in providing integrated rural transport to ensure that older people could best link up with local services (Rural Community Network 2003).

Irish Rural Link (IRL), a national network of organizations and individuals lobbying for sustainable rural development in RoI and Europe, has recently put forward a submission to the Department of Transport to develop a National Rural

Transport Office to link and support the development of rural transport within the overall auspices of developing public transport within the Republic of Ireland. IRL recognizes the benefits delivered by the RTI; however, it contends that the separation between RTI and mainstream transport has accentuated issues of social exclusion. It also supports the availability of flexible services in terms of bus, car, or hackney, depending on local needs and geographical coverage.

### Housing and amenities

The quality of housing in RoI has generally improved over the years. However, a recent survey of housing quality in the Republic of Ireland found that those aged 65 years and over were twice as likely as the average to report a major problem with their dwelling in relation to dampness, food preparation facilities, sanitary facilities or ventilation (Watson and Williams 2003). Older people in rural areas are more likely to be without basic amenities such as bath/shower, indoor toilet, central heating, hot water and washing machines. While the absolute number of older people suffering housing deprivation is small in rural areas, they are still higher than in urban areas, with significant differences for some amenities (Fahey et al. 2004; Fahey et al. 2007). For example, one in five older people living in rural areas lack central heating compared to one in twelve in urban areas. Older people are also likely to be more vulnerable to fuel poverty than other people, as they are likely to spend longer in their homes. A recent study by Healy (2004) identified deprivation of heat in the home as a major problem for many older households (both rural and urban) in the Republic of Ireland.

Most people prefer to live at home even if the quality of accommodation is less than optimal. Therefore, initiatives to improve the standard of housing are of utmost importance for people living in rural areas. Indeed, the objective of housing policy in the Republic of Ireland, as stated by the Department of the Environment, Heritage and Local Government, is to enable every household to have available an affordable dwelling of good quality, suited to its needs, in a good environment and as far as possible at the tenure of its choice. Although there is no specific rural housing policy, there have been a number of grants – most notably to help pay for central heating and essential repairs – introduced to help older people living in poor conditions to improve their homes (NESF 2005). Under a Scheme of Community Support for Older People, funding is also available to enable older people improve the security of their homes. These schemes are critical in allowing people to remain in their own homes as they age but they are sometime administered in isolation, without reference to each other, or to the costs and benefits associated with 'ageing in place' within rural communities.

On-going evaluation and assessment of the impact of these schemes is required, therefore, focusing in particular on efficiency, outcomes and equity. Despite evidence of a genuine effort to improve housing conditions and amenities, there are questions concerning the extent to which existing schemes dovetail each other. This raises further issues about the efficiency, relevance, timeliness and distribution of the schemes. Similarly, housing has been seen as an end in itself rather than a contribution to health and well-being for older people. Linking the various housing programmes that exist to explicit local and national health and quality of life targets would allow for much greater targeting in respect of any new investment in social housing. Housing for older dependent people must be integrated with appropriate levels and types of service provision, up to hospital-level care, where required. A continuum of housing types and options are necessary as the population ages. Ageing in place options should range from remaining in a long-term family home through to specially-designed independent units and supported accommodation in the community.

## Technology and rural ageing

The use of technology to deliver 'best practice' in service provision to older people living in rural areas is increasingly being recognized. Older people have a preference for 'ageing in place' and technology contains the potential to allow this to happen. The term used to describe technologies that have been designed to enable older people and carers to live more independent lifestyles is called 'Assistive Technology' (AT). It is a broad term that describes devices or systems that have been developed to 'allow an individual to perform a task that they would be otherwise unable to do, or increases the ease and safety with which the task can be performed' (McCreadie and Tinker 2005). This improves the quality of life both for the individual using the technology and for their carers who are provided with greater peace of mind (Reeves and Brown 2006).

*Telemedicine* refers to the application of medical practice by telematic means and incorporates the whole range of telematic health care technology. Telecare activity may involve some or all of the following: community alarm services, sensory technology, home health monitoring, care provider technology, rehabilitation technology, health promotion and home therapy. *Smart housing* is the term used to describe the electronic and computer controlled integration of many activities and devices within the home. Within smart homes, technology can control a variety of functions, allow for monitoring of health and well-being and provide communication to the outside world.

There are a number of issues associated with the use of technology in the care of older people both at home and in long-stay care. While older people are confronted with an ever-changing technological environment, there may be a lack of appreciation of the potential benefits of technology for older people and an unwillingness to embrace new technologies easily. There may also be urban-rural differences in relation to the uptake of new technologies among older people (Marcellini et al. 2006). While differential usage is likely to be exacerbated by income and education variables, the absence of a person-centred model of care for older people is also a contributory factor in explaining differences in technology uptake. We do not know what older people want with respect to technology because we rarely ask them directly. There is also an information deficit as to the potential gains from investment in technology for older people. This is linked to an absence of much evaluation of existing schemes and a failure to disseminate existing evaluations as to the potential of technology, particularly in relation to its role in keeping older people out of residential care (Tinker 2003).

It is no surprise, therefore, that some new technologies do not deliver what older people want in a form that they can use. New technologies may be too complex, or designed in such a way as to alienate older users. Older people may not always accept assistive technology, particularly if they have not been involved in the design and testing of devices (McCreadie and Tinker 2005). Older people may also be reluctant to embrace new technologies, even where they exist, due to the stigma attached. The use of technology may infer physical dependence and loss of autonomy that some older people may resist. New technologies may also raise fundamental ethical questions about surveillance and possible loss of privacy and autonomy, particularly for people with dementia and related cognitive impairments.

Notwithstanding all of these problems, technology can be an important resource and support for 'ageing in place' within rural communities; technology can also help to reduce pressure on long stay care and acute hospitals through keeping more older people living in their own homes (Sterns 2005). We need to develop a broadband infrastructure that will facilitate the supply and adoption of new technologies and services in rural areas on an all-island basis. On the demand side, new technology must be developed in collaboration with users, carers and health care providers. Without such collaboration, technology will not achieve its considerable potential. Support networks must also be developed to help users adopt the new technology into their day-to-day lives and home environments. Technology to support rural living must be seen as a social project, requiring integration with existing social care structures and the mobilization of considerable social resources for successful implementation.

## Social entrepreneurship

There is sometimes a stigmatized view that older people are dependent and therefore unwilling and/or unable to contribute to local communities. Indeed it is easier to find evidence of what older people living in rural communities lack or need than what they contribute to society. The reality is that older people make significant and varied contributions to economic, social and civic life; older people 'constitute a dynamic and flexible social and economic resource' (Le Mesurier 2004). Many older people, far from retiring, remain active, and are willing to work and engage in 'lifelong learning'. As many older people are 'ageing in place' they are more likely to understand and identify the needs of their local community. They may be able to find ways and means to address problems through using under-utilized resources – people, buildings, equipment, and finance – and putting these to good use to solve local problems. There is a need therefore to recognize older people as a key resource when implementing policies to encourage social entrepreneurship on the island of Ireland.

Leadbeater (1997) describes the core assets of social entrepreneurs in terms of social capital, meaningful relationships, networks, trust and co-operation, which gives them access to physical and financial capital. Social capital is associated with shared values, trust, networking and co-operation (Putnam 1993), all of which are necessary to transform blighted social landscapes into dynamic living social organisms. The more extensive the social contacts, the more complex the networking, the greater the co-operation, then the more likely it is that social needs can be addressed in an effective way. Social entrepreneurs harness and develop social capital for social productive purposes. Older people are likely to have the skills, experience, wisdom and established social networks necessary to harness economic and social activity in local areas. There is evidence, for example, of the role of social entrepreneurship in empowering older people and transforming care and support networks within Summerhill in County Meath through the establishment of vibrant and inclusive social care communities in the locality (Walsh and O'Shea 2007).

Unfortunately, social entrepreneurs are in scarce supply in rural areas, particularly the variant that can transform communities through an innovative supply of social services. This is not to deny that many communities have people who care about social problems, and who try to do something about these problems. The reality is, however, that many communities do not get beyond the documentation of problems, and, even when they do, their response is often too isolated, too narrowly focused and too poorly funded. Enduring social entrepreneurship is about creating what Leadbeater (1997) calls the virtuous cycle of social capital, whereby physical capital, financial capital, human capital, and organizational capital all grow rapidly to the point of generating social dividends which can be used to create more social capital, and the cycle begins again. Anecdotal evidence from voluntary groups, sports clubs and political organizations suggest that virtuous cycles of social capital are now the exception rather than the rule. The main reasons for this are the absence of a comprehensive and consistent public policy for the creation of social entrepreneurs, the lack of an institutional framework to support social entrepreneurs, and a poor general understanding of the concept of social capital and its importance to both economic and social development.

There is an extensive literature on what makes an economic entrepreneur, but not much on the nature of social entrepreneurs, particularly in rural areas. The promotion of a volunteering ethos in society may be an important pre-requisite for the development of social entrepreneurs. If voluntary effort and community action are not seen as important, or are not highly valued by society, then it is unlikely that entrepreneurs will be attracted to social production. Voluntary effort is still highly valued in Ireland, especially among older people, so a general framework does exist for social entrepreneurship. Paradoxically, the existence of a voluntary ethos may also undermine an enterprise culture among potential social entrepreneurs. Voluntary effort has largely been associated with un-paid, in-kind, often religiously motivated provision of service, and this legacy can make it difficult to generate new forms of innovation linked to quasi-commercial objectives within a social economy framework. Training in management and business oriented skills may be necessary to encourage potential entrepreneurs and embryonic community groups to think in social enterprise terms and ultimately to develop sustainable socially oriented projects.

## Policy and practice

In this section we explore the policy and practice issues associated with the development of rural communities, beginning with the need to set explicit social objectives alongside conventional economic goals. There is significant unmet personal and social need in rural communities. Economies of scale arguments are most frequently invoked to explain spatial differences in social services provision between urban and rural areas. These arguments have not been contested in the Irish context, because the debate has never moved beyond an economic efficiency framework. But clearly social objectives are important, and for fundamental changes to occur there will have to be a radical reassessment of the relative weighting given to efficiency and equity in public policy-making. The visible hand of moral leadership has too often been absent as a counter-balance to the invisible hand of the market in public policy-making.

The enhancement of citizenship and solidarity requires that the social and distributional consequences of public policy be given much higher priority than is currently the case. Social objectives must be explicitly considered if effective strategies to meeting social need in rural areas are to be devised. Social indicators must be developed and used to determine progress in the social arena. Benchmarks need to be established and progress measured with respect to social services provision, transport, housing, technology and general quality of life issues for people living in rural communities. The use and application of indicators and targets have contributed to the alleviation of poverty through focusing attention on policy instruments, programmes and outcomes. The same strategy could serve an equally important function in addressing overall rural deprivation and social care provision for older people living in rural areas. Rural proofing remains an important part of any approach to addressing the needs of older people living in rural communities.

Social entrepreneurship is key to the future development of social progress in rural areas. The development of social entrepreneurship does not have to start from scratch as there already exists a pool of potential entrepreneurs in many parts of the country, many of them older people who wish to remain working post-retirement. However, more will have to be done by Government to encourage and support social entrepreneurship within rural communities. This means the introduction of seed capital and start-up grants for social production, using similar schemes to those currently available to economic entrepreneurs. Entrepreneurs should be given support in identifying commercial social opportunities and generating realistic business plans that match economic imperatives with the realities of social economy provision. Part of the concern of health professionals with the development of the social economy, particularly in respect of social care provision in rural areas, is the potential for inferior quality of care to be delivered to consumers. Training programmes and the development of an appropriate regulatory structure will ensure that quality of care is fundamental to any new initiatives in the social economy sector.

The maintenance of rural communities and the protection of both culture and way of life should be a fundamental goal for the Irish Government. Up to now, the only approach to maintaining rural communities has been the strategy of promoting various types of rural economic development. This is a necessary but not a sufficient condition for the maintenance of a vibrant rural tradition. Economic development means little if the quality of life of rural inhabitants is poor because social services are weak. There must be a dual approach to development that recognizes the importance of the economic and the social in the lives of the people. For that reason, a new development agency is necessary, a National Office for Rural Transformation, to initiate and lead dialogue and interventions that are required to promote major innovation in social production in rural areas.

Specifically, the National Office for Rural Transformation could provide leadership in the following areas:

- establishment of values and goal-setting with respect to social progress in rural communities;
- measurement of social progress and social gain in rural communities through an annual social audit of quality of life;
- integration of economic and social programmes and projects in rural areas;
- identification and nurturing of social entrepreneurs within rural communities, building upon the existing network of volunteer providers; and
- provision of basic seed and start-up capital grants for rural projects meeting specific social economy criteria.

Some people might argue that many of these activities should, more properly, fall under the control of existing statutory agencies, like the HSE or the local authorities, but that would be to miss the point. Social production should be an integral part of overall development policy for rural areas. Despite the existence of a myriad of national organizations and agencies social decline continues to happen in rural areas. It is now time for one agency to oversee the response to social need in rural communities. To have a new National Office for Rural Transformation involved in first of all nurturing, then supporting, and ultimately funding social enterprise and production in rural areas would avoid piecemeal and ad hoc responses to the sort of rural problems we have been discussing. The new Office would serve to integrate economic, social and health goals for rural communities. It would also help to protect the rural way of life that contributes so much to our understanding of both Irish culture and identity.

## Conclusion

The key element in the sustainability and re-generation of rural communities is a broader vision of the potential of older people within rural society. Development has been too narrowly defined in the past to encompass only economic effects. People living and ageing in rural communities experience life in economic terms certainly, but also in social, health and cultural terms. There is no doubt but that the social and broad health needs of older people have been allowed to lag behind economic imperatives within rural communities. This must change if quality of life and wellbeing are to be enhanced for all people living in rural areas. A case has been made in this chapter for the establishment of a new National Office for Rural Transformation, whose brief would be to develop the rural economy and rural society through social production as much as economic production, thereby playing an important role in maintaining and enhancing the quality of life for all rural dwellers. A vibrant social economy dedicated to meeting the social and cultural needs of the community would allow people living in rural areas to move into a future where they might fully realize their human potential and have a greater stake in the communities in which they live and age.

## References

Age Action Ireland (2007) *Older People Count: Pre-Budget Submission 2007.* Dublin: Age Action Ireland.

Ansello, E. (1980) 'Special Considerations in Rural Ageing', *Educational Gerontology* 5, 4: 343–54.

Banister, D. and A. Bowling (2004) 'Quality of Life for the Elderly: The Transport Dimension', *Transport Policy* 11: 105–15.

Benyamini, Y. and E. Leventhal (2003) 'Elderly People's Ratings of the Importance of Health-related Factors to their Self-assessments of Health', *Social Science & Medicine* 56: 1661–7.

Blazer, D. (2003) 'Depression in Late Life: Review and Commentary', *The Journals of Gerontology Series* 58: 49–65.

Eales, J., N. Keating and J, Rozanaova (2006) *Caring Contexts for Rural Seniors: A Case Study of Diversity among Adults in Rural Communities.* Alberta: Research on Aging: Policies and Practice (RAPP), University of Alberta, Canada.

Fahey, T., B. Maitre, B. Nolan and C. T. Whelan (2007) *A Social Portrait of Older People in Ireland.* Dublin: Stationery Office.

Fahey, T., B. Nolan, and B. Maitre (2004) *Housing, Poverty and Wealth in Ireland.* Dublin: Combat Poverty Agency.

Fitzpatrick Associates Economic Consultants (2004) *External Evaluation of the Rural Transport Initiative.* Dublin: Area Development Management Ltd.

Gilhooly, M., K. Hamilton and M. O'Neil (2003) *Transport and Ageing: Extending Quality of Life for Older People via Public and Private Transport.* ESRC Growing Older Programme 16. Sheffield: University of Sheffield.

Healy J. (2004) *Fuel Poverty and Policy in Ireland and the European Union.* Dublin: The Policy Institute TCD.

Higgs, G. and S. D. White (1997) 'Changes in Service Provision in Rural Areas. Part 1: The use of GIS in Analysing Accessibility to Services in Rural Deprivation Research', *Journal of Rural Studies* 13, 4: 441–50. http://www.adm.ie/Pages/rti/overview.htm.

Kim, P. K. H. (1980) 'Public Policies for the Rural Elderly', *Aging and Public Policy: The Politics of growing old in America.* Westport, CT, Greenwood Press.

Leadbeater, C. (1997) *The Rise of the Social Entrepreneur.* London: Demos.

Le Mesurier, N. (2004) 'Older People's Involvement in Rural Communities'. Paper presented to Ageing and Countryside Conference, March 2004.

Lynch, S. (2001) *Older People – the Cycle of Deprivation*, Network News. Cookstown: Rural Community Network.

Marcellini, F., C. Giuli, C. Gagliardi and R. Papa (2006) 'Aging in Italy: Urban-Rural Differences', *Archives of Gerontology and Geriatrics*, doi: 10.1016/J.Archger.2006.05.04.

McCreadie, C. and A. Tinker (2005) 'The Acceptability of Assistive Technology to Older People', *Ageing and Society* 25: 91–110.

McGee, H., R. Garavan and R. Winder (2001) *Health and Social Services for Older People (HeSSOP).* Dublin: National Council of Ageing and Older People (Report No. 64).

McGee, H., A. O'Hanlon, M. Barker, A. Hickey, R. Garavan, R. Conroy, R. Layte, E. Shelley, F. Horgan, V. Crawford, R. Stout and D. O'Neill (2005) *One Island-Two Systems: A Comparison of Health Status and Health and Social Service Use by Community Dwelling Older People in the Republic of Ireland and Northern Ireland.* Dublin: Institute of Public Health in Ireland.

National Economic and Social Forum (2005) *Care for Older People.* Dublin: Stationery Office (NESF Report No. 32).

O'Shea, E, M. Keane and M. O'Connor (1998) *Addressing Social Need in the Gaeltacht: The Role and Potential of Social Entrepreneurship.* Galway: Údarás na Gaeltachta.

Organisation for Economic Co-operation and Development (OECD) (1993) *What Future for our Countryside? A Rural Development Policy.* OECD: Paris.

Palmore, E. (1983) 'Health Care Needs of the Rural Elderly', *International Journal of Ageing and Human Development* 18: 39–45.

Ponzetti, J. (2003) 'Growing Old in Rural Communities: A Visual Methodology for Studying Place Attachment', *Journal of Rural Community Psychology*, 6:1.

Putnam, D. (1993) *Making Democracy Work: Civic Traditions in Modern Italy*. Princeton: Princeton University Press.

Reeves, A. and S. J. Brown (2006) *Remotely Supporting Care Provision for Older Adults*. IEEE Computer Society.

Rural Community Network (2003) *Ageing and Rural Poverty*. Cookstown: Rural Community Network (NI).

Rural Development Council (2003) A *Picture of Rural Change*. Cookstown: Rural Development Council.

Sterns, A. (2005) 'Curriculum Design and Program to Train Older Adults to Use Personal Digital Assistants', *The Gerontologist* 45: 828–34.

Tinker, A. (2003) 'Assistive Technology and its Role in Housing Policies for Older People', *Quality in Ageing: Policy, Practice and Research* 4: 12–14.

Walsh, K. and E. O'Shea (2007). *Third Age Foundation: An Assessment of Role and Contributions*. Galway: Irish Centre for Social Gerontology, NUI Galway.

Watson, D. and J. Williams (2003) *Irish National Survey of Housing Quality 2001–2002*. Dublin: ESRI.

Wenger, C. (2001) 'Myths and Realities of Ageing in Rural Britain', *Ageing and Society* 21: 117–30.

# Chapter 16

# Gender and Sustainability in Rural Ireland

## Sally Shortall and Anne Byrne

## Introduction

This chapter considers if and how gender is relevant for the sustainability of rural Ireland. When we refer to rural sustainability we mean the continuation of the economic, social, institutional and environmental components of rural life. There are many ways in which we could approach a chapter on gender and rural sustainability. Mobility, education, employment, social class, health care and practically every social structure impacts on gender and the sustainability of rural areas. As these topics are covered in other chapters in this book, we have chosen to focus on gender relations and the sustainability of agriculture and rural development programmes. We review the existing body of research on these topics and consider what they tell us about rural sustainability. The literature review demonstrates how initially research reported gender differences but did not analyse them in any depth. The next phase saw scholars starting to examine the role of women on farms and latterly the role of women in rural development programmes. More recently, scholars have turned their attention to the implications for men of changing gender roles in rural areas. It is clear that any renegotiation of women's roles has implications for men's roles, and vice versa. Much of the research we will review focuses on whether a particular construction of a gender role negatively impacts on another. Our rationale is that a good quality of life for men and women seems central to the sustainability of rural living. We conclude by identifying contemporary considerations regarding gender and rural sustainability.

## Gender, sustainability, and early sociological studies

Arensberg and Kimball (2001 [1940]) are credited with the provision of the first anthropological account of the main social and economic conditions of rural Ireland. Prior to this, accounts had been presented only through literary or political commentary and controversy. Their documentation of the social and economic conditions of Co. Clare in the 1930s which they consider to have been representative of Ireland, marks the initiation of rural Irish sociological research. Arensberg and Kimball's work sparked considerable controversy, both for its claims that it is representative of all of Ireland, and for their rigid underlying conceptual model of structural functionalism. As they explicitly indicate in their introduction, their work aimed to be both ethnographic, and to advance the explanatory power of structural functional analysis (Arensberg and Kimball, pp. xxv–xxvii; Hannan 1982). Hannan (1972) and Byrne et al. (2001) conclusively argue that despite theoretical shortcomings, Arensberg and Kimball's work is an important ethnographic account of the social and cultural system characteristic of small scale farming communities in the west of Ireland in the 1930s.

Arensberg and Kimball describe the patrilineal system which existed:

> The Irish family is patrilocal and patronymic, and farm, house, and most of the household goods descend from father to son with the patronym. The father is dominant within the family. He comes to stand for the group which he heads; the farm is known by his name, and the wife and children bear his name (p. 80).

Arensberg and Kimball frequently use 'father, husband and farm-owner' as synonyms (pp. 46/47). They detail the father's 'controlling role' (p. 46), and attribute it to his status as landowner: 'The old man abdicates his controlling position with his transference of the farm to his son' (p. 121). They recount the social standing and precedence accorded to the old fellows, 'the men of full status who head farms and farm – working corporations of sons, those who have turned or are about to turn over their control to a younger generation' (pp. 170–74).

Arensberg and Kimball give us a detailed description of the work carried out by women, which suggests the work performed by women is arduous and time-consuming; 'The first duty of the day falls to the woman. She rakes up the fire and gets it going, and starts getting the breakfast ready' (p. 35). At about 5 o'clock, the work of the men is over for the day but that of the woman goes on (p. 39). She must prepare, serve and clean up after the tea, milk the cows, help the children with school lessons if necessary, and put them to bed. If she returns to join the men at the fire she continues with knitting and baking. When the whole family has gone to bed she closes up the house and slakes the fire in the hearth. They say too that 'the woman's hands are never idle' and the work of women is as important in farm economy as men's work (p. 63).

Even though Arensberg and Kimball focus on reciprocity and complimentary roles, they do outline the different status and prestige vested in each role. They speak of the farmwife and mother 'who serves her men' (p. 35), who, as they eat stands ready to refill their plates (p. 37) and who does not seat herself to eat until the men have finished. Men's status as land owners also gave them access to wider social structures; it is because of their position as farm heads and owners that the old fellows, the men of full status, come to 'represent the interests of the community before priest, schoolmaster, merchant, cattleman and government official' (pp. 170–74).

From the perspective of their structural functionalist framework, Arensberg and Kimball describe a rural society that is sustainable because of complementary gender roles. They present the gender-related division of labour as a functional development within the society. The 'duties of male and female are complementary'

(p. 195), and the division of labour between the sexes simply represents the separation of human activity into male and female spheres. They describe the division of labour between the sexes as one that arises within a field of larger interests and obligations. It is 'part of the behaviour expected reciprocally of husband and wife, it is a functional element of their relationship within the family' (p. 48). They describe the dominant position of the father within the family, and alongside this, provide an anomalous account of reciprocity within the family. This is a charge levied against Arensberg and Kimball by their critics; their concern to assert the importance of structural functionalism means that they provide many descriptions of observed relationships that do not conform to their theoretical model (Hannan 1972). The questions of property, power, women and complementary work roles are clear examples.

McNabb's (1964) study of rural Limerick describes a social structure similar to that described by Arensberg and Kimball. He describes the authority and social standing of farm men, while the lives of women are 'one of unrelieved monotony' (p. 234). McNabb does consider the sustainability of the rural life he describes. He states that traditional society is well established 'because it controls the means of production' (p. 244). Changes regarding the increased availability of education actually serve to maintain traditional norms regarding property and the role of the father, because it becomes more legitimate to have one heir, and educate the other children (p. 244). McNabb maintains that the state also contributes to the sustainability of this structure; These 'being paternalistic, are part of the traditional framework. . .they do not change it' (p. 245). He says too that 'the chief institutions . . .are so organized and related to each other as to guarantee the authority of the father and the conservation of property' (p. 243).

Eithne Viney (1968) described the tough life of women on small farms and labouring families in the 1950s and 1960s. She recounted an unending cycle of hard, physical labour, poor spousal and familial relations, large numbers of children to care for, poor health and little material comforts. This provides an insight into why women might become disenchanted with rural life and she argued that mothers were encouraging daughters to marry farmers hoping that education would provide an opportunity for daughters to make their way in the world away from family farming (p. 338).

John Messenger's anthropological study (1969) is based on research he carried out on an island in the Irish Gaeltacht, which he identifies by the pseudonym 'Inisbeag'. He and his wife spent most of a year there in 1959/60, and they returned eight times between 1961–1966. Like Viney's study, he signalled gender relations as a threat to the sustainability of the rural structure. Messenger's account details the exclusion of women from social structures and practices, and also a measure of discontent with their situation. Women confided to his wife that they were unhappy about being forced to remain at home, minding children, and performing tedious household chores. They were resentful of their husband's greater freedom, and their involvement in numerous social activities 'forbidden by custom to women' (p. 77). Women expressed concern about the pressure of informal controls, particularly with regard to having children (p. 77). Similar to Brody's (1973) analysis of Inishkillane, Messenger reasoned that many of the girls (*sic*) who emigrated did so because they were dissatisfied with the lot of married women on the island (p. 125). Gender dissatisfaction with rural life is presented as a potential threat to the social structure.

Brody's (1973) famous study is based on participant observation carried out in five communities in the West of Ireland between 1966–1971. He lived and worked in the communities as a visitor or additional hand, but never as an investigator. Brody believed that demoralization was rampant in the West of Ireland; 'it is the breakdown of the communities, the devaluation of the traditional mores, the weakening hold of the older conceptions over the minds of young people in particular, to which every chapter will return' (p. 2). Brody maintained that unlike Arensberg and Kimball (2001 [1940]), he is not about to describe a harmonious and self-maintaining system, but rather one in which the people are demoralized and have lost belief in the social advantages and moral worth of their small society (p. 16). Brody presents emigration as a means of escape from a disintegrating society for disenchanted young people, and outlines the differing rates of emigration for men and women. Women leave when they are younger and they leave in larger numbers. Brody describes a way of life that he does not think is sustainable, and nor does he think it should be. Because of women's lack of material possessions it is easier for them to leave, and he also states that that 'country girls have refused to marry into local farms' (p. 98). It is a bleak image of an unsustainable rural life that this research presents.

Hannan and Katsiaouni's (1977) study holds a position of importance in the chronology of research on Irish farm life. Their report makes the leap from the anthropological studies we have considered, to the analysis of a modernized, commercial type of farming. Hannan and Katsiaouni state that their study is an attempt to provide some information on nuclear family interaction patterns in Ireland (p. 11). Their main aim is to identify the principle characteristics of farm family interaction, explain variances in interaction, and examine how and why farm family interaction patterns have changed in Ireland since the 1930s (p. 2). While this farm family structure existed and was suitable within a particular context, this context has changed dramatically and significant changes within the family structure are also to be expected. They identify two crucial processes which are accountable for this: the first is the commercialization of farm production, and the second is the massive expansion of mass communication and modern transport. Hannan and Katsiaouni (ibid.) maintain that these forces combined are likely to lead to changes in people's beliefs and values 'as people begin to take on the perspective of prestigious urban reference groups' (p. 26), and definite adaptations in family task and decision-making patterns will have to be made as a purely circumstantial response to the changing farm and household economy.

The traditional farm family and the modern urban middle-class model are the two anchor points for Hannan and Katsiaouni's study, and they set out to show and explain variations in farm family interaction patterns along a continuum between these anchors. Hannan and Katsiaouni assume it is natural for their 'modern urban middle-class model' to develop in rural communities. They say that the direction of change is 'almost inevitably' (p. 16) towards such a model. The summarized description of this model, developed by Elizabeth Bott (1971), recounts the main features as being minimal or no spousal segregation in housekeeping and childrearing roles, similarly power or authority gradients between spouses and between parents and children are minimized, with decision making being largely a joint consultative process. The greater openness of all interpersonal relationships within the family means that maternal specialization in emotionally supportive functions is no longer obvious or necessary.

The basic economic provider role is still predominantly male, and Hannan and Katsiaouni feel this will be particularly so on farms. They say that 'although the degree of participation by the husband in household and child-rearing tasks is limited by his economic role as provider, what is important is that the norms have changed' (p. 27). They are describing patterns of social interaction that have not significantly changed but are sustainable because the source of legitimization has changed. There has been a reinterpretation of the old pattern which allows it to remain acceptable. Hannan and Katsiaouni identify how the survival of a given system relies on the belief that it is legitimate. They say of the traditional farm structure that such an overall system could only remain intact so long as it continued to be legitimized by the consensual sets of beliefs and values of the community. This legitimizing ideology remains effectively isolated from contending ideals of family organization which hold in external prestigious groups (p. 20). They argue this traditional society no longer exists, and has now moved towards the modern urban middle-class model. They present this shift as contributing to the sustainability of family farming.

## Feminism, farming and rural life

During the 1980s there was an upsurge in feminist studies of the role of women on farms. Irish sociological studies mirror international developments in scholarship. Early research focused on women's farm work, essential to the sustainability of the farm but rarely accounted for in agricultural statistics (Fahey 1990; Shortall 1992; O'Hara 1994). The patriarchal nature of farming and the power relations within the farm family were studied (Higgins 1983; O'Hara 1998; Shortall 1999). Women's agency and resistance within farming structures were also considered (O'Hara 1998; Kelly and Shortall 2002). Recent research has focused on how women's off-farm work impacts on the construction of gender relations within the farm family (Hanrahan 2006; Gorman 2006; Shortall 2006). While women's off-farm employment is now central to the sustainability of farming, the renegotiation of gender roles has led to a sophisticated analysis of the implications for men, and what that means for the sustainability of the sector (Ní Laoire 2001, 2002; Kelly and Shortall 2002). We consider these theoretical developments in turn.[1]

An early and continuing focus of research for sociologists, geographers and economists is the 'invisibility' of farm women's work and theoretical analyses of why it is so (Fahey 1990; Shortall 1992; O'Hara 1994; Heenan and Birrell 1997). This body of work borrows heavily from Marxist debates, particularly notions of petty commodity production, and the separation of productive and reproductive work on the farm (Reimer 1986). Feminist scholars argue that narrow definitions of productive farm work meant that much of the reproductive work carried out by farm women is unacknowledged (Bouquet 1982; Whatmore 1991; Brandth 2002; Little and Panelli 2003). Feminists identify the many ways in which women's farm work is essential to the farm business (Gasson 1992). Attempts are made to bring farm women 'out of the shadows' (O'Hara 1994) of the family farm to illustrate the unequal gender relations within the family and the different status of work carried out by different family members. This research follows broader feminist trends by noting it is not the nature of women's work that leads to lack of recognition, but rather women's position within a patriarchal household (Oakley 1974; Walby 1990; Whatmore 1991; Delphy and Leonard 1992). Whatmore's theory of patriarchal gender relations remains the most sophisticated analysis to date of women's farm work (Whatmore 1991). Her concept of domestic political ideology is developed from the recognition that home and work share the same location on a farm, and production, reproduction, family and economy must be analysed in an integrated rather than a fragmented fashion. Through this approach an understanding of the exploitation of women as farm housewives and unpaid farm labourers is advanced.

Early studies on power relations focused on relations within the farm family and the situations and circumstances that influenced women's involvement in farm decision making (Bokemeier and Garkovich 1987; Hannan and Katsiaouni 1987; Oldrup 1999). The reasons why men and women occupy different positions within the family farm is a central component of empirical and theoretical analysis. Of particular interest is that women enter and engage in farming through specific kinship relations, as wives, daughters, mothers and widows. Given the patrilineal nature of land transfer from father to son, women marry into farming, and thus enter the occupation through marriage rather than through occupational choice (Breen 1984; Kennedy 1991; O'Hara 1998; Shortall 1999). Their husband is already established as the 'farmer', and he is also (at least initially) the owner of the capital resources necessary to farm. This position impacts on the valuation of women's work and on their public place in farming.

A great deal of subsequent research examines the power relations embedded in the public representation of women and the under-representation of women in farming organizations (Shortall 2001). There are very few women in the Irish Farmers' Association or in the Ulster Farmers' Union. Other work explores the

---

1    There are two excellent reviews of the international literature on this topic: Brandth (2002) and Little and Panelli (2003).

stereotypical representations of men and women in the farming media which reinforce the conception of farm work as masculine (Duggan 1987). Farm women on the other hand are portrayed primarily as mothers and housewives by way of their domestic functions. Other research examines how agricultural education and training also reflects and reinforces power relations through gendered provisions of training (O'Hara 1994; Shortall 1996). Theoretical analysis of why these power differentials exist and persist, leads to the critical factor that allows men to hold the occupational position of farmer and to occupy the public face of farming; land ownership. The prevalent patrilineal line of inheritance means that women rarely own farms in their own right. This is central to gendered power divisions within farm families (Shortall 1999). While the centrality of land ownership to gender power differentials in farming is debated (Silvasti 1999; Brandth 2002), it is likely that even though there are situations where women are land owners, the pervasiveness of male land ownership is a key component lending weight to the ideology that positions men at the heart of farming.

The initial focus of research on farm women sought to illuminate women's farm work which had previously been eclipsed, and to understand the different gender and power relations within the farm family. It tended to present subordinate women and dominant men as static and homogeneous categories, and sought structural and causal explanations. Research in the 1970s and 1980s is described as occupying 'the rural women's subordination category' (Berg 2004). More recently, choice, agency, resistance and the altering of gender identities over time have become more prominent in the research agenda. Research examines how women on farms are not simply accepting victims of patriarchal relations, but rather they are active agents, constructing and shaping their roles within farming (O'Hara 1997, 1998; Gorman 2006; Hanrahan 2006). Indeed O'Hara's research found strategies of resistance that threatened the sustainability of farming families; her research found that women had left agricultural areas in order to avoid the types of lives their mothers had led, and she also found that mothers are encouraging their daughters to leave rural areas. It is a similar act of resistance to the one that Viney and Messenger had reported almost forty years earlier, but this time mothers are actively participating in their daughters' exit strategies. Other research identifies off-farm work, and the subsequent financial independence, as an expression of women's agency and resistance (O'Hara 1998; Hanrahan 2006).

The growing emphasis on agency, choice and resistance is a necessary counterbalance to explanations that seemed to lean towards structural determinism. Hoggart (2004) cautions that there is still a tendency for research to focus on women's 'subordination', and there is more scope for 'celebratory explorations' of women in rural societies (p. 2). While there is merit in this assertion, it remains the case, in farming at least, that women's options in terms of resistance and choice are ones that have not greatly diminished the patriarchal nature of agricultural institutions.

The research debate has tended to present research as sitting in one of two opposing camps; on one side is research that focuses on 'patriarchy and the subordination' of women, and on the other is research that focuses on the 'resistance and agency' of farm women. However both structural constraints and strategies of resistance co-exist. Women are agents who make choices and engage in both strategies of resistance and co-operation in farming and within the farm household. It is also the case that the patriarchal nature of farming and the farming industry persists despite resistance and a changing society. Gorman's recent work (2006) neatly combines a structural focus with strategies of agency. She examines how farm household livelihoods are influenced by evolving gender relations within farm families. She considers how gender roles and relations impact on the process of livelihood decision making, and whether individuals within the farm family pursue off-farm employment or strategies of farm diversification. Most importantly Gorman's work contributes to the complex debate on how to combine research on individual behaviour and action within the farm family, alongside collective household strategies.

The extent to which off-farm earnings alter gender roles and positions within the farm family has been a research question of interest for some time and classifications or models of farm women have long included the category of 'women working off the farm' or 'women in paid work' (O'Hara 1998). Women's unequal status within the farm family is seen as tied to their subordinate economic position in relation to the male breadwinner. With the generally declining income of the agricultural industry, women's off-farm work is increasingly subsidizing the farm, and women on farms are now more likely to be the primary breadwinner (Kelly and Shortall 2002; Gorman 2006; Hanrahan 2006). In many cases the decision to work off-farm is motivated by the dire need for more income. Other research found the decision to work off the farm and increase independent earnings to have been a positive choice (O'Hara 1998).

Regardless of the motivating factors, this represents a fundamental change in women's economic status within the family farm, and could potentially have significant implications for gender relations. However, it is not necessarily increased resource contributions that lead to renegotiated domestic work and gender roles, but rather gender ideologies (Layte 1998; Shortall 2006). Agrarian gender ideology is such that even though women may have an independent source of income off the farm, the fact of living on a farm means they continue to be positioned as farm women and traditional gender roles remain pervasive. It is also the case that for farm women, an individualistic approach confuses the fact that women not only act as individuals but also as members of farm households (Wheelock and Oughton 1996). Women's off-farm labour is often part of a farm household survival strategy to maintain the farm and men's occupation as the farmer. Any analysis of the likely impact of women's off-farm earnings on gender roles within the farm family must also take account of the historical context, power and established gender relations in the farm family. But in terms of the sustainability of family farming, there is no doubt regarding the contribution of farm women's off-farm income to the continuation of this social structure in Ireland.

With modernization and globalization, the economic position and social status of traditional rural professions weaken. Farmers have difficulty providing for their families and see their identity as the breadwinner disappearing (Bock 2006). This has led to a difficult reconstruction of gender roles for farm men. For a very long time the Irish rural man was seen as financially independent, a property owner, and having a romantic way of life. Men feel a sense of failure about not being the primary breadwinner on farms, and research has also shown the problems of mental illness, alcoholism, isolation and loneliness amongst both farming and non-farming rural men. We have described early anthropological and sociological studies that demonstrate the extent to which Irish farming/ rural masculinity was tied to land ownership, control of property, being the breadwinner, and being the 'head' of the farm family (Martin 1997). Ní Laoire (2001) demonstrates how the reconstruction of masculine identity negatively affects men's well-being.

Similarly Kelly and Shortall (2002) found that men in Northern Ireland had difficulties with their changed economic status and felt a sense of personal failure about no longer filling the breadwinner role. They suggest that women try to protect men's mental well-being and sense of self-esteem by maintaining this image of farming men as breadwinners. It is likely that a sustainable agricultural sector will continue to rely on off-farm incomes. Ní Laoire (2002) argues that the increased competitiveness and rationalization of agriculture threaten institutions and values that are at the core of masculine farming identities such as the patriarchal family farm, and the prestige of land ownership. Kelly and Shortall (2002) also found demoralization linked to changing masculine identity. The way in which gender roles will be negotiated and structured within a sustainable structure of agriculture requires further research.

## Gender and rural development

For the last couple of decades, rural studies have moved from an almost exclusive focus on agriculture to an extensive engagement with debates on rural development and the most appropriate way to ensure the sustainability of rural areas. The Irish sociological study of rural development has examined different questions over time. Earlier work focuses on increasing participation in rural development initiatives, their holistic nature, representativeness and what was meant by 'community', governance, partnerships and social inclusion (Cuddy 1992; O'Malley 1992; Commins and Keane 1994; Shortall 1994a). While rural development programmes aim to achieve sustainability, they also aim to enhance participatory democracy and the legitimacy of sustainability initiatives. Scholars note that while there is a considerable body of research on Irish farm women, there is a less well developed body of research on Irish rural women (McNerney and Gillmor 2005). Most of the research that has emerged considers the role of women in rural development structures and programmes (Owens 1992; O'Hara 1994; Byrne 1995; O'Connor 1995; Byrne and Owens 1998; Shortall 2002).

While women in farming have had to contend with the ideological and cultural barriers of a very masculine industry, feminists view participatory forms of rural development as providing considerable potential to include women in political structures in a way that has not previously been achieved. Rural women had always been active in community and voluntary activities (O'Hara 1994; Byrne 1995; Byrne and Owens 1998; Shortall 1994b). However, it quickly became obvious that few women were participating within new rural structures of governance. In Northern Ireland the Department of Agriculture and Rural Development (DARD) regularly state a commitment to engaging women in their Rural Development Programmes, but how this will be done is never explicitly stated. In addition, no data on gendered participation in rural development initiatives exists, although qualitative research demonstrates women's under-representation in development initiatives (Shortall and Kelly 2001).

In the Republic, women are better represented in the community and voluntary sector as participants and staff members, while continuing to be under-represented in the formal political sphere or in leadership positions in the statutory sector. This is despite a Government commitment to achieve a 40/60 per cent gender balance in respect of nominees to state boards (Department of Agriculture, Food and Rural Development 2000; National Women's Council of Ireland 2002; Department of Justice, Equality and Law Reform 2003). O'Connor (1995) argued that the few women in senior positions in rural development initiatives receive a disproportionate amount of media coverage which distorts public perceptions about the level of women's engagement.

A recent review of women's participation in decision making in national and local politics, in regional authorities, on state boards and representation on National Development Plan monitoring committees points to women's continuing under-representation and exclusion from decision-making (National Women's Council of Ireland 2002). Women's marginal access to political power shows 'the deep and persistent inequality between women and men in Irish society... (raising) fundamental questions about the representative nature of decision-making in this country' (National Women's Council of Ireland 2002, p. 5). The report of the Advisory Committee on the Role of Women in Agriculture (2000) comments that the involvement of women in decision-making is necessary to ensure that the broader social perspective of rural development is fully realized (p. 24). Engaging women in decision-making is clearly one important aspect of sustainable rural development.

Similarly to the position of women in agriculture, the debate around the participation of women in rural development follows lines of agency versus structural constraints. It is argued by some involved in rural development that the initiatives are there if women choose to get involved, as Shortall and Kelly (2001) reported from their research. A certain amount of rural development funding has focused on capacity building and empowerment programmes for women. On the other hand, while globalization processes are linked to the sub-national structures that have developed, it remains the state that hollows itself out (Jessop 1994; Rankin 2001), thereby reproducing the power struggles, contradictions and dominant ideologies of the state at sub-national levels (Rankin 2001). The ideological perspective adopted by rural development initiatives may well embrace traditional gender ideologies. It is for this reason that the more recent research on women's role in rural development is arguing for a more critical and complex analysis of the construction and management of rural development initiatives and the gender ideology they suppose (O'Connor 1995; Byrne and Owens 1998; Kelly and Shortall 2002). The Northern Ireland Women's Resource and Development Agency reports that women's networks have to work doubly hard to demonstrate that they are development bodies rather than women's groups. It is noted that involving women and addressing gender are two very different matters. Indeed it is suggested that a focus on the number of women involved in rural development initiatives is invidious because it detracts from an examination of the gender relations that underpin rural policies, while giving the impression that gender inequalities are being addressed. Rural development programmes have assumed the male norm and women must adopt this pattern of behaviour to participate. An ideological perspective that accepts the male norm may persist unabated if the focus is solely on women. When the focus moves from involving women to addressing gender relations, it becomes apparent that women's under-participation cannot be addressed by focusing on women alone.

The North Leitrim Men's Group's report (2001) explores the issues for men with a changing sense of masculinity. They consider the farming community but also rural men more generally. They report the far higher rate of single men in North Leitrim, and also the problems of mental illness, alcoholism, social isolation and depression. They recount the difficulties of getting support and assistance to many of these men because of the 'hard man image' which prevents rural men from admitting their problems or seeking help. Again it is clear that rural studies need to consider the implications for men and women of renegotiated gender roles. A sustainable way of rural life needs both men and women.

Research from the UK has examined the gendered nature of the construction of rural development policy. Studies of rural policy tend to avoid references to the relationship between policy and the construction of gender identities, or the operation of gender relations (Little and Jones 2000). From their research, Little and Jones (2000) argue that male control of the rural development policy process sustains patriarchal gender relations. They argue that greater attention needs to be given to the construction of rural policy and the priorities assumed. They contend that male power is reinforced within the policy making process, favouring particular masculine working practices and values (p. 637). The very projects that appear inclusive and transformative may turn out to be supportive of a status quo that is highly inequitable to women and to the diversity of women now living in rural Ireland. The approach to rural development may have changed but a particular gendered ideology persists to the detriment of women. The 'weak version of equality' adopted by the state falls short in promoting an egalitarian culture, in which the renegotiation of gender roles may be possible (Connolly 1999; Kirby 2002). Further Irish research on this topic would be useful.

The exclusion of women from emerging structures of rural governance is linked to its management through the two Departments of agriculture on the island, and rural development organizations, all predominantly 'masculine' in culture. Both Government departments have equality strategies. In the Republic, gender mainstreaming is a stated goal though critics are sceptical of its capacity to anticipate and deeply engage with the chronic problems of gender inequality (McGauran 2005). Including 'new' people in policy making, increasing the proportion of women in decision making positions, grassroots pressure from women's groups, political will and a commitment to changing organizational culture are some of the changes McGauran identifies if gender mainstreaming is to become real as opposed to ideal (p. 11). In the North, all Government departments are obliged to have an equality strategy following the 1998 Northern Ireland Act. Northern Ireland's approach to gender mainstreaming has received some favourable reviews in policy circles both for its insistence on a participatory-democratic model and the relative sophistication of the model of equality impact assessment to be used (Beveridge et al. 2000). While this legislation has put gender equality onto the DARD agenda, a lack of gendered baseline information, lack of a gender equality ethos, and lack of expertise have contributed to a limited impact to date (Donaghy and Kelly 2001). Favourable equality legislation exists North and South. However there is a time lag between structural change and change at the agency level. Vigilance is needed to ensure unfavourable gender relations are not incorporated in rural policies in a new guise.

## Conclusions

It is immediately obvious that employment, transport, migration patterns and education are essential to rural sustainability. This chapter demonstrates that gender roles, and the re-negotiation of gender roles are central to rural sustainability. Here we have examined how gender interacts with a sustainable way of rural living. We focused specifically on farming and rural development programmes. The literature reviewed shows that after the foundation of the State there was a rigid patriarchal gender order that led to a sustainable way of rural life regarded as legitimate and beyond question. Messenger's study in the 1960s is one of the first times that mention is made of gender roles threatening the sustainability of rural life; he reports that young women leaving the island because of dissatisfaction with the types of lives the older women had to endure. O'Hara (1998) also identified women's flight as a threat to the sustainability of farming. More recent research has shown that men's difficulties with the restructuring of rural male gender roles poses a potential threat to the sustainability of rural areas (Ní Laoire 2002, 2005; Kelly and Shortall 2002). Arguably this may also present another opportunity to de-legitimize patriarchal ideologies in a changing society.

The exclusion of women from rural governance structures has been identified as a problem at a policy level by both relevant Government departments. Including women in high-level decision-making is important as is reviewing the dominant but light version of gender equality in operation. This chapter demonstrates that gender roles, and the negotiation of gender roles is central to rural sustainability. Rural sustainability is more likely to occur with gender equality.

# References

Arensberg, C. M. and S. T. Kimball (2001 [1940]) *Family and Community in Ireland.* 3rd edition. Ennis: CLASP Press.

Berg, N. G. (2004) 'Discourses on Rurality and Gender in Norwegian Rural Studies', in H. J. Goverde, H. de Haan and M. Baylina (eds), *Power and Gender in European Rural Development.* Aldershot: Ashgate, pp. 127–44.

Bock, B. (2006) 'Rurality and Gender Identity', in B. Bock and S. Shortall (eds), *Rural Gender Relations: Issues and Case Studies.* Oxfordshire: CAB International, pp. 279–88.

Bokemeier, J. and L. Garkovich (1987) 'Assessing the Influence of Farm Women's Self Identity on Task Allocation and Decision Making', *Rural Sociology*, 52: 13–36.

Bouquet, M. (1982) 'Production and Reproduction of Family Farms in South West England', *Sociologia Ruralis*, 22, 3: 227–44.

Brandth, B. (2002) 'Gender Identity in European Family Farming: A Literature Review', *Sociologia Ruralis*, 42, 3: 181–201.

Breen, R. (1984) 'Dowry Payments and the Irish Case', *Comparative Studies in Society and History*, 26, 2: 280–96.

Brody, H. (1973) *Inishkillane – Change and Decline in the West of Ireland.* London: The Penguin Press.

Byrne, A. (1995) 'Making Development Work for Women', *Women's Studies Centre Review*, 3: 201–13.

Byrne, A. and M. Owens (1998) 'Gendering Rural Development', *Administration* 46, 3: 37–52.

Byrne, A., R. Edmondson and T. Varley (2001) 'Arensberg and Kimball and Anthropological Research in Ireland: Introduction to the Third Edition', in C. Arensberg and S. Kimball, *Family and Community in Ireland.* Ennis: CLASP Press, pp. 1–101.

Commins, P. and M. Keane (1994) 'Developing the Rural Economy – Problems, Programmes, and Prospects', *New Approaches to Rural Development* NESC Report No 97. Dublin: NESC.

Connolly, E. (1999) 'The Republic of Ireland's Equality Contract: Women and Public Policy', in Y. Galligan, E. Ward and R. Wilford (eds), *Contesting Politics: Women in Ireland, North and South.* USA: Westview/PSAI, pp. 74–89.

Cuddy, M. (1992) 'Rural Development: The Broader Context', in M. Ó Cinnéide and M. Cuddy (eds), *Perspectives on Rural Development in Advanced Economies.* University College Galway: Social Sciences Research Centre, pp. 65–77.

Delphy, C. and D. Leonard (1992) *Familiar Exploitation*. Cambridge: Polity Press.

Department of Agriculture, Food and Rural Development (2000) *Report of the Advisory Committee on the Role of Women in Agriculture*. Dublin: Department of Agriculture, Food and Rural Development.

Department of Justice, Equality and Law Reform (2003) *Assessment of the Main Gaps in Existing Information on Women in Agriculture*. Dublin: Department of Justice, Equality and Law Reform.

Duggan, C. (1987) 'Farming Women or Farmers' Wives? Women in the Farming Press', in C. Curtin, P. Jackson and B. O'Connor (eds), *Gender in Irish Society*. Galway: Galway University Press, pp. 54–70.

Fahey, T. (1990) 'Measuring the Female Labour Supply: Conceptual and Procedural Problems in Irish Official Statistics', *Economic and Social Review*, 21, 2: 163–91.

Gasson, R. (1992) 'Farmers' Wives: Their Contribution to the Farm Business', *Journal of Agricultural Economics*, 43: 74–87.

Gorman, M. (2006) 'Gender Relations and Livelihood Strategies', in B. Bock and S. Shortall (eds), *Rural Gender Relations: Issues and Case Studies*. Oxfordshire: CAB International, pp. 27–47.

Hannan, D. (1972) 'Kinship, Neighbourhood and Social Change in Irish Rural Communities', *The Economic and Social Review*, 3, 2: 163–89.

Hannan, D. F. and L. A. Katsiaouni (1977) *Traditional Families? From Culturally Prescribed to Negotiated Roles in Farm Families*. Dublin: Economic and Social Research Institute.

Hanrahan, S. (2006) *Gender Relations and Women's Off-farm Employment: A Critical Analysis of Discourses. Women in Agriculture*. Dublin: Teagasc.

Heenan, D. and D. Birrell (1997) 'Farm Wives in Northern Ireland and the Gendered Division of Labour', in A. Byrne and M. Leonard (eds), *Women and Irish Society: A Sociological Reader*. Belfast: Beyond the Pale Publications, pp. 377–94.

Higgins, J. (1983) *A Study of Part-time Farmers in the Republic of Ireland*. Dublin: An Foras Talúntais.

Hoggart, K. (2004) 'Structures, Cultures, Personalities, Places, Policies: Frameworks for Uneven Development', in H. Buller and K. Hoggart (eds), *Women in the European Countryside*. Aldershot: Ashgate, pp. 1–14.

Jessop, B. (1994) 'Post-Fordism and the State', in A. Amin (ed.), *Post-Fordism: A Reader*. Cambridge: Blackwell, pp. 251–79.

Kelly, R. and S. Shortall (2002) 'Farmers' Wives': Women who are Off-farm Breadwinners and the Implications for On-farm Gender Relations', *Journal of Sociology*, 38, 4: 327–43.

Kennedy, L. (1991) 'Farm Succession in Modern Ireland: Elements of a Theory of Inheritance', *Economic History Review*, XLIV, 3: 477–99.

Kirby, P. (2002) 'Contested Pedigrees of the Celtic Tiger', in P. Kirby and M. Cronin (eds), *Reinventing Ireland: Culture, Society and the Global Economy*. London: Pluto Press, pp. 21–37.

Layte, R. (1998) 'Gendered Equity: Comparing Explanations of Women's Satisfaction with the Domestic Division of Labour', *Work, Employment and Society*, 12, 3: 511–32.

Little, J. and O. Jones (2000) 'Masculinity, Gender and Rural Policy', *Rural Sociology*, 65, 4: 621–39.

Little, J. and R. Panelli (2003) 'Gender Research in Rural Geography', *Gender, Place and Culture*, 10, 3: 281–9.

Martin, A. (1997) 'The Practice of Identity and an Irish Sense of Place', *Gender, Place and Culture*, 5: 277–300.

McGauran, A. M. (2005) *Plus ca change? Gender Mainstreaming of the Irish National Development Plan.* Dublin: The Policy Institute, TCD.

McNabb, P. (1964) 'Social Structure', in J. Newman (ed.), *The Limerick Rural Survey 1958–1964.* Ireland: Muintir na Tire Rural Publications, pp. 193–247.

McNerney, C. and D. Gillmor (2005) 'Experiences and Perceptions of Rural Women in the Republic of Ireland: Studies in the Border Region', *Irish Geography*, 38, 1: 44–56.

Messenger, J. C. (1969) *Inis Beag – Isle of Ireland.* New York: Holt, Rinehart and Winston.

National Women's Council of Ireland (2002) *Irish Politics – Jobs for the Boys! Recommendations on Increasing the Number of Women in Decision-making.* Dublin: NWCI.

Ní Laoire, C. (2001) 'A Matter of Life and Death? Men, Masculinities and Staying behind in Rural Ireland', *Sociologia Ruralis*, 41, 2: 220–36.

Ní Laoire, C. (2002) 'Young Farmers, Masculinities and Change in Rural Ireland', *Irish Geography*, 35, 1: 16–27.

Ní Laoire, C. (2005) '"You're not a Man at all!" Masculinity, Responsibility, and Staying on the Land in Contemporary Ireland', *Irish Journal of Sociology*, 14, 2: 94–114.

The North Leitrim Men's Group (2001) *Men in North Leitrim.* Published by the North Leitrim's Men's Group.

O'Connor, P. (1995) 'Tourism and Development: Women's Business?' *The Economic and Social Review*, 26, 4: 369–401.

O'Hara, P. (1994) 'Out of the Shadows: Women and Family Farms and their Contribution to Agriculture and Rural Development', in M. Van de Burg and M. Endeveld (eds), *Women on Family Farms: Gender Research, EC Policies and New Perspectives.* Wageningen: Wageningen University Holland, pp. 39–49.

O'Hara, P. (1997) 'Interfering Women: Farm Families and the Reproduction of Family Farming', *Economic and Social Review*, 28, 2: 135–56.

O'Hara, P. (1998) *Partners in Production, Women, Farm and Family in Ireland.* New York and Oxford: Berghahn.

O'Malley, E. (1992) *The Pilot Programme for Integrated Rural Development 1988–1990.* Dublin: The Economic and Social Research Institute.

Oakley, A. (1974) *The Sociology of Housework.* Oxford: Martin Robertson.

Oakley, A. (1998) 'A Brief History of Gender', in A. Oakley and J. Mitchell (eds), *Who's Afraid of Feminism: Seeing through the Backlash.* Harmondsworth: Penguin, pp. 22–55.

Oldrup, H. (1999) 'Women Working off the Farm: Reconstructing Gender Identity in Danish Agriculture', *Sociologia Ruralis*, 39, 3: 343–58.

Owens, M. (1992) 'Women in Rural Development: A Hit or Miss Affair', *Women's Studies Centre Review*, 1: 15–19.

Owens, M. and A. Byrne (1996) 'Family, Work and Community – Rural Women's Lives', *Women's Studies Centre Review*, 4: 77–94.

Rankin, K. (2001) 'Governing Development: Neoliberalism, Microcredit, and Rational Economic Woman', *Economy and Society*, 30, 1: 18–37.

Reimer, B. (1986) 'Women as Farm Labour', *Rural Sociology*, 51, 2: 143–55.

Shortall, S. (1992) 'Power Analysis and Farm Wives and Power – An Empirical Study of the Power Relationships Affecting Women on Irish Farms', *Sociologia Ruralis*, 32, 4: 431–51.

Shortall, S. (1994a) 'The Irish Rural Development Paradigm – an Exploratory Analysis', *The Economic and Social Review*, 25, 3: 233–61.

Shortall, S. (1994b) 'Farm Women's Groups: Feminist or Farming or Community Groups or New Social Movements?' *Sociology*, 28, 1: 279–91.

Shortall, S. (1996) 'Training to be Farmers or Wives? Agricultural Training for Women in Northern Ireland', *Sociologia Ruralis*, 36, 3: 269–85.

Shortall, S. (1999) *Women and Farming: Property and Power.* London/New York: Macmillan.

Shortall, S. (2002) 'Gendered Agricultural and Rural Restructuring: A Case Study of Northern Ireland', *Sociologia Ruralis*, 42, 2: 160–76.

Shortall, S. (2006) 'Economic Status and Gender Roles', in B. Bock and S. Shortall (eds), *Rural Gender Relations: Issues and Case Studies.* Oxfordshire: CAB International, pp. 303–17.

Shortall, S. and R. Kelly (2001) *Gender Proofing CAP Reforms.* Cookstown, Tyrone: Rural Community Network.

Silvasti, T. (1999) 'Farm Women, Women Farmers and Daughters-in-law – Women's Position in the Traditional Peasant Script in Finland'. Unpublished paper. Helsinki, University of Helsinki.

Viney, E. (1968) 'Women in Rural Ireland', *Christus Rex*, xxlx, 333–42.

Walby, S. (1990) *Theorizing Patriarchy.* Oxford: Blackwell.

Whatmore, S. (1991) *Farming Women: Gender, Work and Family Enterprise.* London: Macmillan.

Wheelock, J. and E. Oughton (1996) 'The Household as a Focus for Research', *Journal of Economic Issues*, 30, 1: 143–59.

Chapter 17

# The Irish Language and the Future of the Gaeltacht Regions of Ireland

Seosamh Mac Donnacha and Conchúr Ó Giollagáin

## Introduction

The Irish language planning process has been successful in nurturing positive attitudes towards the language among the population at large. It has also succeeded in developing an educational process which can replicate the reproduction of second language Irish-speakers on an intergenerational basis. What it has not succeeded in achieving is sustainable language planning outcomes – an increase in Irish language usage among the populace in general and, more importantly, intergenerational increases in the number of first language Irish-speakers, as a result of parents who have learned Irish as a second language and who opt to bring their own children up through the medium of Irish. Thus, despite a significant number of people living outside the traditional Irish-speaking (Gaeltacht) districts being able to speak Irish to various levels of competence, Irish has not taken hold again as a social and community language in any place outside of the Gaeltacht. Because of this, the future of Irish as a living community language is inextricably linked to the sustainability of Irish-speaking communities in the Gaeltacht regions of the Republic of Ireland (RoI).

## Irish in the Republic of Ireland

With the establishment of the Irish Free State in 1922, the Irish language was accorded the status of the 'national language' (with English being recognized as an 'official language') under Article 4 of the Constitution of the Irish Free State (Saorstát Éireann) Act, 1922. A new constitution, Bunreacht na hÉireann, enacted in 1937, declared that the 'Irish language as the national language is the first official language', and that English had the status of being 'a second official language'. But how much Irish was spoken? By 1926, Irish speakers in the Irish Free State numbered 543,511 (18 per cent), out of a total population of 2,971,992. A year before Bunreacht na hÉireann was enacted in 1937, the number of Irish speakers among those aged three years and over had increased to 666,601 (24 per cent) (Census 2002).

This divergence between the favoured constitutional status of Irish and its *de facto* position as the lesser-used language in the RoI is, perhaps, best explained in a reference from a 1934 court case in which Justice Ó Cinnéide, referring to the constitutional status of Irish in the 1922 constitution, stated:

> The declaration by the Constitution that the national language of Saorstát Éireann is the Irish language does not mean that the Irish language is, or was at that historical moment, universally spoken by the people of the Saorstát, which would be untrue in fact, but it did mean that it is the historic distinctive speech of the Irish people, that it is to rank as such in the nation and, by implication, that the State is bound to do everything within its sphere of action ... to establish and maintain it in its status as the national language and to recognize it for all official purposes as the national language (*Ó Foghludha v McClean* (1934) IR 469 68 ILTR 189 [1934]).

Subsequently, the Irish state did invest heavily in efforts to maintain Irish as a living language in those areas where it is still spoken, and to revive it in the rest of the state. In some respects, this investment in language planning has achieved positive outcomes in relation to language attitudes and language ability. A 1993 survey reported that most Irish people were favorably disposed towards the Irish language, mainly because they saw it as an integral element of their own identity as Irish people in an Irish nation (Ó Riagáin and Ó Gliasáin 1994). Language ability among the population, when defined as the number of people who can speak the language to some degree of competence, has increased steadily, from 18 per cent in 1926 to 42 per cent in 2002 (Census 2002). Of course, census figures relating to language ability – while useful for inter-census comparative purposes – can be somewhat misleading unless cognizance is taken of their composition and of the exact question being asked and answered on the census form.

The 2002 census reveals that the proportion of Irish speakers in the school-age cohorts peaks at 68 per cent in the 10–14 age group and drops slightly to 66 per cent in the 15–19 age group. It drops to 51 per cent in the 20–24 age group and is in the region of 31–39 per cent for all age groups older than 24 years. The census data relating to language usage suggests that 9 per cent of the population use Irish on a daily basis. Again, however, when the school age cohorts are eliminated, it reveals that less than 3 per cent of the population use Irish on a daily basis. Even among that portion of the population reported in the census as being able to speak Irish, the percentage reported as using Irish on a daily basis drops from 50 per cent in the school age cohorts to 7 per cent in the 20+ age group (Census 2002). This suggests that although the language planning process in the RoI is able to replicate the reproduction of second language Irish-speakers on an inter-generational basis, it is not succeeding in achieving sustainable outcomes. In the vast majority of cases, second language Irish speakers revert to their first language once their active engagement with the language during their school years ceases.

## Irish in Northern Ireland

Data from the 2001 Census of Northern Ireland show that 75,125 (4.6 per cent) of the population aged three years and over reported that they were able to 'speak, read, write and understand Irish', with a further 5.7 per cent having a more limited knowledge of the language (Northern Ireland Statistics and Research Agency 2007). While this data suggests that active support for Irish is very limited among the Northern Ireland population in general, cognizance has to be taken of the very different political context that exists in Northern Ireland and of the way in which this impacts on efforts to promote the sustainable development of the Irish language there.

While efforts to promote the survival of Irish in the Republic of Ireland are broadly supported among the community in general and among all shades of political opinion, the position of Irish in Northern Ireland is more complex. O'Reilly (1999) suggests that much of the debate on the Irish language in Northern Ireland has been shaped by three types of discourse, namely 'decolonizing discourse' in which Irish is seen as an inherent part of the political struggle; 'cultural discourse' which sees Irish as a cultural activity which should be kept separate from politics; and 'rights discourse' which tries to place the Irish language debate in the context of human rights and minority language rights. In addition, because of the political context that exists in Northern Ireland, Irish has not had the same level of Government and institutional support as has been the case in the Republic of Ireland, with the result that its continued promotion and survival has been more dependent on the efforts of committed individuals and community organizations and networks.

The establishment of the Shaw's Road Urban Gaeltacht in Belfast and its impact on other Irish language initiatives in the city in subsequent years is a case in point. Established in 1969 by a group of committed Irish speakers the Gaeltacht enclave initially consisted of a core of five young families (Nig Uidhir 2006, p. 138). However, its impact on the Irish language in Northern Ireland in subsequent years was much more significant than this initial beginning would suggest. Nig Uidhir (ibid., p. 142) says that 'the impact of the Shaw's Road Community on the fortunes of the language [in Northern Ireland] on a much larger scale has been realized in two ways: (a) the leadership and supportive role played by members of the Shaw's Road Community in a range of significant social and educational developments and (b) the generation of a growing strategic network of Irish-medium schools and relevant services and infrastructure that have been established indirectly or directly as an outcome of the pioneering work of the Shaw's Road Community in those areas'. Established by the Shaw's Road parents in 1971 to provide Irish-medium education for their own children, Bunscoil Phobal Feirste operated without Government support until 1984 and became the catalyst that led to the emergence of the Irish-medium education sector in Northern Ireland, which currently consists of 'around 65 educational sites across the three phases: nursery, primary and secondary, with around 3,300 children; 165 teachers and 35 nursery directors' (Comhairle na Gaelscolaíochta 2004, cited in Mac Corraidh 2006, p. 181).

As a result of the provision made for the Irish language in the Northern Ireland Peace Agreement, 1998, several bodies have been established in the intervening years with a statutory role in supporting Irish in Northern Ireland. These include Foras na Gaeilge and Comhairle na Gaelscolaíochta (Council for Irish-medium Education) established in 1999 and 2000 respectively. While these developments suggest that official support for the Irish language may be more forthcoming in the future, it remains to be seen whether such support will be at a level and delivered in a way that increases the likelihood that the Shaw's Road Gaeltacht will become a sustainable language community and whether similar initiatives can be successfully undertaken in other areas of Northern Ireland.

## An Ghaeltacht

### Defining the Gaeltacht

The first attempt to provide an 'official' definition of the geographical extent of the Gaeltacht and to estimate the number of Irish speakers within it was made by Coimisiún na Gaeltachta (Gaeltacht Commission) which was established by the State in 1925.

The Commission recommended that two categories of Gaeltacht be recognized: districts in which 80 per cent or more of the community could speak Irish should be regarded as a Fíor-Ghaeltacht (Irish-speaking district) and districts in which between 25 and 79 per cent of the community could speak Irish should be regarded as a Breac-Ghaeltacht (partly Irish-speaking district). The district electoral divisions (DEDs) were the spatial units the Commission used to define the Gaeltacht. The Commission reported that Fíor-Ghaeltacht districts were to be found in the counties of Donegal, Mayo, Galway, Clare, Kerry, Cork and Waterford. Breac-Ghaeltacht districts were also found in the above counties, as well as in the counties of Sligo and Tipperary (Coimisiún na Gaeltachta 1926).

In the years between 1926 and 1956 Government departments remained inconsistent in their definition of the Gaeltacht. The degree of inconsistency in the definitions of the Gaeltacht proposed during this period is clear from the spatial scope of the definitions used – the Housing (Gaeltacht) Act, 1929, includes a total of 658 DEDs (in 14 counties) in its definition of the Gaeltacht, while the School Meals (Gaeltacht) Act, 1930, limits its definition of the Gaeltacht to 122 DEDs in five counties.

This inconsistency was rectified in 1956 with the enactment of the Ministers and Secretaries (Amendment) Act, which established the Department of the Gaeltacht and included a provision which allowed the Minister for the Gaeltacht to place a parliamentary order before the Houses of the Oireachtas defining the extent of the Gaeltacht and to make changes to this definition if necessary. The first such order, the Gaeltacht Areas Order, 1956, designated 85 DEDs in full, and parts of a further 57 DEDs in the counties of Donegal, Mayo, Galway, Kerry, Cork

**Figure 17.1    The Gaeltacht as Currently Defined Under the Provisions of
the Ministers and Secretaries Act 1956**

and Waterford as being in the Gaeltacht. Since 1956, three further parliamentary
orders have been placed before the Houses of the Oireachtas which have extended
the boundaries of the Gaeltacht. The most significant of these was the 1967 order
which included the Gaeltacht colonies of Ráth Cairn and Baile Ghib in County
Meath as part of the officially designated Gaeltacht. No provision was made in the
1956 Act for distinguishing between Fíor-Ghaeltacht and Breac-Ghaeltacht areas
as recommended by Coimisiún na Gaeltachta's 1926 Report.

No changes have been made to the boundaries of the Gaeltacht since 1982,
and it is now estimated that the Gaeltacht as designated under the provisions of

the 1956 Act (see Figure 17.1) constitutes approximately 7 per cent of the state (Commins 1988). In the years subsequent to 1956, all new legislative enactments which have included a provision for the Gaeltacht have relied on the definition of the Gaeltacht as provided for under the provisions of the Ministers and Secretaries (Amendment) Act, 1956.

Despite these changes, it is generally accepted that the 1956 boundaries exaggerated the true size of the Gaeltacht as it then stood. Based on an analysis of earlier research on the Gaeltacht in the early 1970s (Mac Aodh 1971; Ó Riagáin 1971) and the Committee on Irish Language Attitudes Research (CILAR) Report (1975), Ó Riagáin (1982; 1997, p. 77) concluded that only:

> 30 per cent of Gaeltacht communities were predominantly Irish-speaking and stable and another 25 per cent were almost entirely English-speaking. The remainder were bilingual but were unstable and showing evidence of a shift towards English. It was only in the two largest Gaeltacht areas of Donegal and Galway that bilingual core areas of any significant size were to be found. Adjacent to these core areas was an intermediate zone of limited bilingualism. Beyond this zone and moving to the margins of the official Gaeltacht, was a more or less completely anglicized zone.

Subsequently, however, Ó Gliasáin's analysis of the Department of the Gaeltacht's data for the period 1974–1984 in relation to the number of children qualifying as Irish-speaking children, under one of its language support schemes, suggested that the process of language shift occurring in the Gaeltacht was such that 'even the core areas were becoming unstable' (Ó Gliasáin 1990, p. 74).

*Language shift in the Gaeltacht*

CILAR's (1975) analysis of the language shift occurring in the Gaeltacht suggested 'that this was a trend beginning with the introduction of English in the form of new contexts or "neo-events" (i.e. non-traditional contexts or persons for which no established associations exist within the local community) and with its progressive intrusion into the more traditional contexts' (p. 267). English was being introduced in this way by 'new social actors, new participants in Gaeltacht life or new types of transactions' (ibid., p. 348). In addition, CILAR's social network analysis of particular Gaeltacht areas suggested that the 'domains in which Irish is not used generally involve the presence of persons who are already "mapped" in terms of competence, linguistic repertoire, language using habits, attitude and commitment and adjudged to be not disposed to use Irish' (ibid., p. 350).

Against this background CILAR suggested (ibid., p. 254) that the sustainability of the Gaeltacht as a bilingual or monolingual Irish language entity was dependent not only on 'the transmission of competence to use the language, i.e., the production of new native speakers in the home, but the transmission of the propensity for use of the language as well'. CILAR further maintained that the sustainability of the

Gaeltacht as a bilingual entity was dependent largely on 'maintaining any existing stability in diglossic usage in the face of (a) new institutional penetrations (e.g., media) from outside the traditional Gaeltacht social system, or (b) the erosion of the traditional Gaeltacht communities as they adopt the economic and social characteristics of the commercial economy (e.g. by migration)' (ibid., p. 257). In view of this they argued that a key issue was the question of 'how the "external" agencies, personnel, or institutions articulate with the traditional Gaeltacht community, particularly whether they are actively supportive of the language, passively acquiescent in existing linguistic trends, represent an obstacle to the maintenance of Irish, or deliberately hasten its demise' (ibid., p. 258).

Broadly speaking a sociolinguistic study of the Corca Dhuibhne Gaeltacht in Co. Kerry in 1983 (Ó Riagáin 1997, pp. 79–142), supports the CILAR thesis. Ó Riagáin's study gives us a more in-depth picture of the dynamics surrounding language shift in the Gaeltacht, and, in particular, of the link between language shift and the type of interactions that happen *within* local networks and *between* local networks and regional and national networks. Ó Riagáin studied the degree to which small Gaeltacht communities are linked through various social, economic, educational and emigration networks with the broader regional, national and international (and mainly English-speaking) world. He concluded that these:

> ...interaction systems ... shifted very considerably in the period since 1960, due to changes in demographic, occupational, educational, retailing and car ownership patterns. Up to this point, social networks tended to be localized. Since 1960, they have become increasingly more extensive and differentiated. Formerly, in rural areas work, school, shop, and church networks tended to coincide and operate within the same locality. Nowadays, people may only reside in rural areas, but work, attend school, and shop elsewhere (1997, p. 141).

A key element in these interactions is the impact of migration flows on the stability of the bilingual nature of the Gaeltacht:

> Even in periods of high emigration there was always a certain, though limited, movement into rural areas and some former emigrants returned. As population trends assume more favourable patterns, the balance between these different kinds of movements can be expected to change. Out-migration becomes less pronounced and, for a variety of reasons, in-migration and return migration become more significant. These population movements, and the many variations possible within each type, can carry very considerable implications for language patterns. For example ... out-migrants may leave in disproportionate numbers from Irish-speaking areas, in-migrants may not be fluent Irish speakers; and returned migrants may have married English-speaking spouses or may otherwise have changed their attitude towards Irish (Ó Riagáin 1997, pp. 116–17).

Ó Riagáin's own survey of the Corca Dhuibhne Gaeltacht in 1983 reported that 15 per cent of respondents could be classified as in-migrants. These were reported to be 'younger and more likely to be women than men. As they are also more likely (80 per cent) to be married, it would seem that a significant proportion of in-migrants are women marrying into Corca Dhuibhne families from outside' (ibid., p. 117). He also noted (p. 119) 'evidence of a connection between in-migration and return migration'; in all, 'one quarter of returned migrants married while outside of Corca Dhuibhne', suggesting that 'an increase in return migration will almost certainly imply an increase in in-migration'. Ó Riagáin also found that 'although in-migrants and internal migrants form relatively small proportions of the total sample, they form a large proportion of married respondents. Thus their influence on the language composition of household networks is potentially considerable and growing' (ibid., p. 119).

In his study of the Gaeltacht, Hindley (1990) concluded that by the early 1980s less than ten thousand 'native Irish speakers [were] living in communities with sufficient attachment to Irish to transmit it to a substantial majority of their children as the language of home and community. They alone are living in circumstances in which continued transmission seems possible or even probable in the light of experience' (p. 251). Hindley also found that the major causes of decline were related 'primarily to economic forces which have promoted the modernization of the Gaeltacht economy and the mobilization of its people, involving them intimately in much wider and constant social and economic relationships' (ibid., p. 248).

These interactions and their impact on language usage patterns at a community, household and individual level were studied by Ó Giollagáin who presented his results in two reports (Ó Giollagáin 2002 and Ó Giollagáin 2005). The first of these considered the linguistic composition of the Ráth Cairn Gaeltacht in Co. Meath and the second looks at the community of Ros Muc in the Gaeltacht of Co. Galway. The Ráth Cairn Gaeltacht is the more unusual of the two communities. It was established in 1935, when over a period of two years a total of 40 families (333 persons in all) were transferred, under the auspices of the Land Commission, from Gaeltacht districts in Co. Galway to land that had been acquired by the Land Commission in Co. Meath.

Ó Giollagáin's research in Ráth Cairn, which was conducted in 2001–2002, looks at the linguistic composition of the community, at both the individual and household levels. He illustrates how the language dynamic of the Ráth Cairn community is evenly balanced on either end of the spectrum, with 33 per cent of the community being native speakers of Irish and a similar percentage being English speakers. A further 25 per cent were categorized as second-language speakers, with the rest being semi-native speakers or learners of Irish. More importantly, Ó Giollagáin's analysis of language competence among the younger age groups suggests that only 38 per cent of those in the pre-school and primary school age groups were acquiring competence in Irish through the medium of the home with the other 62 per cent being dependent on institutional support – primarily the education system – to acquire such competence. When this analysis was replicated at the household level it was revealed that while 47 per cent of the total number of households were Irish-speaking, the number of Irish-speaking households containing young families (19 per cent) was much less than the number of English-speaking households containing young families (34 per cent).

Ó Giollagáin's analysis of these younger families suggests that where the parents are both native speakers of Irish the resulting offspring are usually brought up through the medium of Irish. This is a very positive conclusion, and indicates that all other things being equal, Gaeltacht parents are predisposed to speaking Irish as a primary home language to their own children. However, in the Ráth Cairn context it can only be viewed as a minor contribution to the sustainability of the Irish-speaking community as only 9 per cent of families belonged to this category. A similar percentage of Irish-speaking families in the Ráth Cairn Gaeltacht come from a non-Gaeltacht background (i.e. parents who have learned Irish as a second language who have decided to bring their children up through the medium of Irish). In younger families where a female native-speaker was married to a male from outside Ráth Cairn some effort was made to create semi-native speakers in 75 per cent of cases, whereas when the roles were reversed i.e. where the native speaker in a linguistically mixed family was male, no such instances were found.

Ó Giollagáin's initial analysis of the Irish-speaking community of Ros Muc, which lies 35 miles west of Galway City, indicates that the community is in a much stronger position *vis-à-vis* the language than is so in Ráth Cairn. In Ros Muc 93 per cent of the community were regarded as having a high level of competence in Irish and 82 per cent of them were native Irish speakers. However, when the figures are analysed by age group they clearly reveal that English is quickly becoming the home language of a significant percentage of young families, with 44 per cent of preschool children in the area being categorized as 'English speakers'. Similarly, when analysed at the household level it was found that, overall, 82 per cent of households were Irish-speaking and 18 per cent were English-speaking. However, the ratio of English-speaking to Irish-speaking households increases significantly among the younger families in the area. In addition, Ros Muc faces further challenges due to its peripheral location and falling population. Ó Giollagáin's research clearly shows that the population of the area, which fell from 1,452 in 1911 to 461 in 2002, is still under pressure, with the younger age groups being under-represented in the population as a whole.

Ó Giollagáin found, as was the case in Ráth Cairn, that in the vast majority of Ros Muc families, where both parents were 'native Irish speakers' the resulting offspring are usually brought up through the medium of Irish; but that in cases where only one parent was a native Irish speaker the language of the household tends to be English. This suggests that the introduction of any complexity into the linguistic mix of Gaeltacht households reduces considerably the likelihood that the next generation of children will be Irish-speaking.

Mac Donnacha et al. (2005), in a baseline study of primary and second level schools in the Gaeltacht, provide data on the language competence of children attending Gaeltacht schools. In order to allow for the fact that the Gaeltacht as currently defined contains some areas that are now mainly English-speaking and other areas that are still mainly Irish-speaking, schools were grouped into three categories as follows, using data from Census 2002:

- Category A: Schools located in areas in which 70 per cent+ of the population speak Irish on a daily basis. Some 30 per cent of the schools and 28 per cent of the pupils surveyed fell into this category.
- Category B: Schools located in areas in which between 40 per cent and 69 per cent of the population speak Irish on a daily basis. Some 16 per cent of the schools and 16 per cent of the pupils surveyed fell into this category.
- Category C: Schools located in areas in which less than 40 per cent of the population speak Irish on a daily basis. Some 53 per cent of the schools and 56 per cent of the pupils surveyed fell into this category.

On this basis the analysis of the language competence and usage patterns of Gaeltacht school children provides a clearer picture of the degree to which the official boundaries of the Gaeltacht have been overestimated, and suggests that the patterns of language shift reported by Ó Giollagáin (2002 and 2005) for Ráth Cairn and Ros Muc are being replicated in all of the still mainly Irish-speaking Gaeltacht communities. If we take the percentage of school children currently attending a Category C school, as defined above, to be an approximate indication of the percentage of the Gaeltacht children currently living in communities which are mainly English-speaking, it is clear that well over half of the current Gaeltacht population live in areas which are little different from the rest of the country in linguistic terms.

What the research shows here is that of the children attending junior infants classes in these schools, only 2 per cent were reported as having 'Gaeilge líofa' (Fluent Irish). Even more worrying, however, was the data relating to the language competence of junior infant pupils in Category A schools, i.e. the schools located in the linguistically strongest Gaeltacht communities. These figures showed that only 43 per cent of these pupils had 'Gaeilge líofa'. These figures suggest that less than half of the households in such communities were now using Irish as a home language, thus supporting Ó Gliasáin's (1990) thesis that the core areas of the Gaeltacht have become unstable and that the language shift to English has made considerable progress even in the strongest Irish-speaking communities.

Mac Donnacha et al. (2005) also provide data on the number of school children attending Gaeltacht schools who were born outside the Gaeltacht or who lived outside the Gaeltacht for a period of time before attending school in the Gaeltacht – 26 per cent in the case of primary pupils (p. 28) and 23 per cent in the case of second level pupils (p. 78). The finding here supports Ó Riagáin's (1997) thesis, mentioned above, that in-migrants are likely to have a considerable influence on the language composition of younger households. In addition 18 per cent of the students currently attending second-level schools in the Gaeltacht live in English-speaking areas outside the Gaeltacht (ibid., p. 79). As these pupils receive their primary education through the medium of English, it becomes very unlikely that the second level schools in question could operate effectively through the medium of Irish. The effect of this mixed linguistic intake of pupils on language usage patterns in Gaeltacht schools is clear, with only 46 per cent of primary schools and 26 per cent of secondary schools reporting that their pupils used 'more Irish than English' in their everyday social interactions while in the environs of the school (Mac Donnacha et al. 2005, pp. 46 and 97).

Harris (1984) and Harris and Murtagh (1987, 1988) (discussed in Harris 2006, pp. 8–9) found evidence of a 'distinctive improvement in the Irish achievement of Gaeltacht pupils between second grade and sixth grade' (Harris 2006, p. 9). Nonetheless, Harris (2006) in a major study of the level of achievement of pupils in the Irish language in ordinary, all-Irish and Gaeltacht schools found that pupils in all-Irish schools consistently outperformed pupils in Gaeltacht schools on the relevant standardized listening, reading, and speaking tests. Mac Donnacha et al. (2005, p. 124) concluded that although their study also indicated that the school system in the Gaeltacht plays a pivotal role in providing the institutional support necessary to assist children coming from a non-Irish speaking background to achieve a significant increase in their general level of competence in Irish, they were not succeeding in transmitting, what CILAR (1975) refers to as 'the propensity for use of the language'.

What this most recent research suggests is that the current generation of Gaeltacht children participating in the education process are entering a school-based socialization process that is predisposed towards the use of English. The effect of this socialization process is likely to be reinforced by a reliance on English as a language necessary to achieve academic success, as a result of the scarcity of Irish-medium teaching resources at both education levels. Such a problem is particularly acute at Leaving Certificate level where it was reported that Irish language textbooks were used in only 18 per cent of Leaving Certificate class sessions in the second level schools situated in the strongest remaining Gaeltacht areas (Mac Donnacha et al. 2005, p. 109).

This analysis suggests that even when Gaeltacht parents opt to bring their children up through the medium of Irish, the likelihood is that once these children enter the school system they are entering a socialization process which operates partly in English at primary level and mainly in English at second level. Thus as young adults they are more likely to be predisposed to speaking English within the social networks comprised mainly of their school acquaintances, even if Irish remains their primary home language. One can only conclude from this that the Gaeltacht education system itself plays a significant role in facilitating the process of language shift towards English.

## Conclusion

Although the current official definition of the Gaeltacht, as prescribed under the provisions of the Ministers and Secretaries (Amendment) Act, 1956, represents a more realistic picture of the true geographical extent of the Gaeltacht than that proposed by the Gaeltacht Commission of 1926, it is clear from the research carried out since the early 1970s that the boundaries as currently set out still seriously overestimate the overall size of the Gaeltacht. It is also clear from the work of CILAR (1975), Ó Gliasáin (1990), Ó Riagáin (1997), Ó Giollagáin (2002; 2005) and Mac Donnacha et al. (2005) that the process of language shift from Irish to English has continued in Gaeltacht areas even after the current Gaeltacht boundaries were established in 1956, and that this process now threatens the future sustainability of the core surviving Gaeltacht communities.

While the process of language shift in previous centuries may have been driven by the politics and economics of the era, it is clear that the process in recent decades has been driven by the introduction of 'new social actors, new participants in Gaeltacht life [and] new types of transactions' (CILAR 1975, p. 348) as well as changes in the range and frequency of the interactions that Gaeltacht communities have with other regional, national and international and mainly English-speaking networks (Ó Riagáin 1997). There has been an increase in the number of in-migrants through marriage and as a result of Gaeltacht migrants returning home and bringing with them English-speaking family members. The number of English speakers living in Gaeltacht areas contiguous to larger urban areas (such as Galway city) has increased as a result of the overflow of the urban English-speaking population into the rural Gaeltacht hinterland. Gaeltacht areas in more scenic rural districts have witnessed an increase in recent years in the number of people buying short- and long-term holiday homes in their area. And the economic and industrial development that has taken place in the Gaeltacht over the last three decades has led to an increase in the number of English speakers moving into the Gaeltacht to take up employment opportunities available to them there. The interactions between these factors, by altering the linguistic composition of the Gaeltacht population, have helped destabilize the position of Irish as the main community language of the Gaeltacht.

An exacerbating factor in the changing linguistic composition of the Gaeltacht is the fact that English-speaking in-migrants tend to form a larger proportion of younger families and households (Ó Riagáin 1997; Ó Giollagáin 2002, 2005). Mac Donnacha et al. (2005) have shown that *circa* a quarter of the current generation of Gaeltacht school children come from a non-Gaeltacht background and the likelihood is that the percentage of such in-migrants and return migrants in the Gaeltacht population in general is substantial, given the accumulation of such persons as a proportion of the population over several generations. The research of Ó Giollagáin (2002, 2005) and Mac Donnacha et al. (2005) suggests that where English-speakers participate in the normal socialization processes of Gaeltacht society, through the school system or through inter-marriage with native Irish speakers, the usual outcome is an increase in the propensity to use English as the language of normal communication.

Although this increased propensity for the use of English emanates chiefly from changes in the linguistic composition of the community, it could also be argued that it is being encouraged, however unintentionally, by the education system in the Gaeltacht which, to paraphrase CILAR (1975, p. 258), is 'passively acquiescent [in supporting] existing linguistic trends'. Encouragement also comes from an educational policy that ignores the linguistic socialization process that occurs as part of the overall education process, and which, when left to its own devices, always favours the dominant language in a society. Thus, as English-speaking children enter the Gaeltacht school system and the socialization process therein, their effect on language norms within schools is such that native Irish speakers, both children and teachers, tend to switch to English to accommodate them. In this school setting native speakers of Irish encounter a peer-group socialization process which is dominated by the use of English. Thus, young people living in the Gaeltacht of today, even those whose home language is Irish, are reliant to a predominant degree on English as their normal language of communication with their peers. To this can be added their reliance on English to access the main sources of text-based educational knowledge available to them (i.e., the English-medium textbooks predominantly in use in Gaeltacht schools), and the constant presence of English in their normal everyday life through their interactions with modern broadcasting media, the internet and other computer-based activities.

The consequence of all this is that the use of the Irish language by young people in the Gaeltacht is becoming increasingly restricted to a limited number of social settings. Young people's use of Irish differs from that of their parents' generation in that it is largely restricted to the home and to the formal aspects of school settings, whilst English tends to dominate as the medium of communication in social networks which the young have established in their own age group. In other words, they increasingly associate the use of Irish as a social medium in settings where the communication is established and maintained by the presence of adults (Ó Giollagáin et al. 2007). This conclusion has very serious implications for the future sustainability of the Gaeltacht as a linguistic entity, as young adults who use English as their primary language of interaction with their peers are unlikely, except in exceptional cases, to switch to Irish as their primary language of communication with their partners and children on becoming young parents.

Clearly the patterns revealed by all the analysis calls for major changes in the state's language planning policy and strategy for the Gaeltacht, which has heretofore focused mainly on the economic and industrial development aspects of Gaeltacht life, with only a limited number of language maintenance initiatives (Ó Cinnéide et al. 2001, pp. 139–55). The state has taken some tentative steps in more recent years which have been designed to provide support and secure the position of Irish in the Gaeltacht. These include the establishment of An Chomhairle um Oideachas Gaeltachta agus Gaelscolaíochta in 2002, which has a broad advisory and support brief in relation to Gaeltacht and Irish-medium education; Section 10 (2) (m) of the Planning and Development Act, 2000 which allows local authorities to take the Irish language and the linguistic stability of the Gaeltacht into account in deciding on planning applications which refer to Gaeltacht areas; and the passing of the Official Languages Act in 2003.

These developments clearly indicate that state agencies are taking the threat to the future sustainability of the Gaeltacht seriously. However, as yet no clear strategy has emerged which is designed to assist Gaeltacht children who have been raised through English to socialize with the Irish speakers of their peer group in settings which are designed to increase the 'propensity for the use of Irish' among them, and so counter the normal linguistic processes which lead bilingual native Irish speakers to switch to English to accommodate (initially) monolingual English speakers. In addition, no statutory arrangements have been made that would allow state agencies to differentiate between the different types of linguistic community currently contained within the official boundaries of the Gaeltacht and to adopt their language planning and economic development strategies accordingly.*

It is evident that basic questions remain as to whether an overall strategy for the Gaeltacht exists, and if so who is responsible for it. This weakness in the overall planning process for the Gaeltacht was identified by Ó Cinnéide et al. (2001, p. 141), and before that by CILAR (1975), who suggested that the lack of a co-ordinated and planned approach among the many state agencies operating in the Gaeltacht resulted in 'many of the agencies of central and local government or of other statutory-linked functions [being] associated with [an] anglicizing impact on the Gaeltacht' (CILAR 1975, p. 349). CILAR (1975) and Ó Cinnéide et al. (2001), also refer to the fact that no one agency has overarching responsibility for language planning and maintenance issues in the Gaeltacht; in particular CILAR clearly stated its view that:

> without a greater degree of unified executive autonomy within the Gaeltacht areas, the frequent expressions of well-meaning concern for Irish-speaking communities on the part of individual administrative bodies are unlikely to result in a co-ordinated strategy to conserve Irish-using populations. Only an Authority with powers to direct the activities of the most relevant departments of state within the Gaeltacht can hope to offer a real solution to the problem (1975, p. 354).

If and when such an Authority is established, or such authority is given to one of the existing government agencies with responsibility for the Gaeltacht, it will undoubtedly face a daunting challenge.* However, there seems to be no alternative to this new departure because the existing approaches, however well intended, are failing to stem the ongoing language shift from Irish to English.

---

\*    At the time of going to press a Cabinet Committee, under the chairmanship of An Taoiseach, Brian Cowen T.D., is considering these issues in response to the Comprehensive Sociolinguistic Study of the Use of Irish in the Gaeltacht prepared by Ó Giollagáin et al. 2007.

## References

Census of Population (2002) *The Irish Language. Volume 11.* Dublin: Central Statistics Office.

CILAR (Committee on Irish Language Attitudes Research (1975) *Report.* Dublin: Government Publications Office.

Coimisiún na Gaeltachta (1926) *Report.* Dublin: Government Publications Office.

Comhairle na Gaelscolaíochta (2004) *Bunachar Sonraí Gaelscoileanna: 2004-2004.* Belfast: Comhairle na Gaelscolaíochta.

Commins, P. (1988) 'Socioeconomic Development and Language Maintenance in the Gaeltacht'. *International Journal of the Sociology of Language*, 70: 11–28.

Harris, J. (1984) *Spoken Irish in Primary Schools.* Dublin: Institiúid Teangeolaíochta Éireann.

Harris, J. (2006) *Irish in Primary Schools: Long-Term National Trends in Achievement.* Dublin: Department of Education and Science.

Harris, J. and L. Murtagh (1987) 'Irish and English in Gaeltacht Primary Schools', in G. Mac Eoin, A. Ahlqvist, and D. Ó hAodha (eds), *Third International Conference on Minority Languages: Celtic Papers.* Clevedon: Multilingual Matters, pp. 104–24.

Harris, J. and L. Murtagh (1988) *Ability and Communication in Learning Irish.* Unpublished Report. Dublin: Institiúid Teangeolaíochta Éireann.

Hindley, R. (1990) *The Death of the Irish Language: A Qualified Obituary.* London and New York: Routledge.

Mac Aodh, B. (1971) *Galway Gaeltacht Survey.* Galway: Social Sciences Research Centre, University College Galway.

Mac Corraidh, S. (2006) 'Irish-medium Education in Belfast', in F. de Brún (ed.), *Belfast and the Irish Language.* Dublin: Four Courts Press, pp. 177–83.

Mac Donnacha, S., F. Ní Chualáin, A. Ní Shéaghdha and T. Ní Mhainín (2005) *Staid Reatha na Scoileanna Gaeltachta.* Dublin: An Chomhairle um Oideachas Gaeltachta agus Gaelscolaíochta.

Nig Uidhir, G. (2006) 'The Shaw's Road Urban Gaeltacht: Role and Impact', in F. de Brún (ed.), *Belfast and the Irish Language.* Dublin: Four Courts Press, pp. 136–46.

Northern Ireland Statistics and Research Agency (2007) *Northern Ireland Census Access: Cultural Profile for Northern Ireland.* Internet Address: http://www.nicensus2001.gov.uk/nica/browser/profile.jsp?profile=Cultural&mainArea=&mainLevel= (17/02/07).

Ó Cinnéide, M., S. Mac Donnacha and S. Ní Chonghaile (2001) *Polasaithe agus Cleachtais Eagraíochtaí Éagsúla le Feidhm sa Ghaeltacht.* Dublin: An Roinn Gnóthaí Pobail Tuaithe agus Gaeltachta.

Ó Giollagáin, C. (2002) 'Scagadh ar rannú cainteoirí comhaimseartha Gaeltachta: gnéithe d'antraipeolaíocht teangeolaíochta phobal Ráth Cairn', *The Irish Journal of Anthropology*, 6: 25–56.

Ó Giollagáin, C. (2005) 'Gnéithe d'antraipeolaíocht theangeolaíoch Phobal Ros Muc, Co. na Gaillimhe', in J. Kirk and D. Ó Baoill (eds), *Legislation, Literature and Sociolinguistics: Northern Ireland, the Republic of Ireland, and Scotland*. Béal Feirste: Cló Ollscoil na Banríona, pp. 138–62.

Ó Giollagáin, C., S. Mac Donnacha, F. Ní Chualáin and A. Ní Shéaghdha (2007) *Staidéar Teangeolaíoch ar Úsáid na Gaeilge sa nGaeltacht/Comprehensive Linguistic Study of the Use of Irish in the Gaeltacht*. Dublin: An Roinn Gnóthaí Pobail, Tuaithe agus Gaeltachta.

Ó Gliasáin, M. (1990) *Language Shift among Schoolchildren in Gaeltacht Areas 1974–1984: An Analysis of the Distribution of £10 Grant Qualifiers*. Dublin: Institiúid Teangelaíochta Éireann.

O'Reilly, C. C. (1999) *The Irish Language in Northern Ireland: The Politics of Culture and Identity*. New York: Palgrave.

Ó Riagáin, P. (1971) *The Gaeltacht Studies: A Development Plan for the Gaeltacht*. Dublin: An Foras Forbartha.

Ó Riagáin, P. (1982) 'Athrú agus Buanú Teanga sa Ghaeltacht', *Taighde Sochtheangeolaíochta agus Teangeolaíochta sa Ghaeltacht*. Dublin: Institiúid Teangeolaíochta Éireann, pp. 3–28.

Ó Riagáin, P. (1997) *Language Policy and Social Reproduction: Ireland 1893–1993*. Oxford: Clarendon Press.

Ó Riagáin, P. and M. Ó Gliasáin (1994) *National Survey on Languages 1993: Preliminary Report*. Dublin: Institiúid Teangeolaíochta Éireann.

### Acts of the Oireachtas and other Statutory Instruments

Constitution of the Irish Free State (Saor Stát Éireann) Act, 1922.
Gaeltacht Areas Order, 1956.
Housing (Gaeltacht) Act, 1929.
Ministers and Secretaries (Amendment) Act, 1956.
Official Languages Act, 2003.
School Meals (Gaeltacht) Act, 1930.
*Source: www.irishstatutebook.ie*

# PART V
# Sustainability and Civil Society

# Chapter 18

# Environmental Movements in Ireland: North and South

John Barry and Peter Doran

## Introduction

The island of Ireland is marked as much by peripherality as by its self-consciously marketed image as the 'Emerald Isle' of rolling green hills and valleys, rugged coastlines, misty topped mountains and wild bogs and wetlands. Whether peripherality is interpreted literally in the sense of Ireland being to the west of mainland Europe, or politically, culturally and historically in terms of its colonial experience and post-colonial legacies which continue to shape and mark it, the island of Ireland has, until relatively recently, neglected environmental and sustainable development issues. This is particularly so in the case of the Irish countryside where the politics of sustainable development can cover a wide range of positions. Farmers and rural communities can be viewed, for instance, as being ecological stewards of the land and as being the biggest threat to sustainable development in the countryside.

In this chapter we are particularly keen to explore the thesis that there is a marked difference in style, intent and strategic political action between a focus on the 'environment' and 'environmental protection' as opposed to 'sustainable development'. We will offer the view that adopting an explicit sustainable development focus to analysing the environmental movement provides a much more challenging and potentially radical political discourse and guide to action than does an orthodox 'environmental' focus – typically expressed through the domesticated prism of 'environmental protection' – though the two of course can overlap.

The 'triple bottom line' conception of sustainable development that we will draw on concerns itself with the ecological, economic and social dimensions of development. In keeping with this conception one cannot understand the green/environmental movement (distinguished below, but combined together here) without understanding the political economy of *unsustainable* development and the pursuit of orthodox economic growth. It is this that has caused environmental degradation, social inequality and decreasing levels of economic (and energy) security.

We will further contend that the 'path dependence' of the island cannot be discussed and analysed without incorporating its colonial and post-colonial legacy and dynamics. It is for this reason that the localized campaigns that have typified the Irish environmental movement's myriad of mobilizations against specific state-backed industrial and infrastructural projects are best explained by using an 'environmental justice' and social movement framework. Such a framework, as we will see, is quite different from the Eurocentric 'post-materialist' and middle class-centred explanation often given to account for the appearance of environmental campaigning and movement formation.

## The environmental movement in Ireland

Although precise figures are not available, the Sustainable Ireland Sustainability Directory lists over 1,100 groups, campaigns and organizations in the Republic of Ireland (Sustainable Ireland 2006). On the Northern Ireland Environment Link website we find over 50 environmental and sustainable development organizations listed (Northern Ireland Environment Link 2006). These groups and organizations range from large internationally known ones such as Friends of the Earth and the World Wildlife Fund to small locally based conservation groups and campaigns. These lists of environmental groups do not include more radical groups such as Reclaim the Streets or groups engaged in the anti-globalization/global justice movement.

A glance at the Irish Indymedia website will also testify to the variety and debate within the Irish environmental movement, and the connections being made between 'new' social movements such as the environmental one, and 'old' social movements such as anti-capitalist/socialist and anarchist and feminist movements and groups (Indymedia 2006). An interesting recent development for the Irish environmental movement was an announcement in January 2006 of a new umbrella body of the Irish environmental movement entitled *The Irish Environmental Forum* bringing together 20 of the leading environmental groups in the Republic of Ireland (RoI). This was established to rectify the weakness of the Irish environmental movement as a national movement in making its voice heard in European environmental lobbying and policy-making, particularly at the level of the European Commission. Such pooling of resources by environmental groups is one of the most significant developments for the environmental movement, alongside, as this chapter hopes to demonstrate, the 'politicization' and 'radicalization' of the environmental movement though embracing sustainable development rather than a narrow 'environmental protection' agenda.

## The political economy of unsustainable development in Ireland

The green movement/s in Ireland (north and south) has not had it easy. Whether measured by membership or influence on policy and politics, it emerges as one of the weakest in Europe. Attitude surveys from the 1980s onwards show the public in the Republic and Northern Ireland consistently placing environmental concerns below other concerns, especially orthodox economic growth, security and employment (Whiteman 1990; Devine and Lloyd 2000). Environmental concerns, as measured by Euro barometer studies, have traditionally been lower in both jurisdictions than in other EU countries.

In this section we will sketch the main contours of the underlying causes of unsustainable development which the environmental movement on the island of Ireland must contend with. On either side of the border the Irish and British states have prioritized an orthodox view of economic growth as the state's main goal (though in Northern Ireland security had long been the state's primary interest before the recent 'peace process' and power-sharing executive in March 2007). Across the island the environmental costs of pursuing economic growth are visible in the excessive use of nitrogen and other fertilisers in industrialized agriculture, the pollution of inland waterways from agricultural, industrial and domestic sources, the loss of biodiversity and habitats, the unsustainable increases in carbon dioxide emissions from burning fossil fuels, in patterns of land use and urban and suburban development which each year decrease green spaces, and in the congestion and pollution associated with an explosion of privatised car transport in a context where the road and transport infrastructure are seriously inadequate.

From a sustainable development point of view there are also many 'non-environmental' costs associated with the neo-liberal pursuit of economic growth and wealth creation. The Republic of Ireland, according to the 2005 UN Human Development Report, with over 15 per cent of its population living in poverty is second only to the USA in income inequality, while Northern Ireland has some of the highest rates of environmentally-linked childhood asthma and respiratory problems.

In both parts of the island, the governance and political structures for sustainable development are marked by weak or absent accountability processes. Unlike market actors, interests and imperatives, citizens have often been discouraged from participating in policy decision-making processes. While lip service is paid to 'joined up thinking', sustainable development continues largely to be confined to the 'policy ghetto' of 'the environment' and not linked to economic policy for example. Until it is taken as an overarching, integrated policy approach the potential for sustainable development to redefine economic development will, in our view, not be realized on the island.

Both sides of the border have witnessed the pursuit of export-led development strategies, based around the attraction of FDI (foreign direct investment) usually through state subsidies and other incentives, in particular lower environmental standards which reduce costs and the burden of compliance. As Yearley has put it:

> Some foreign companies have seen Ireland as a country in which they can locate processes which have been rendered uneconomical or outlawed by changing environmental regulations in their home countries. The Irish Republic has had more lax environmental controls and has been so keen to attract investment that operations effectively exiled from the USA have turned up there (Yearley 1995, p. 659).

The desire to attract inward investment of whatever kind reminds us of Baker's (1990, p. 47) observation of the 'developing world' character of Ireland's political economy. The pharmaceutical and chemical industries, despite the 'troubles' and the continuing legacy of the conflict, have similarly established a presence in Northern Ireland. In short, both jurisdictions have shared a similar narrow view of economic development in which jobs at any cost typify state policy.

Criticism and resistance to the growth model of development on environmental (or social) grounds have routinely been dismissed by both business interests and state representatives. Pádraic White, former head of the Industrial Development Authority in the Republic, is on record as claiming that local opposition to pharmaceutical development in Cork was made up of 'small undemocratic groups' intent on blocking industrial development (Allen and Jones 1990, p. 1; Allen 2004). A more recent example of the same discourse was the attempts by the Progressive Democrats in the May 2007 election to claim the policies of the Green Party would destroy the 'Celtic Tiger' economy. In the ideological battle over the meaning and realization of 'development' and 'progress', local opposition to proposed developments – whether incinerators, motorway expansion, genetically modified crops or new power stations – is often deliberately misrepresented as 'anti-development' when in fact what often motivates opposition are alternative models of development.

The popularity and official endorsement of the discourse of 'sustainable development' shows that opposition to particular forms of 'development' are in fact better understood as involving competing (and often mutually exclusive) understandings of 'sustainable' and 'unsustainable' development. Thus opposition to building more roadways – whether in particular ecologically and historically significant sites such as the Glen of the Downs or the Hill of Tara – is in reality to be interpreted as part of a wider debate about whether building more roads, encouraging more car use and the consequent rise in the burning of fossil fuels can be regarded as 'sustainable'.

From the viewpoint of environmental groups and their approach to 'sustainable development', the issue then is whether more motorways, car use and greater dependency on oil are to be judged as helping or hindering ecological protection (locally, in terms of air and water pollution as well as globally in terms of climate change), economic competitiveness and long-run prosperity (given scarce, insecure and costly carbon fuel sources) and social health (in view of the growth of commuter culture). We should therefore view the analysis and perspective of the environmental movement as always potentially challenging the orthodox model of political economy (whether particular groups explicitly articulate their concerns in this manner or not). Within the discourses and rationales which motivate and orientate the Irish environmental movement, North and South, one can increasingly discern a common critique of this orthodox model and its 'autistic' and dogmatic belief that like Achilles' lance in Greek mythology, more economic growth can heal the wounds and damage caused by the growth process itself (Barry 1999).

In all this the Irish environmental movement – particularly if we include within it the Green Party (which exists in both jurisdictions and exists as one party on the island since 2006), environmental groups such as Friends of the Earth, and green think tanks such as Feasta – has much in common with other non-ecologically based critiques of this orthodox 'neo-liberal' political economy. A common rejection of the social injustice and inequality which is structurally part of the orthodox growth model (Barry 2006) is what enables (parts of) the environmental movement to be grouped with a variety of interest groups that would include the Conference of Religious in Ireland (CORI) and its influential Justice Commission, left-leaning and socialist-orientated parties, groups and campaigns and the trade union movement.

We can for example find in CORI's 2006 submission on Ireland's *Rural Development Strategy Plan 2007–2013* policy recommendations which are not only consistent with the integrated 'triple bottom line' of sustainable development but which echo the views of more radical environmental/green economic arguments. Commitments to a basic income, decentralization of services and the promotion of the social economy have been long-standing policy aims of the Irish Green Party. The encouraging of organic farming and balancing housing development with environmental protection are other positions which all environmental groups would support. Much the same can be said of the Irish trade union movement's view of the downsides of globalization, its support for fair trade as oppose to the neo-liberal vision of 'free'.

What is also evident is that some local campaigns can develop in a way that moves towards the integrated 'triple bottom line' conception of sustainable development. The current 'Shell to Sea' campaign provides a good illustration here. Its campaign has gone beyond the specific issue of the siting of a pipe line to bring gas from the Corrib gas field to embrace a wide set of issues that include anxieties about the role of multinational corporations within Ireland. The issues raised by the 'Shell to Sea' campaign, by touching on the undemocratic manner in which the Irish state makes infrastructural and economic decisions, and by giving voice to nationalistic and social justice concerns about the selling off of 'Irish' oil cheaply to foreign multinational corporations with little benefit to the people of Ireland, are deeply central to sustainable development.

The 'Shell to Sea' campaign shows how the Irish environmental/green movement has the capacity to push conceptions of 'sustainable development' beyond narrowly conceived 'environmental' issues and the protection and/or conservation of nature. By so doing it is also able to make links with a wide variety of other justice-minded social movements and interests. It is of course true that there groups and campaigns exist which do have a specific environmental focus and it needs to be acknowledged that not all such groups would consider themselves as part of a wider 'green' movement for sustainable development. It is therefore analytically useful to distinguish 'environmental' groups and campaigns whose sole or main concern is the protection of the natural or physical environment or species or particular habitats or spaces (urban or rural) and a wider 'green' movement which moves in the direction of embracing the expansive 'triple bottom line' conception of sustainable development. To explore this distinction further we will now turn to the growing 'Environmental Justice' movement.

## Materialism and post-materialism: The environmental justice movement in Ireland

The 'environmentalism of the poor' (Martinez-Alier 2002), and the associated 'environmental justice' literature, are not often applied in analyses of the environmental movement in European countries. While the latter has been extensively used and applied in the United States in relation to environmental racism and the systematic injustice inflicted on minorities and communities of colour through the siting of toxic waste facilities or dumping of waste (Schlosberg 2002; Bullard 2005) and other environmental 'bads', only recently has it begun to be used in analysing the distribution of environmental bads and costs in Europe (Boardman, Bullock and McLaren 1999). If we accept that Ireland shares more with other post-colonial countries than with the developed nations of Europe or North America, the adoption of an environmental justice frame is particularly apt in the Irish/Northern Irish case.

Using Inglehart's well-known 'post-materialist' thesis, the standard sociological explanation for the rise of environmental concern in the developed world has traditionally drawn attention to the rise of a 'new middle class' of knowledge-based workers (often in the state and private sectors of education/training, social work, caring and social welfare) whose interests differ both from either the wage and employment concerns of organized labour and from the economic growth and accumulation concerns of the capitalist class. The political interests of this new class are with 'post-materialist' concerns of personal autonomy, environmental amenity, identity and quality of life.

Yearley uses this 'post-materialist' analysis to describe the dynamics of the Irish environmental movement (Yearley 1995), and while it can certainly explain some aspects of it, we find that the ways in which Ireland differs from other developed countries necessitates recourse to explanations other than the 'post-materialist' one. Tovey (1992, p. 285) offers a salutary reminder to us of the peculiarities of the Irish case when she observes that the values espoused in a lot of Irish anti-pollution campaigns are those of family, community, locality and tradition rather than those of the international green movement.

A recent report by Kelly, Kennedy, Faughnan and Tovey (2003) points out how:

> The international literature would lead us to expect that those with post-materialist values would be more likely than their counterparts to have pro-environmental attitudes and to act in a manner that protects the environment. However, in Ireland, only a small percentage of people can be said to hold post-materialist values...When it comes to understanding differences between Irish people's attitudes and behaviours, the evidence suggests that post-materialism is not of much help...those with post-materialist values are more likely than those who do not hold such values to be involved in indirect methods of promoting the environment as a social and political issue...As such, then, post-materialism is

not a particularly useful perspective in understanding differences between Irish people's environmental attitudes and behaviours (p. 52).

Equally, unlike the UK, membership of national environmental groups and organizations is markedly lower in Ireland, and there are few strong national environment organizations in the Republic. Across the island, on the other hand, there are and have been numerous local environmental campaigns motivated by 'materialist' concerns. These give rise to a pattern which leaves the Irish environmental movement with more in common in terms of its origins with the 'environmental justice' movement and the 'environmentalism of the poor' as outlined by Martinez-Alier. This is not to say that the 'post-materialist' explanation does not apply at all. There is certainty some evidence for a 'new middle class' supporting a green social movement, perhaps best exemplified by the socio-economic background of members and voters of the Green Party (Garry forthcoming; McWilliams 2005, p. 22).

In reality the environmental movement in Ireland is marked by both 'materialist' and 'post-materialist' impulses; it is also understandable in terms of forms of mobilizations for 'environmental justice'. Writing in 1989, Baker noted the connection between local exploitation of lignite resources in Co. Antrim and unaccountable and undemocratic bargaining between the British state and Northern Ireland agencies and foreign multi-national corporations, a common theme in many local environmental campaigns (see also Just Say No 2005; Magerity 2007). According to her:

> Access to, and participation in the decision-making process and, ultimately, some degree of control over that process by the different groups that comprise society, are at stake here. The involvement of MNCs in the mining operations raises this question in a direct manner. Foreign-owned (and this with their power centre located outside the area in question), such companies are rarely, if ever, subject to public accountability (Baker 1989, p. 65).

According to Schlosberg (2002), a key aim of the environmental justice movement is for 'voice' and recognition which cannot, either in theory or practice, be separated from 'community empowerment' (Foreman, cited in Leonard 2006a, p. 196). The denial of voice to local interests in resource use or infrastructural decision-making processes is something that one sees throughout the island; it is a main cause for local community mobilizations on environmental and often public health and safety grounds.

## Anti-incinerator campaigns

Another good example of a local environmental justice campaign is CHASE (Cork Harbour Alliance for a Safe Environment) which began in October 2001 and has campaigned generally '…for a safe environment, and most urgently, for preventing construction of a hazardous waste incinerator at Ringaskiddy' (CHASE 2008). The Government, however, seems determined to proceed with the project and sees incineration as a key part of the solution to tackling the Republic's waste crisis. It is significant that the anti-incinerator campaign in Cork has links to other anti-incinerator campaigns in Ireland, indicating as it does a broadening out of the movement beyond specific local contexts and in ways which can lead to more developed forms of analysis and protest (Leonard 2006a). Most of these anti- incinerator campaigns have also non-environmental concerns based on the commercialization and privatization of waste management and the 'crowding out' of community-based and social economy waste initiatives (Davies 2005). 'Essentially', as Leonard (2006a) points out, 'anti-incinerator campaigns in the Irish case have mobilized communities and experts against the state's waste policy by exploiting the combination of rural sentiment and democratic deficit that has surfaced in Ireland in the recent post-scandal tribunal era' (p. 177). Leonard's analysis demonstrates that it is not just environmental and health concerns that mobilize local communities; other 'bottom lines' built around social, political, democratic and economic concerns are relevant as well.

While Northern Ireland currently does not have large- scale incinerators for municipal waste, it is clear that the Government is determined to build incinerators as part of its waste management strategy. In the press release announcing the launch of the Department's waste management strategy in March 2006, Lord Rooker, the then direct rule Minister with responsibility for the environment, noted that he '[recognized] that energy from waste will be a *necessary component* of the mix of technologies required, particularly in light of the urgent need to develop energy from renewable sources' (DOENI 2006, emphasis added). This description of incinerators using the common euphemism of 'energy from waste' (other such terms include 'energy recovery') is significant for what it says not just about the politics of waste and incinerators, but indeed about the politics of sustainable development as a whole.

Not alone is the environmental movement on the island proposing or opposing specific projects or policies; it is also engaged in a battle over discourse and language to define the terms of political debate. We see this when anti-incinerator groups challenge the 'newspeak' of 'energy from waste' and environmental activists defend a view of 'sustainable development' against corporate or state definitions that reduce it to 'business as usual' which boil to policies for enhanced 'eco-efficiency' measures. In these ideological and discursive struggles campaign aims can be attached symbolically – a good current example is the Shell to Sea campaign – to notions of defending the 'nation', 'the people' or 'rural Ireland'.

Many rural environmental campaigns, as Leonard (2006a, p. 41) points out, have elements of an 'agrarian nationalism' which can trace its roots back to Michael Davitt and the Land League of the 19th century (Leonard 2006a, pp. 41, 247–8). While this 'agrarian nationalism' can influence forms of 'environmental nationalism' and 'populist environmentalism' (Tovey 1992; Leonard 2006a, pp. 244–5; Leonard

2006b), it is not necessarily true of all aspects of the environmental movement. If we consider the Green Party, it is significant that a survey of Irish Green Party members found that they were less nationalistic that either Green Party voters or the population as a whole (Garry forthcoming), something which can perhaps be explained by a commitment to *internationalist* principles. However, in terms of policies, the Irish Green Party does expresses clear (if non-strident) nationalist elements (support for the Irish language, criticism of the UK Government's policy in Northern Ireland); it is also capable of using nationalist rhetoric in relation to specific campaigns (such as the prominent support by the Green Party leadership for the Shell-to-Sea campaigners or the use/abuse of Shannon airport by the US military as a violation of Irish neutrality). Thus a link can be established between the discourses, mobilization frames and practices of post-colonialism and de-colonialism (throughout the island, but particularly in Northern Ireland) and the politics of sustainable development in general and the ideology and political objectives of the green movement in particular (Doran 1994; Barry 2004).

The sharing of experience, expertise and the widening out of localized campaigns is something that typifies most green campaigning activity on the island of Ireland. It is therefore a gross misrepresentation to view these campaigns as simply NIMBYist (Not in My Back Yard). Far from 'NIMBY' being a value-free designation, it is best seen as an ideologically loaded one, which has its origins in struggles between state and market-imposed development projects and local communities. Typically it is used as a pejorative term to devalue and undermine local campaigns as part of a deliberative effort to portray protest mobilizations as irrational, anti-progress, selfish and endangering the economic competitiveness of the national or local economy.

While obviously not true of each and every campaign (or at least of each and every campaign in its initial stages), there is a discernable pattern shared across many Irish environmental campaigns on the island of Ireland that sees local issues as a symptom or effect of deeper socio-economic and political causes. For example, local anti-incinerator mobilizations are not campaigning for incinerators to be located somewhere else on the island. Rather, their campaigns share a common agenda to reject incinerators per se as part of the solution to dealing with the Irish waste crisis in Ireland. By the same token, campaigns against genetically modified crops would not regard it as a success that Ireland was 'GM Free' while the rest of Europe was not.

## The environmental movement, rural communities and farming interests

One part of the environmental movement (certainly in the past) has been animated by a (typically urban-based) romantic view of the 'rural' qua 'Arcadian or ecological idyll' (Rennie-Short 1999). Within this romantic-cum-environmentalist perspective 'the countryside is seen as the last remnant of a golden age...the nostalgic past, providing a glimpse of a simpler, purer age...[a] refuge from modernity' (1991, pp. 31, 34). Such non-productivist, 'external' visions of the countryside have of course been rejected by most farming communities and have from time to time been the source of tension and suspicion between environmentalists and farming interests and communities.

In debates between farming groups (both North and South) and the two Governments around the implementation of EU directives, particularly the Nitrates Directive, farming interest groups present farmers as the 'natural' (and indeed 'national') stewards of the land who should be trusted (and left alone) to take care of the countryside. Particularly when in conflict with environmental groups, who criticize farmers for polluting rivers, over-using fertiliser and pesticides, destroying natural habitats and so on, the exchange between farming and rural interest groups (for example, Countryside Alliance Ireland, or Hare Coursing) is not simply about specific issues, such as the Nitrates Directive or cross-compliance, but above all it is about the meaning and understanding of the 'land', 'countryside' and 'rurality'. What is fundamentally in contention is the 'rural' itself, who should manage it, how and in what manner and for what purpose.

Leonard, in an excellent review of the Irish environmental movement and the relationship between rurality, rural sentiment and environmentalism, has noted that:

> With the onset of a technologically driven agri-business sector, mass production and scientization drove a wedge between rural dwellers and their hinterland. Farming would become synonymous with over production, fish kills derived from slurry spillages and images of EU subsidies for non-production in the wake of the 'butter mountains' and 'gravy lakes' which stemmed from unsustainable practices. In the era of globalized production local production for local markets came to be dismissed as small-minded thinking. The damage caused to local interactions between communities and hinterlands was significant (Leonard 2006a, p. 40).

The often antagonistic relationship between the environmental movement and farming communities and organizations (not of course synonymous), can therefore be viewed in part as a reflection of larger processes of 'modernization', which have transformed 'agriculture' into an 'industry' (i.e. 'agribusiness') and brought dramatic changes in land-use and production practices.

Productivist agriculture, of course, has not had it all its own way. One could say that the modest influx of new (often non-Irish) members to rural communities – particularly organic farmers, traditional craftspeople and so on – has inspired some attempts to integrate ecological visions of rurality with tradition productivist conceptions and practices. There are in addition some clear signs of more 'common ground' and co-operation between environmental groups and rural communities, particularly in relation to various crises within the rural community, particularly on issues arising from infrastructural and other controversial economic developments. The Green Party's 'reaching out' to rural Ireland in the run-up to the March 2007 campaign, through the promotion of bio-diesel as a 'win-win' for post-CAP farming communities and ensuring renewable and secure energy supplies, is a case in point.

## The ecosystem of political action

Under the impetus of EU environmental and other directives as well as concerted pressure from environmental and other civil society groups, the state, in both Northern Ireland and the Republic, has made some progress in dealing with environmental degradation through measures addressing water quality issues, waste minimization and recycling, energy efficiency and biodiversity preservation.

There is, unlike the situation in Northern Ireland, an independent Environmental Protection Agency (EPA) in the Republic. The nearest equivalent to this in the North is the Environment and Heritage Service, a non-departmental public body located within the Department of Environment (Macrory 2003). However, with the Review of Environmental Governance currently taking place, one prompted by the co-ordinated lobbying of seven large environmental NGOs, there is a good chance that an independent Environmental Protection Agency will materialize in Northern Ireland.

It is fair to say that neither jurisdiction is noteworthy for state innovation or leadership in the area of environmental protection, never mind sustainable development. State action in regards to environmental issues across the island can be described as slow, minimal and reluctant; what has appeared has largely been the result of EU Directives and the threat of EU infraction fines (Yearley 1995; Fagan, O'Hearn, McCann and Murray 2001; Turner 2006a, 2006b). The Republic has the highest rates of non-transcription of EU environmental legislation of any member state, while Northern Ireland has EU infraction fines totalling millions of euro, proportionately higher than any other region of the UK (Turner 2006a).

Another notable feature of the evolving policy regime is the manner in which the state, both North and South, has sought to 'support' and enter into partnership arrangements with different environmental groups and organizations. Largely effected through state funding arrangements, both jurisdictions are characterized by a blurring of the separation of civil society environmental interests and those of the state, such that we can now talk of 'environmental governance' (Macrory 2003; Turner 2006b) or 'governance for sustainable development' (Lafferty 2004).

## The ecologically modernizing state on the island

Across the island as a whole, to the extent that ecological considerations have been taken into account in devising and implementing state and business economic policy these have been couched within a weak 'ecological modernization' framework (Barry 2003). The origins and subsequent development of the Irish EPA provide a case in point. At the outset, the then Minister for Environment, Mary Harney, was at pains to make sure that the EPA would not be seen as 'anti-industry' or as a threat to business competitiveness (Harney 1991, p. 31). This imperative to 'balance' orthodox economic growth with environmental protection is a classic 'ecological modernization' strategy in which international competitiveness, Foreign Direct

Investment and export-led growth are not up for negotiation or serious amendment. In 'weak' understandings of ecological modernization (Barry 2003), the limits of environmental protection are set not by the natural world or ecosystem limits, but by the non-negotiable limits of a capitalist organized economy. In Northern Ireland, ecological modernization, while present in aspects of Government policy, and in the economic policy of environmental groups such as Friends of the Earth and the World Wildlife Fund, can hardly be said to be on the same level as the policy debate and practical developments within sectors of the economy in the rest of the UK (Barry and Paterson 2004).

Within both jurisdictions there is no political leadership committed to making sustainable development a strategic objective of state policy as a whole rather than a discrete 'policy area' with little or no connection to mainstream policy. Across all European states a clear 'hierarchy' of policy areas can be found within which environmental interests always rank lower than those of finance, economic matters, foreign affairs and defence. This pattern is particularly noticeable in Ireland where sustainable development has been defined (and we would say, confined) in terms of 'the environment'. Environment and sustainable development are thus seen as policy areas on the bottom rung of the ladder for Irish Governments and for ambitious politicians and policy-makers.

Clearly there is insufficient cooperation and co-ordination of activities on an island-wide basis between the two jurisdictions. Despite the Good Friday Agreement specifying the environment as one of the areas of cross-border co-operation, there has been a lack of political will or imagination to use the Agreement institutions, whether one looks at the North-South bodies or the East-West ones such as the British Irish Council, to explicitly make sustainable development a key policy objective.

**Market-based dynamics**

There are two aspects to the analysis of the role of market interests and actors in respect to the issues of environment and sustainable development in Ireland. One concern the ways in which powerful market actors have shaped, influenced and constrained state policy and have attempted to neutralize or co-opt community/ civil society resistance to unsustainable development initiatives. Here we find the familiar pattern in which a capitalist economic system, particularly under conditions of neo-liberal globalization (acutely so in the case of the Republic of Ireland), undermines and degrades the environmental basis upon which its wealth-creation is premised. As well as degrading the natural environment, the unfettered pursuit of economic growth also degrades bonds of social solidarity and community, creates socio-economic inequalities, lowers the collective quality of life through increasing the individual and collective stress brought by a '24/7' work ethic and associated competitiveness and productivity rationalities.

The other, more positive, but not unproblematic, aspect of market activity concerns the rise of what we may call the 'green' business sector, in which we include the following three types of economic activity and actor:

1. Those private companies engaged in environmental activities such as recycling or providing environmental services and consultancy.
2. Those sectors of the private, profit-making economy which are actively 'greening' their production, research and design, distribution and marketing.
3. Those parts of the social economy that explicitly aim to fulfil the 'triple bottom line' of sustainable development.

This growing 'green' part of the economy we view as the outcome of an 'ecological modernization' strategy in which the state together with leading economic actors and organizations (particularly those drawn from sectors at the cutting edge of technological innovation in the chemicals, life sciences and some service provision and manufacturing sectors), together with the support of sections of the green movement, attempts to 'steer' and 'encourage' the emergence of more resource efficient and less environmentally damaging economic practices (Barry 2002; Barry and Paterson 2004).

Whether to include this green business sector within the broad environmental/ green movement is a moot point. The ambivalent (if not negative) attitude of the orthodox business community on both sides of the border can be gauged from various statements on sustainable development made by members of business interest groups such as the Irish Business and Employers Confederation. What these convey is that the business community sees sustainable development not as a new paradigm for doing business and re-defining the Irish economy for the 21st century, but as a negotiable and purely environmental/resource 'side-constraint' that can be tolerated so long as it does not undermine orthodox economic growth and international competitiveness.

Here it is interesting to reflect how much lobby groups and initiatives to educate and push business in a greener director are noticeably absent within the Republic of Ireland. There is nothing like the UK and NI based organizations such as the *Arena Network* of the *Business in the Community* organization, *Forum for the Future* or *Green Alliance*, which work with business to develop sustainable development strategies. However, while *Arena Network* does have a presence in Northern Ireland, its role is more concerned with information provision than with active dialogue and building capacity within industry to move in a more sustainable direction.

In contrast it is possible to make a much stronger case for the inclusion of 'green' social economy organizations within the broad environmental/green movement. While not often explicitly political, such organizations range from charities such as *Bryson House* in Northern Ireland, which works with individuals with learning and physical disabilities in recycling and energy efficiency schemes, to the *Sunflower*

*Project* in Dublin which focuses on combating long-term unemployment through an inner city recycling business.

The renewable energy industry on the island is one of the clearest examples of the new green economy which represents a paradigm shift not only within the energy sector, but also represents a wider shift of the economy onto a more sustainable development path. To the extent that the energy sector is increasingly made up of renewable energy companies sitting alongside fossil fuel based energy providers, the more we can say the economy is moving in a sustainable direction. What the renewable energy sector represents is the 'sustainable' dimension of business in Ireland, at least in terms of the environmental bottom line. While it does not, as is the case in Denmark, pursue a community ownership approach, the renewable energy sector in Ireland does consciously present itself as helping to decrease the $CO_2$ emissions of both the Republic and Northern Ireland as part of both jurisdictions' commitment to $CO_2$ reductions under the Kyoto Protocol.

A notable feature of the evolution of the renewable energy sector on the island, and one shared with other parts of Europe, is the resistance to the siting of wind turbines and other forms of renewable energy plant such as anaerobic digesters. Whether its local government initiated mobilization against the siting of off-shore wind turbines in North Antrim (Barry, Ellis and Robinson 2008) or on-shore proposals in other parts of the island, the transition to a post-carbon, renewable energy society is not simply a technological issue. The debate between those representing social forces, movements and mobilizations pushing for a more sustainable economy and society and those whose primary aim is to maintain the status quo is as much (if not more) about competing ideas, values and principles as it is about concrete issues of contention.

### The biotechnology industry

State support for biotechnology and life science based technological innovation has become another source of debate and controversy in Ireland. Science Foundation Ireland (SFI) was established in 2000 with the aim of pump priming knowledge-based industrial innovation in the biotechnology and ICT areas. Its budget of €646 million between 2000–2006 is clear evidence of the stress and importance the Irish state has placed on this sector of the economy, as is SFI's contribution to the Sustaining Progress Partnership agreement, in which 'efficient use of resources' is top of its priorities (SFI 2005).

A similar pattern, in part motivated by the 'environmental' demands of resource efficiency and the need to find less polluting forms of industrial production, is observable in Northern Ireland. For example, Invest Northern Ireland's Green Technology Initiative offers established businesses up to £50,000stg in interest free loans for fitting environmental technologies and there are well-established and growing linkages between biotechnology research at Queens University and the University of Ulster and the biotechnology industry (Invest Northern Ireland 2004).

While state support for biotechnology can be viewed as part of an ecological modernization agenda on the island of Ireland as a whole, most environmental groups and organizations would reject, or at least find counter-intuitive, the idea that biotechnology can be seen as part of a sustainable development path. The reasons for this would reflect concerns over biotechnology's environmental and health impacts and its contribution to propping up the 'treadmill of production'. What sustainable development requires is a re-definition and re-calibration of the economy, not finding more ingenious ways to keep the orthodox economic growth system going (Barry 1999, 2006).

However, some farming groups and others representing rural interests would not necessarily share such concerns. Although there is some support for 'branding' Irish and Northern Irish agricultural produce on the basis of 'GM Free Ireland', the position of farming organizations (with the exception of those groups promoting organic production such as the Soil Association in Northern Ireland and the Irish Organic Farmers' and Growers' Association in the Republic) is different. These have expressed no principled objection to following the lead of countries like America, Argentina and China in growing GM food crops, provided Irish consumers and the 'market' find GM foods attractive and there is a demand for them.

At the same time, there is considerable interest in initiatives to encourage farmers to switch from food production towards producing energy crops, particularly in light of concerns about energy security, climate change, 'peak oil' and the fact that the island as a whole is one of the most vulnerable to oil shocks givens its high dependence on imported fossil fuels. There is a something of a consensus among farming organizations on both sides of the border (among the Irish Farmers Association and the Ulster Farmers Union) about the need for the agricultural sector to diversify into this new and profitable area. It is also a policy proposal promoted heavily by the Green Party, despite some environmental groups, such as Friends of the Earth, raising serious concerns. This shift towards integrating an energy production aspect to farming and rural communities, while at a very early stage of development, does promise the possibility of a less oppositional relationship between the environmental movement and farming interests, communities and organizations.

## Conclusion

As Andrew Rowell puts it in his book *Green Backlash*, 'The tide is turning against the environmental movement worldwide. Environmental activists are increasingly being scapegoated by the triple engines of the political Right, corporations and the state. The backlash has one simple aim: to nullify environmentalists and environmentalism' (Rowell 1996, p. v). It is clear from the experience of the environmental movement on the island of Ireland that there is indeed a struggle between its values and policies and those of major state and business/corporate interests, and, certainly in the past, between the environmental movement and farming interests and some rural local community interests.

The backlash that Rowell writes about is particularly encouraged by a number of circumstances in Ireland: the current neo-liberal political climate on both sides of the border; resource wars, such as the one in Iraq; continuing climate change and concerns about energy security for the island; ongoing degradation of the island's natural and built environments by industrial, housing, infrastructural and agricultural demands; rising levels of economic insecurity and income inequality; and declining perceptions of quality of life as well as democratic participation and accountability on both sides of the border. To the extent that environmental groups can be said to challenge or disrupt the smooth functioning of the state's accumulation and/or legitimation imperatives (Offe 1983; Dryzek, Downes, Hunold and Schlosberg 2003), their continued (if fitful and not always successful) 'politicization' and broadening of campaigns to include issues around 'sustainable development', 'environmental justice', 'anti-capitalism/global justice' can be expected.

As such, perceptive analysts of Irish social movements as Cox (2006) and Leonard (2006a) have noted, there is a growing impatience among many social movement activists with the accepted mechanisms of participation and inclusion, whether this is through the formal employment market or the social partnership process in the Republic of Ireland. While not all aspects of the environmental movement exhibit a broader anti-government or anti-capitalist agenda, it is clear that increasingly a politicization and radicalization is happening. Clear examples of this are found in the 'Shell to Sea' campaign (Leonard 2006a, Chapter 11), in anti-infrastructural project protests (such as the opposition to the building of motorways near heritage sites at the Hill of Tara (Leonard 2006a, pp. 226–31) and in anti-incinerator protests at the Battle of the Boyne site.

What is further evident is that even such 'reformist' environmental groups as the Royal Society for the Protection of Birds (in Northern Ireland) or An Taisce (in the Republic) have become progressively politicized over the last decade, and have moved away (if partially and not without tensions and contradictions) from a narrow 'environmental' or 'heritage' protectionism towards a the wider and more challenging agenda compatible with the transition to sustainable development. What can be said to be happening to the environmental movement in Ireland is an uneven shift towards embracing the politics of sustainable development and the political economy of the transition to a sustainable Ireland. Of course, as Cox (2006) points out, opposition to the dominant state and corporate model of development in the Republic of Ireland – moulded in the image of 'Boston' rather than 'Berlin' to use the Progressive Democrats' former leader Mary Harney's phrase – can result in state coercion – witness the Gardaí violence against the Glen of the Downs campaigners in 1997 and the 'Reclaim the Streets' protesters in May 2002. The jailing of the Rossport Five in 2005 became a crucial milestone in the Shell-to-Sea campaign. One might even expect that as the threat from IRA and other terrorist organizations decreases, more attention will be focused on redeploying police and intelligence services against 'eco-terrorism' (as has been the case in the United States where the 'Ecoterrorism Prevention Act of 2004' specifies 'eco-terrorism' as one aspect of the 'war on terror') and environmentally motivated opposition to economic, development and infrastructural projects.

What seems certain is that as the neo-liberal political economy of post-Celtic Tiger Ireland intensifies, with its associated environmental degradation, continuing socio-economic inequality and competitive 'race to the bottom', the environmental movement will increasing find itself pitted against these neo-liberal political forces. Equally in Northern Ireland, it is clear that the economic vision outlined for it by the British Government (and local business and political interests) is of a broadly similar one to that within the Republic, thus effectively making the island of Ireland one neo-liberal economic area. Tackling the dominance of neo-liberalism will force more and more parts of the environmental movement to politicize themselves and to make alliances with other social movements and forces to fulfil their objectives.

If the environmental movement wishes to deal with the *causes* of ecological destruction for example, rather than simply dealing with its *effects*, we can expect to see a greater degree of critiquing, challenging and proposing alternatives to the underlying political economy of the island as part of the transition to a more sustainable Ireland. In particular with 'peak oil' looming and Ireland (both North and South) being so heavily dependent on this imported, non-renewable energy source, a serious debate around energy security has started in which the transition to a post-carbon economy now pits renewable, clean energy against nuclear power, a technology whose introduction Irish environmentalists thought they had defeated in the late 1970s. Battles the movement had won in the past will now have to be re-fought in much more testing times, and against a coalition of state and business interests determined to find a technological fix for our energy-hungry economy. All the indications are that the proponents of the neo-liberal order will try to maintain the status quo rather than use the energy crisis as an opportunity to plan a transition to a more sustainable and different type of *society* and democratic system.

## References

Allen, R. and T. Jones (1990) *Guests of the Nation: People of Ireland versus the Multinationals*. London: Earthscan.

Allen, R. and T. Jones (2004) *No Global: People of Ireland versus the Multinationals*. London: Zed Press.

Asthma Northern Ireland (2005) *A Moving Picture: Asthma in Northern Ireland Today*. London: Asthma UK.

Baker, S. (1989) 'Community Survival and Lignite Mining in Ireland', *The Ecologist*, 19, 2: 63–7.

Baker, S. (1990) 'The Evolution of the Irish Ecology Movement', in W. Rudig (ed.), *Green Politics One*. Edinburgh: Edinburgh University Press, pp. 47–81.

Barry, J. (1999) *Rethinking Green Politics: Nature, Virtue and Progress*. London: Sage Publications.

Barry, J. (2003) 'Ecological Modernisation', in J. Proops and E. Page (eds), *Environmental Thought*. Cheltenham: Edward Elgar, pp. 191–214.

Barry, J. (2004) 'Where the Islands Meet', *Fortnight*, 11–12.

Barry, J. (2006) 'Towards a Concrete Utopian Model of Green Political Economy: From Economic Growth and Ecological Modernisation to Economic Security', *Post-Autistic Economics Review*, 36, 5–25.

Barry, J., G. Ellis and C. Robinson (2008) 'Cool Rationalities and Hot Air: A Rhetorical Approach to Understanding Debates on Renewable Energy', *Global Environmental Politics*, 8, 2, 67–99.

Barry, J. and M. Paterson (2004) 'Globalisation, Ecological Modernisation, and New Labour', *Political Studies*, 4, 54: 767–85.

Berger, P. (1987) *The Capitalist Revolution*. Aldershot: Wildwood House.

Biotechnology Ireland (2005) 'About Us', http://www.biotechnologyireland.com/bfora/systems/xmlviewer/default.asp?arg=DS_BIR_ABOUTART_14/_firsttitle.xsl/2 (accessed 22/11/05).

Boardman, B., S. Bullock and D. McLaren (1999) *Equity and the Environment: Guidelines for Green and Socially Just Government*. Edinburgh: Catalyst/Friends of the Earth Scotland.

Bullard, R. (ed.) (2005) *The Quest For Environmental Justice: Human Rights and the Politics of Pollution*. Berkeley: University of California Press.

CHASE (Cork Harbour Alliance for a Safe Environment) (2008) 'About Chase', http://www.chaseireland.org/about.htm (accessed 3/09/08).

Conference of Religious in Ireland Justice Commission (2006) *Submission to the Department of Community, Rural & Gaeltacht Affairs and the Department of Agriculture & Food on preparation for Ireland's Rural Development Strategy Plan 2007–2013*, http://www.cori.ie/justice/soc_issues/spec_issues/rural_submission0106.htm (accessed 23/3/06).

Cox, L. (1999) 'Structure, Routine and Transformation: Movements from below at the End of the Century', in C. Barker and M. Tyldesley (eds), *Alternative Futures and Popular Protest 5: A Selection of Papers from the Conference*. Manchester: Manchester Metropolitan University, pp. 56–81.

Cox, L. (2003) 'Global Social Movements', Talk to West Papua Festival, NUI Maynooth, 22/3/03 http://www.iol.ie/~mazzoldi/toolsforchange/rev/papua.html (accessed 20/9/06).

Cox, L. (2006a) 'News from Nowhere: The Movement of Movements in Ireland', in L. Connolly and N. Hourigan (eds), *Social Movements and Ireland*. Manchester: Manchester University Press, pp. 210–29.

Cox, L. (2006b) Personal Communication (18/9/06).

Davies, A. R. (2005) 'Incinerator Politics and the Governance of Waste', *Environment and Planning C: Government and Policy*, 23, 3: 375–98.

Davies, A. R. (2007) 'A Wasted Opportunity?: Civil Society and Waste Management in Ireland', *Environmental Politics*, 16, 1, 52–72.

Department of Environment, Northern Ireland (2006) 'New Strategy for Waste Management', http://www.nics.gov.uk/press/env/060330d-env.htm (accessed 20/3/06).

Devine, P. and K. Lloyd (2000) 'Being Green: Attitudes to the Environment', http://www.ark.ac.uk/publications/updates/update15.pdf (accessed 2/1/04).

Doran, P. (1994) '1968 was a Fine Year', *The Rowan Tree*.

Dryzek, J., D. Downes, C. Hunold and D. Schlosberg (2003) *Green States and Social Movements*. Oxford: Oxford University Press.

Fagan, H., D. O'Hearn, G. McCann and M. Murray (2001) *Waste Management Strategy: A Cross-Border Perspective*, National Institute for Regional and Spatial Analysis, Working Paper, 2a Dec 01.

Foyle Ethical Investment Campaign (2006) 'History', http://www.feiconline.org/feic.htm (accessed 6/4/06).

Garry, J. (forthcoming) 'Irish Green Party Members', in Rudig, W. (ed.), *Green Party Members*. Boston: MIT Press.

Harney, M. (1991) 'The Irish Environmental Protection Agency', in J. Feehan (ed.), *Environment and Development in Ireland*. Dublin: The Environment Institute, pp. 52–67.

Hunter, S. and K.M. Leyden (1995) 'Beyond NIMBY: Explaining Opposition to Hazardous Waste Facilities', *Policy Studies Journal*, 23, 4: 601–19.

Invest Northern Ireland (2004) *Winter 2004/05 Newsletter*, http://www.investni.com/environment_and_energy_newsletter_winter04.pdf (accessed 22/11/05)

Irish Business and Employers Confederation (2004) 'Companies concerned at the level of Regulation', http://www.ibec.ie/ibec/press/presspublicationsdoclib3.nsf/wvPCICCC/C3B642DE0A5A3AE480256F420058CC88?OpenDocument (accessed 22/11/05).

Just Say No (2005) *Newsletter*, http://justsaynotolignite.org/ (accessed 3/4/06).

Kelly, M., F. Kennedy, P. Faughnan and H. Tovey (2004) 'Cultural Sources of Support upon which Environmental Attitudes Draw', 2nd report of National Survey Data, http://www.ucd.ie/environ/reports/envirattitudessecondrept.pdf#search='environmental%20attitudes%20in%20Ireland (accessed 6/04/06).

Lafferty, W. (ed.) (2004) *Governance for Sustainable Development: The Challenge of Adapting Form to Function*. Cheltenham: Edward Elgar.

Leonard, L. (2006a) *Green Nation: The Irish Environmental Movement from Carnsore Point to the Rossport Five*. Dundalk: Choice Publishing.

Leonard, L. (2006b) Personal Communication (15/9/06).

Macrory, R. (2004) *Transparency and Trust: Reshaping Environmental Governance in Northern Ireland*. London: University College London.

McWilliams, D. (2005) *The Pope's Children: Ireland's New Elite*. Dublin: Gill and Macmillan.

Magerity, E. (2007) 'Social Capital and Environmental Campaigns in Northern Ireland' (Unpublished PhD). Belfast: Queens University Belfast.

Martinez-Alier, J. (2002) *Environmentalism of the Poor: A Study of Ecological Conflicts and Valuation*. Cheltenham: Edward Elgar.

Northern Ireland Environment Link (2006) *Members* (accessed 3/4/06).

Offe, C. (1983) *Contradictions of the Welfare State*. London: Heinemann.

Rennie-Short, J. (1991) *Imagined Country: Society, Culture and Environment*. London: Routledge.

Rowell, A. (1996) *Green Backlash: Global Subversion of the Environmental Movement*. London: Routledge.

Schlosberg, D. (2002) *Environmental Justice and the New Pluralism*. Oxford: Oxford University Press.

Science Foundation Ireland (2005) 'Sustaining Progress', http://www.sfi.ie/content/content.asp?section_id=489&language_id=1 (accessed 22/11/05).

Shell to Sea (2006) http://www.corribsos.com/ (accessed 3/4/06).

Sustainable Ireland (2006) 'Directory', http://www.sustainable.ie/directory/index.php (accessed 3/4/06).

Tovey, H. (1992) 'Environmentalism in Ireland: Modernisation and Identity', in P. Clancy, M. Kelly, R. Zolaniecki and J. Wiatr (eds), *Ireland and Poland: Comparative Perspectives*. Dublin: University College Dublin, pp. 275–87.

Turner, S. (2006a) 'Transforming Environmental Governance in Northern Ireland: Part One – The Process of Policy Renewal', *Journal of Environmental Law*, 18, 1: 55–87.

Turner, S. (2006b) 'Transforming Environmental Governance in Northern Ireland: Part Two – The Case of Environmental Regulation', *Journal of Environmental Law*, 18, 2: 245–75.

United Nations Development Program (2005) *Human Development Report 2005*, http://hdr.undp.org/reports/global/2005/ (accessed 23/3/06).

Whiteman, D. (1990) 'The Progress and Potential of the Green Party in Ireland', *Irish Political Studies*, 5, 1: 45–58.

Yearley, S. (1995) 'The Social Shaping of the Environmental Movement in Ireland', in P. Clancy, S. Drudy, K. Lynch and K. O'Dowd (eds), *Irish Society: Sociological Perspectives*. Dublin: Institute of Public Administration, pp. 652–75.

# Chapter 19

# Populism and the Politics of Community Survival in Rural Ireland

Tony Varley

## Introduction

Over the past two centuries inequality, conflict and social and economic transformation have been enduring themes in the rural history of modern Ireland. In the 19th century and beyond much sporadic or continuing conflict surrounded access to and the distribution of Irish land (Lee 1980; Nairn 1998, pp. 107–8; O'Dowd 2005, p. 93); and from these land struggles some clear 'winners' and 'losers' would emerge. We know that the challenge to landlord power of the more substantial tenant farmers partly depended on the decline of their old class adversaries, the agricultural labourers (Hoppen 1984, p. 103). If the farm-workers and landlords were the big losers historically, then the larger tenant or 'strong' farmers were the big winners (Larkin 1984, pp. xiv–xv).

The rise of this strong farmer class would become one plank of a rather conservative social order partly founded on the 'dominance of bourgeois values in the Irish countryside' (Kissane 2002, p. 74). What is also evident, as we move deeper into the 20th century, is that the position of the strong farming class would become more advantaged vis-à-vis struggling smallholders (Crowley 2006, pp. 32–5). Striking as this pattern may be it conceals a number of complexities. One of these is that in the market and political turmoil of the inter-war period (especially after Fianna Fáil's rise to power in the 1930s) the economic and political fortunes of the strong farmers would suffer a traumatic (albeit temporary) reverse (Cronin 1997, p. 112).

The turbulent 1930s would see the appearance of Muintir na Tíre (People of the Land), the earliest of two community organizations to be examined in this chapter. Remarkably, on the basis of reinventing itself a number of times, Muintir na Tíre (hereafter Muintir) would survive to continue as a community movement to the present day. Relatively speaking our other case – the Rural Community Network (RCN) which appeared in Northern Ireland (NI) in 1991 – is but a relatively recent arrival on the scene.

Apart from a common focus on using collective action to promote the welfare of rural communities, why these two cases have been chosen for discussion reflects their attempts to achieve a 'national' coverage for themselves (albeit in two separate jurisdictions) and their common engagement with the challenges of community survival in the countryside. On the surface level the differences between our two cases are evident in their origins, aims and in their conceptions of what rural communities need to do to sustain themselves and survive in the near to medium term. Other differences are apparent in their abilities to mobilize resources and exploit opportunities, abilities that some have taken to be the *sine qua non* of social movement effectiveness more generally (Klandermans 2004, p. 281); and in the actual headway they have made in realizing their aims. In their many differences our two cases illustrate what some have seen to be a characteristic general feature of Irish community development – its diversity (Ó Cinnéide and Walsh 1990).

It is, of course, possible to make too much of surface differences. Groups and tendencies may differ on the surface only to be found to be contiguous at a deeper level. One contiguity, evoked in the literature that links collective action on the part of disadvantaged groups with the phenomenon of 'empowerment' (Friedmann 1992, pp. 31–4; Hanna and Robinson 1994; Gaventa et al. 1995), is that it is possible to construe community-based collective action as attempting to reject the perceived powerlessness that springs from the imbalanced development that makes for rural decline.

In a sense there is nothing new in this recent linking of community-based collective action with 'empowerment'. Such a linkage was anticipated by those versions of populism that have viewed rural interests to be in a structurally disadvantaged and relatively powerless position in urbanized societies (Lipton 1977; Kitching 1989); and that have looked on collective action on the part of relatively powerless rural interests as one potential means of generating forms of countervailing power (Varley 2003, 2006).

For our purposes populism has the advantage not only of consciously focusing on the declining, 'losing' or underdog interest, but of forcing us to see 'power' as central in any discussion of the politics of community 'survival' and 'sustainability' in the countryside. Viewed in terms of the 'power over' and 'power to' distinction familiar to students of power (see Morriss 2002, pp. xiii–xiv), the resort to collective action becomes a crucial first step in a process of generating sufficient 'power to' capacities to allow relatively powerless groups assert themselves as organized interests, capable of resisting the 'power over' forces that dominate and exploit them.

But what sort of perceived popular powerlessness has been relevant historically to rural populist-type collective action? Distinguishable conceptions of popular powerlessness can be associated with three specific (though often inter-related and overlapping) constituencies. These are: struggling 'small man' interests (smallholders and small business people present prime examples), beleaguered local communities and those who see themselves being left relatively powerless by virtue of the way they perceive their interests to be misrepresented, ignored or even betrayed by the formal political system. Each of these three constituencies can be shown to produce its own distinctive strand of populism 'from below' – what can be termed the 'small man', the 'communitarian' and the 'representational' respectively.

If these three constituencies form the elements of a populism 'from below', there is also a populism 'from above' that can be associated with anti-colonial nationalism and the postcolonial state (Kitching 1989), features deeply rooted in the Irish historical experience. Fianna Fáil, the most successful party of power in independent Ireland, would consciously identify itself with a small man constitutency in its early period (Mair 1987, pp. 25, 51).

Lying behind the 'communitarian' populism of most concern to us is a commitment to 'community life' and to the welfare and preservation of local communities and community-based interests (Midgley (1995, p. 90). A characteristic communitarian populist suggestion, one that echoes through the academic literature that takes the decline of community as its theme (Etzioni 1995; Putnam 2000), is that the modern world is often hostile to community. Within communities in disadvantaged rural areas, and with but weak local economies to rely on, the challenges of resisting the forces producing decline become all the greater. In the post-war period the theme of the decline of Irish rural communities has spawned an abundant academic literature (Brody 1973; Wilson and Donnan 2006, pp. 23–7).

## Populism, power and sustainable communities

Before turning to our two Irish cases, we will need to look briefly at how populist conceptions of power might impinge on conceptions of sustainable communities and effective collective action. As much as the 'power to' and 'power over' distinction draws our attention to the negotiated or transactional nature of power relationships, it sheds no light on the direction community interests might take or on how far they might go in their conceptions of sustainable communities and in their efforts to turn collective action into forms of countervailing power. To explore these issues, with a view to providing benchmarks for the discussion of our two cases, two ideal-typical scenarios – the 'radical' and the 'pragmatic' – will now be outlined. Since these ideal-typical scenarios (though empirically possible) exaggerate and simplify complex realities they cannot be regarded as empirical descriptions in any real sense. All that can be expected of them is that they can facilitate interpretations and comparisons of real world cases (see Burger 1987, pp. 154–79).

Our ideal-typical radicals and pragmatists will be made to differ in how they conceive of sustainable rural communities, and therefore in whom they seek to represent and in what they look to achieve *via* collective action. They will further be made to differ in their views of the internal resources and external opportunities relevant to turning collective action into forms of countervailing power; and in how they see these resources being mobilized and opportunities exploited to best effect.

The radical conception of 'the community' concerns itself with the plight of the weakest elements of rural society who feel the effect of structural decline most sharply. Radical thinking about what to do, reflecting a normative commitment to equalitarian and communitarian ideals, centres on seeking some alternative to an established order that is built upon 'power over' forces and that has structural decline and inequality as its persisting consequences. In contrast our ideal-typical pragmatists opt to take the 'whole community' (rather than the most powerless segments) as the people to be represented and defended. And, instead of looking to some radical alternative to the *status quo*, they are prepared to settle for an accommodation that seeks either to preserve the *status quo ante* or to deliver incremental change or change at the margins.

An analysis that points to whom they should represent, and to what they should do, becomes a critical resource therefore for both radicals and pragmatists. A related valued resource for radicals is the cultivation of a participative organizational culture that seeks to encourage universal active involvement in challenging collective action among the relatively powerless.

Tactical preferences that speak to how collective actors should relate to the state become another important resource. And here the radical preference is to combine opposition to the state (seen as underpinning the system of imbalanced development and structural inequality) with building campaigning alliances with other structurally disadvantaged community interests. Radicals see a stream of opportunities flowing from the ongoing and progressively more severe crisis conditions generated by structural decline. Given its analysis and tactical preference, the radical tendency is to dismiss as unrealistic any suggestion that state-inspired partnerships might be genuinely interested in bringing about structural change and that community partners might somehow be equal with state partners in any authentic sense.

For their part our ideal-typical pragmatic populists can accept that external large-scale 'power over' forces are well capable of inflicting serious damage on local communities, small-scale producers and other rural interests. Where they would differ from their radical counterparts is in their view that such damage becomes a matter for most concern only when ruptures sporadically occur that significantly and unacceptably worsen the existing imbalance to the further detriment of community interests. Underlying such a view is a pragmatic acceptance that the best local community interests can do is to accommodate themselves to the overwhelming reality of a world in which external large-scale 'power over' forces and interests are massively in the ascendant. Searching for radical alternatives to the *status quo* is deemed to be counter-productive, if not entirely utopian.

As well as a reactive analysis that thinks in terms of restorative and incremental change, a key resource for our ideal-typical pragmatists is a leadership-centred and directive organizational culture in which local notables are expected to lead the way.

Opportunities, in pragmatic eyes, are linked to the sporadically occurring ruptures that threaten a fairly stable system (however imbalanced or unequally constructed) with crisis conditions; and that prompt the state to intervene so as to restore the *status quo ante* or to concede incremental improvements. Pragmatic considerations again dictate that tactics aimed at cultivating close working relations with the state be favoured over forms of oppositionalism. Consistent with this preference for working within the system, our ideal-typical pragmatic populist collective actors would optimistically see partnership-type relations with the state as offering at once valuable opportunities and resources (such as funding and experience) to local community interests intent on advancing their various projects.

A feature of the two ideal-typical scenarios sketched above is the absence of any common conception of sustainable rural communities and of the conditions deemed necessary for effective collective action. Our ideal-typical radicals and pragmatists have been made to differ in their constituencies and aims, in their thinking about relevant resources and opportunities and in how they see resources being mobilized and opportunities exploited so as to generate forms of countervailing power capable of delivering more sustainable futures to rural communities.

With these ideal-typical scenarios as our guides we are now ready to turn to our two Irish cases. In all, four issues will be considered. Early aspirations and aims will be examined to piece together conceptions of sustainable communities. What internal resources and external opportunities (in particular those arising in the state sphere) were considered necessary early on to effective collective action will then be taken up. From here we will ask how well (or poorly) our two cases have fared in practice in mobilising resources and exploiting opportunities, and with what effects. Finally, the experience of our two cases will be compared in the light of our radical and pragmatic ideal-typical populist scenarios.

## Muintir na Tíre

Who did the early Muintir seek to represent? Its early days were spent organizing agricultural interests. When it was clear by 1933 that this approach was making little progress, moves began to build a new rural movement around the Catholic parishes. Even after this shift of focus, the assumption that the rural economy and society were essentially agricultural remained strong till the 1960s.

But why did agricultural interests and parishes require organization in the first place? Here the Muintir idea was that the right form of organization would play a vital part in countering debilitating crisis conditions born of agricultural difficulties and the divided society left behind by syndicalist-inspired class war in the 1917–23 period (O'Connor 1988), and by the short but bloody civil war (1922–3) that had attended the creation of the 26-county Free State.[1]

Inspired by nationalist and Catholic ideas, the parishes – once organized along vocationally in the form of guilds and parish councils (Rynne 1960) – were imagined as potentially dynamic generators of civic patriotism and social Catholicism. The early expectation was that this parish-based community action would help give rural communities a more secure future by turning local people with declining

---

1   The island of Ireland, under the Government of Ireland Act of 1920, had already been partitioned.

prospects and divided loyalties into better neighbours, better patriots and better Catholics. The challenge for Irish patriots, as Muintir's founder (the Tipperary-based priest Fr John Hayes) frequently put it, was no longer to die for Ireland but to live for it.[2] Not alone did localities stand to benefit from active communities, but a vibrant tradition of community action promised to breathe new life both into the Irish nation and Irish Catholicism.

To have any major impact, it was recognized that the Muintir approach would have to spread itself across the country and that the new parish guilds and councils would have to be more than paper entities. Unless they were active and remained so most of the potential benefits of collective self-help effort would never materialize. Something else the top leadership recognized early on was that local level effectiveness would require that Muintir achieve and retain a high profile for itself at the national level. Otherwise it would be more difficult to keep alive a sense of local identification with the national movement.

*Resources*

What resources were at Muintir's disposal or could be created as it set about its work? Apparently Fr Hayes's ideas about civic patriotism were influenced by the nationalist experiences of the Land League and the pre-split Sinn Féin movement (Rynne 1960, p. 103). In *Quadragessimo Anno*, the papal encyclical of 1931, he found a tangible expression of social Catholicism and a means of legitimating the idea of constituting parish councils and guilds along vocational lines. Vocational organization was intended to be inclusive – labourers were to have the same sectional standing (at least numerically) as farmers, shopkeepers and professionals. Sometimes provision was made for recognizing women and youth as distinct sections in their own right (Rynne 1960, pp. 141, 217).

How much of a resource did the Catholic Church, with its presence in every corner of the country, its enormous following and its large pool of parish clergy, provide to Muintir? There is no doubt but that the numerous Catholic clergy along with local teachers proved to be a crucial source of local leadership in the early Muintir. At the same time Fr Hayes had some reservations about his parish councils being 'completely dominated' by the local clergy (Tierney 2004, p. 35). What is clear as well is that nothing like all the Catholic clergy were willing to give Muintir their active support (see O'Leary 2000, pp. 164–5).

*Opportunities*

Acceptance of the Catholic version of the principle of subsidiarity meant that the early Muintir put great store on preserving its own autonomy vis-à-vis the state. Yet this never implied any rejection of the ideal of working closely with state authorities so as to advance the common good and the well-being of local

---

2    Interview, Tom Fitzgerald, Canon Hayes House, Tipperary, 20 August, 1998.

communities. From early on it became clear that many of the opportunities that Muintir, at the national and local levels, would try to exploit would derive from state campaigns to respond to emergencies (such as those inspired by wartime food and fuel shortages or by attacks on the elderly in their homes in the 1980s) or to bring services (such as electricity or running water) to rural areas.

Whatever residual suspicion of state control lingered on would disappear when Muintir's national leadership, under the influence of the international Community Development movement in the late 1950s, began to warm to the ideal of voluntary and state bodies working closely together to advance the common good. By the early 1970s it had come to see Muintir's fortunes as tied inextricably to state funding. Sufficient state funding, it was believed, would clear the way for Muintir to overcome its historic inability to spread its local organization more comprehensively, and to provide a wider range of local services to its constituent councils.

By the 1970s what was found especially appealing was the idea that the state should grant Muintir a monopoly in the organizing and servicing of community councils. A professionally staffed development unit was envisaged that, with adequate state funding, would be able to serve the needs of the ever more numerous set of Muintir-affiliated community councils expected to spring up in urban as well as rural Ireland (Muintir na Tíre 1971, pp. 36–7).

*Outcomes*

How well did the early Muintir deliver on the ambitious task of using local community action to stimulate civic patriotism and social Catholicism in ways whose reverberations would be felt nationally? An ability to adapt and improvize would allow the national organization to survive for over 70 years, even if its existence has frequently been precarious and crisis-ridden. Certainly its spatial spread has always been limited and activity, even in those places where Muintir did put down its deepest roots (in Tipperary and surrounding counties), has tended to be cyclical rather than cumulative in character. Involvement in the post-war rural electrification effort exemplifies the way in which local interest in Muintir would expand and contract as national campaigns (the community policing scheme known as 'Community Alert' is the current one) have come and gone. Inspired by the post-war rural electrification drive, which fortuitously threw Muintir 'a life-line' (Tierney 2004, p. 76), the number of affiliated parish councils rose to 295 in 1951 before peaking at 417 in 1955. Thirty years later (when community councils had replaced parish councils) the number of affiliated community councils stood at 142; and today it stands at around 200.[3]

To make its presence felt nationally and integrate the local activists into the wider national movement, the early Muintir produced its own yearbook and

---

3    Figures provided by Mr. Tom Fitzgerald, Muintir na Tíre, Tipperary. It appears that Fr Hayes was keen that Muintir establish a presence for itself north of the border though, despite some tentative efforts (Rynne 1960, pp. 210, 175–7), nothing came of this.

various other publications; it also held 'rural weeks' that brought the 'Muintir family' together each summer and helped to spread the movement into new counties. Some supra-local but sub-national organization – such as county (and provincial) federations – appeared. Since the 1960s, however, there has been a considerable fall-off in all this activity. The yearbook and other publications faded away and the last rural week was held in 1969. Nowadays only Cork survives as a county federation.

Identified very much with its clerical founder, a big test for Muintir was whether it could survive the death of Canon Hayes in 1957. At this point the international Community Development movement came to Muintir's rescue. And in time the national organization's embrace of 'community development' would allow the 'representative' (or elected) community council take over from the old and increasingly seen as outdated vocational parish councils. Largely thanks to the national (and even international) reputation Fr Hayes had gained for himself and his movement, maintaining a high profile was something that came easier to Muintir in the early years. Today the clerical physical presence, once so strong at both the national and local levels, has all but disappeared. Of course, not all Muintir supporters would lament this radical declericalization. In suggesting that Muintir 'needs a new vision of its role in Irish society', Fr Mark Tierney (2004, p. 187) sees Ireland's developing multi-culturalism as one reason why it would no longer do for Muintir to be perceived 'as a religious organization'.

Tierney's (2004) assessment is that Muintir, for all its trials and tribulations, did contribute significantly to the cause of rural 'modernization'. This it did through its support for and involvement in post-war rural electrification, the Parish Plan for Agriculture (1947–1958), the formation of the early credit unions, the provision of rural water supplies and through its espousal of 'community development' ever since the 1960s. Based on such involvements it is clear that Muintir (at national and local levels) developed a capacity to respond to rural needs, as these were given priority from time to time and became the basis of campaigns of varying duration.

With the move to 'community development' the enervating as well as dynamic tensions and struggles generated by a number of oppositions, evident throughout Muintir's history, were not destined to disappear. One key opposition was between voluntarism and dependence on the state. Since the 1960s much importance has been given to establishing partnership-type relations with the state. Yet the state has so far declined to concede the sort of monopoly Muintir wished to be granted in organizing and servicing community councils.

Such were the circumstances that saw Muintir depend on pilot EC funding (secured in 1973) to advance its long-term aims. Once this funding ran out in 1979 the spurt of activity in had made possible fizzled out. The team of paid organizers Muintir had assembled and trained had to be disbanded, thus making for much disruption and discontinuity in the community animation process. Despite this and other setbacks, restoring partnership-type relationships with the state continues to appeal to Muintir's national leadership as its best chance of advancing its community council-based approach to community development. Easily the most ambitious of

Muintir's new partnership-type initiative has been Community Alert, a form of community policing that involves local groups working in collaboraton with the Gardaí (police) and that dates to a spate of attacks on the rural elderly in 1984.

Other notable oppositions that have generated tensions in Muintir are those between the desire to represent everyone as opposed to the worst off and the choice between having a Christian public persona as against a purely Catholic one. Fr Hayes's personal circumstances – he was born in a Land League hut in east Limerick after his family had been evicted by Lord Cloncurry – had left him with a strong sense of social justice and identification with the rural underdog (see Tierney 2004), especially the impoverished farm labourers. Nonetheless his approach to community organizing was the all-together one that sought to have the socially powerful and powerless of local society pull together in pursuit of the common good. Several accounts of local Muintir councils in operation suggest that, reflecting local class and gender realities, these tended to be leadership- rather than membership-led and that local notables (priests, teachers and strong farmers especially) were heavily over-represented in the local leadership (see Varley 2006, pp. 408–9).

At one point the tension surrounding the choice of public persona was partly resolved by Fr Hayes opting for a Christian rather than an exclusively Catholic identity (Rynne 1960, p. 113). At the same time, Fr Hayes' desire that Muintir project itself as Christian never meant a rejection of its Catholic identity. It would after all be rather difficult for a movement founded by a Catholic priest, inspired by Catholic vocationalism, organized around the Catholic parishes and which (in its early days) relied heavily for its activists on priests to disguise its clericalist and Catholic character.

Today (as often before) Muintir finds itself at a crossroads where its future is concerned. Most of the national organization's energies are being absorbed by the upkeep of the still expanding Community Alert programme for which state funding looks secure in the near to medium term. Yet the benefits of this initiative have not been entirely positive; a paradoxical effect of its success has been to eclipse somewhat the distinctive Muintir approach to community development, based on the elected community councils.

### Rural community network

The experience of the Rural Community Network (RCN) illustrates how the wider social context can profoundly influence how collective and state actors approach the notion of sustainable rural communities. Against the background of a divided society slowly emerging from a prolonged civil war (Murtagh 2003), rural community groups in NI have been presented as seeing their work as building 'a better future which is targeted at promoting reconciliation, economic growth and social inclusion' (Murray and Greer 1999a, p. 9). While a city-based 'community development movement' had emerged in NI in the 1970s (see Robson 2000, chapter 6), an 'increasingly organized rural constituency, which [had] suffered equally from the worst effects of deprivation and violence', was taking shape by the early 1990s (Murray and Greer 1999a, pp. 8–9).

The RCN's appearance in 1991, as the voice of 60 community groups (Fitzduff 2005, p. xiv), was at once a reflection of and a significant stimulus to the organization of this emerging rural movement. In keeping with its origins in the Rural Action Project (RAP) of the Second European Programme to Combat Poverty (1985–9), the RCN set out to be 'the voice of rural communities on issues of poverty and disadvantage and to support community development as an important approach to meeting the needs of rural communities' (Fitzduff 2005, p. xiv). On this basis it sought (as Muintir had tried with its parish councils and later with its Community Alert groups) to represent *both* whole communities and relatively powerless social segments within localities. The struggle, according to the RCN's former director Niall Fitzduff (2005, p. xiv), has always been to 'gain equity for rural communities' and to make sure that any 'comprehensive rural policy for Northern Ireland' would take the community interest into account (Fitzduff 2005, p. xiv).

A feature of this emerging community interest and 'sector' was its dynamism – by 1996 the RCN had 191 affiliated community organizations 'from all parts of Northern Ireland' (Murray and Greer 1999a, p. 19). Besides representing these and being engaged in advocacy across an expanding range of issues, the RCN began building the expertise to help community groups negotiate an 'almost bewildering mosaic of stakeholders and programmes' (ibid., p. 26).

What then, based on whom the RCN sought to represent and tried to do, can we say about its early conception of sustainable rural communities? Given the combination of forces making for rural decline and sectarian conflict, the challenges facing community actors in the northern countryside were formidable to say the least. Whatever chance the rural areas had of achieving a secure future was, as the early RCN leadership saw it, very much linked to the formation of community groups that could engage not merely with job creation but with issues of reconciliation and a broad ranging 'social inclusion' agenda (Fitzduff 2003, p. xxi; Murray and Greer 1999a, p. 26). The wider view taken by the RCN leadership was that 'traditional community development practices should not be dismissed as a precondition, but appreciated as an ongoing requirement for self-sustaining success' (Murray and Greer 1999a, p. 25). For the RCN leadership (as with that of RAP) building close working relations with the state was seen as offering by far the best prospect for community groups if gains were to be made and added to locally (Fitzduff 2003, p. xx).

*Resources*

One important resource that the early RCN leadership could draw on was the experience RAP activists had already acquired. We hear how the RCN 'largely inherited the mantle of the Rural Action Project...' (Fitzduff 2003, p. xx); and a clear family resemblance is to be seen between the RCN's evolving programme and the earlier RAP's 'research strategy'. This strategy had aimed to:

...identify, implement and evaluate alternative approaches to rural development in Northern Ireland. The bottom-up approach to development planning, in which the Community plays an active and direct decision-making role, will be a central theme for investigation. The Rural Action Project will take the initiative to form a coordinating committee consisting of representatives from relevant government departments and agencies, the voluntary sector and project staff. This committee will act as a forum for discussion of state support for community-based development projects; the integration of local community needs and aspirations with regional planning priorities, and the evaluation of the Rural Action Project experiments for future rural development policy (ISG 1987, p. 167).

What also stood to the RCN was the proliferation of community groups and their preparedness to support the alliance in large numbers. It was this broad support that gave the RCN its standing and legitimacy in the eyes of community activists, state officials and politicians. Partly fuelling the growth of the expanding community sector was the deep reservoir of popular discontent with the way a combination of a contracting rural economy and unsympathetic state authorities was perceived as making a bad situation worse (Fitzduff 2003, p. xviii)

The demands imposed by advocacy and serving an expanding membership encouraged professionalization within the RCN; and the use of professionalized planners and community workers became a highly valuable resource in itself (Fitzduff 2005, p. xiv). At the outset, with but one development staff member and an administrator, RCN had little choice but to 'achieve its objectives by working closely with others' (Fitzduff 2003, p. xx). Without increasing professionalization, however, it is hard to imagine the stream of regular newsletters and research reports on pressing issues (such as rural housing, the predicament of farm women, the transport needs of the elderly and the worsening farming crisis) that began to flow from the RCN. And this was but one dimension of the network's activities. An expanding range of new commitments followed from RCN becoming a key partner in NI's rural development regime and from direct involvement in 'project and programme administration' (Fitzduff 2003, p. xxi)

*Opportunities*

Acceptance by state elites that 'locally-based action' was to be the cornerstone a new rural development regime meant that the wider political context was broadly sympathetic to the RCN assuming an expanding and more central role for itself (Murray and Greer 1999b, p. 43). In response to stimulation (some of it originating in the RCN) and the availability of substantially increased funding, 'an energized rural constituency' was showing itself increasingly eager to avail of 'the opportunities held before it' (Murray and Greer 1999a, p. 9).

The ability to speak with a strong and united voice in its dealings with the state came to partly depend on the RCN's adeptness at exploiting the opportunities that came its way. In 1990 an Inter-Departmental Committee on Rural Development

(IDCRD) had recommended the establishment of an independent Rural Development Council (RDC) with a brief for promoting 'community-led rural development' and organizing 'a forum for discussion between communities' (ibid., p. 10). Once instituted, however, the RDC did not move immediately to create the envisaged discussion forum, thus handing the fledgling RCN an important opportunity. Under the emerging division of labour between the RDC and the RCN, the former became entrusted with 'developmental work' and the latter with 'rural issues advocacy' (ibid., p. 13). In the 1995–9 period the RDC's move towards 'job creation', and its 'sudden withdrawal' from promoting community development would fortuitously hand another opportunity to the RCN, this time to work directly with local community groups.

A consequence of the 'ongoing transformation of rural development governance in Northern Ireland' was the emergence of a 'policy community' for rural development that had DANI's Rural Development Division, the RDC and the RCN as its set of members (ibid., pp. 18–19). Ultimately the RCN's standing as a member of this policy community would depend on 'being recognized by the agency [DANI's Rural Development Division] as the natural representative of a given social sector' (ibid., p. 21).

*Outcomes*

How influential has the RCN been in shaping rural policy? The RCN may not have been responsible initially for the state's new community-centred rural development regime, but its presence has clearly been important in grounding and deepening the whole approach. More generally, in its role as intermediary between the community and the state, the RCN would challenge the 'guiding assumption… that all brokerage expertise lay with public officials and that no challenge to this dependency relationship could be brooked' (ibid., p. 13).

Of course the RCN has also been a midwife in the birth of a new type of community work in which community groups 'have been on a steep learning curve during the 1990s, having participated in what could be dubbed an informal curriculum of management by objectives, public administration and partnership processes' (ibid., p. 24). A precondition of this development was the way that very many community groups had by the 1980s 'been absorbed as part of the broader state welfare system, with earlier oppositional stances to public policy tempered by a cooperative engagement based on responsibility and funding' (ibid., p. 8).

In this new environment what community groups do may not be wholly determined by state decisions, but the context in which they operate is one in which the state has achieved considerable influence. There are those who would stress the risk of co-optation for community groups in all this (Rolston 1997). Others, in contrast, are impressed by the opportunities that state funding provides to advance local community development work (Fitzduff 2003, pp. xx–xxi).

Murray and Greer are in no doubt about the wider significance of what is being achieved in rural NI:

> State-community interaction in the sphere of rural revitalization has facilitated
> the emergence of a learning democracy based on participation; it is encouraging
> common discoveries and shared visions of the future from the ground up (1999a,
> p. 26).

From the same source we hear how 'community organizing and development'
are 'reinforcing effective learning and doing, and are transforming civic culture
through localized empowerment, albeit in the difficult circumstances of division
in society' (ibid., p. 26). Fitzduff (2003, p. xxii) takes the view that 'the positive
engagement of rural communities, area-based partnerships and sub-regional
networks is evidence of a stronger civil society'.

We have seen how the RCN has been adept in exploiting new opportunities,
particularly so in areas where the RDC had chosen not to venture or to withdraw from.
Its admittance to a 'policy community' would constitute it as a major player in NI's
evolving rural development regime. But has 'policy community' membership come
at a cost for the RCN? Certainly there is, as is characteristic of policy communities
generally, an observable tendency to consensualism (Murray and Greer 1999a, p.
19). Mutually dependent as RDC and RCN may be, with each one being called on
to make its own 'particular operational contributions', their common reliance upon
DANI 'for substantial funding' has meant that there can be no 'equal distribution of
power' in the policy community within which they are members (ibid., p. 19).[4]

The question also arises whether the RCN is 'helping to legitimize state
intervention by facilitating policy formulation and implementation through its
leadership position and thereby reducing criticism of more unpopular aspects
of policy?' (ibid., p. 22). In considering this question Murray and Greer suggest
that 'RCN has been very mindful of this tension and has been careful not to lose
faith with those it seeks to represent' (ibid., p. 22). As evidence for this they can
point to the RCN's successful struggle to preserve 'the principle and the practice
of community development activities' when these were threatened by the RDC's
adoption of a narrower job creation focus in the late 1990s (ibid., p. 26).

How influential has the RCN been in promoting reconciliation in local
communities? Although perceived as a mainly Catholic and nationalist organization
in some quarters, the RCN has:

> Sought to counter this perception through serious internal discussion and
> promotion of its role in challenging poverty, promoting inclusion and tackling
> sectarianism throughout rural Northern Ireland (Murray and Murtagh 2004, pp.
> 43–4, 106–7).

Within the RCN efforts to bridge the ethnic and religious divide through the
'promoting of dual identities in single-community initiatives' are taken very

---

4   In 1996, for instance, over 60 per cent of the RCN's total income of £106,500 had
DANI's Rural Development Division as its source (Murray and Greer 1999a, p. 19).

seriously (Murray and Greer 1999a, p. 25, 1999b). Reflecting the state's acceptance that peace has to be constructed at the local as well as at the supra-local level, considerable funding has been made available to advance community-based peace and reconciliation initiatives (Fitzduff 2003, pp. xxi–xxii). That said, it is also clear that the challenges facing reconciliation efforts on the ground can sometimes be immense (Crawley 2002; RCN 2003; McCall and O'Dowd 2008, p. 44; O'Dowd 2005, pp. 92–3). Murtagh (2003, p. 187) reminds us of 'the high rates of spatial segregation which exist across Northern Ireland'; and of the depth of division in:

> highly contested areas, where differential rates of development are producing distinctive futures for Protestant and Catholic communities and where social need cannot be disentangled from the precarious ethnic geography of places such as mid-Armagh (p. 182).

## Interpreting the patterns

In considering the contribution of community actors to the politics of sustainability, we began by suggesting that populism can help us organize the discussion around perceptions of imbalance making for powerlessness that collective action, conceived as a potential countervailing and counterbalancing force, can begin to remedy. And to develop the suggestion that populism can have something useful to contribute to the study of community-based attempts to defend relatively powerless rural people from the forces that threaten them with decline and disappearance, two ideal-typical scenarios were sketched. These scenarios provide us with a vantage point from which we can see certain features of the world in a way that is *potentially* useful for research purposes. The question now is how useful has this approach *actually* proved to be? In comparing our two cases as forms of potential countervailing power, we can ask how near or far they fall relative to our ideal-typical radical and pragmatic scenarios.

At a general level both Muintir and the RCN have sought to use the positive power of collective action to counter the negative power that flows from the various forces that destablize rural societies or result in their decline or disappearance. Each one, by trying simultaneously to represent whole communities and disadvantaged categories, falls somewhere between the ideal-typical pragmatic and radical conceptions of the people. Can the same be said of their aims? Did these think in terms of radical alternatives to the *status quo* or of merely restoring the *status quo ante*? Clearly there were features of Ireland's past – its nationalism and Catholicism in particular – that the Muintir leadership was keen on reinterpreting, revitalizing and retaining. To achieve this a new form of community-based collective action was required that would be equal to the crisis conditions of the 1930s and beyond. The Muintir leadership therefore may never have desired a completely radical alternative to the *status quo*, but neither did it want to restore a crisis ridden *status quo ante*.

Nor was the RCN leadership, in a context where rural communities were emerging from a prolonged civil war compounded by a long history of rural decline, interested in restoring a *status quo ante* that was synonymous with strife and decline. The challenge was not to try to reach the future through the past but to build a new future, based on moving forward together with a shared vision centred on acceptance of the ideals of social inclusion, justice and reconciliation.

Any assessment of how participative our two cases have proved to be has to take account of the differences in their respective projects. Muintir has always sought to organize community groups along distinctive Muintir lines. Clearly the RCN has thrown the net wider and has been happy to receive the support of different sorts of community groups, though it has urged its affiliated community groups to adopt a participative culture in which decision-making power would be widely dispersed.

How participative did the Muintir councils prove to be in practice? The early Muintir may have insisted on rural labourers receiving sectional representation on the parish councils, but it is clear that this approach was founded on an acceptance of the prevailing class and gender structures of Irish society. And to the extent that the more powerful social elements of local society exercised a controlling leadership within the parish councils, Muintir moves close to pragmatic populism as regards participation. Indeed the experience of the early Muintir illustrates how, in populist-type community action, local notables can create a resource from projecting the constituency they speak for as relatively powerless. With the move to community councils (and notwithstanding the disappearance of the clergy) elitist tendencies have declined though not disappeared in Muintir.

What is also clear is that the professionalization evident in national Muintir in recent years marks a trend that RCN has carried substantially further. Besides having more professional employees with a wider range of skills nowadays, the RCN has made a point of commissioning independent experts to report on a wide array of rural problems. If the authority of expertise counts for much in the RCN, what does this say about rural leadership? Clearly expertise (if undisputed) carries its own authority, though there is also an acceptance within the RCN that professionals and experts alike have to work in close collaboration with community interests.

Did our two cases opt for integration or opposition in their dealings with the state? Whatever early suspicion of the overly interventionist state there was among Muintir's top leadership did not stop the local councils building close working relations with sections of the state. Under the influence of the international community development movement of the 1950s, and partly inspired by a desire to get state support to fund a new wave of community council-centred community development, national Muintir became an early convert to the ideal of community groups (and their representative associations) building close partnership-type working relations with the state. In all this Muintir gravitates to the pragmatic populist position where tactics are concerned, though matters are complicated somewhat by the way Muintir has occasionally been critical of features of state policy, particularly in response to the closure of rural post offices in recent years.

How do we characterize the RCN's position vis-à-vis the state? We have seen how it became a strong supporter of the ideal of organized community interests working closely with the state; and how it assumed membership of the policy community that was at the centre of NI's new rural development regime of the 1990s. None of this, however, would prevent the RCN continuing to exercise the advocacy role it had cultivated from its earliest days. Not only has it been regularly critical of the trend of state policy, but it has helped orchestrate campaigns of opposition (as, most recently, in response to the closure of rural post offices) against specific state policies. What all this suggests is that, in terms of tactics, the RCN emerges also as a mixed case.

## Conclusion

The mixture of radical and pragmatic tendencies evident in our two cases speaks to the way the worlds inhabited by Muintir and RCN activists are characterized by many shades of grey. At a very general level Muintir and the RCN are comparable in their views of the people, their desire to seek for change that goes beyond the restoration of some *status quo ante*, their increasing reliance on professionalization as a key resource and their common desire to work closely with the state to advance their respective projects. Each has had experience of using community development to counteract the legacy of division and bitterness left behind by civil wars. Each has accepted that local community development, to reach its potential, requires the assistance of supra-local alliances. Each has tried to give voice and coherence to a plethora of community groups separated in space and pursuing different agendas. And each has accepted that the state, to a large degree, has the future of rural communities in its hands.

For all these similarities, our two cases present a fairly sharp contrast in terms of outcomes. Muintir may have endured for more than 70 years and have many achievements to its credit, but it has languished in significant respects in recent years. The decisive breakthrough that would insert it at the centre of the state's rural development regime, in the way rapidly accomplished by the RCN in the 1990s, has remained elusive. How much can this difference in outcomes between our two cases be attributed to differences in their respective projects, to the resources they have inherited or been able to create, to the availability of state-related opportunities and to adeptness at making the most of these?

There are some important differences between our two cases at the level of projects. While Muintir set out to build a community movement, the RCN's concern was to represent, defend and build a rural community movement that already partly existed. Today the RCN's project, in encompassing advocacy, partnership as well as the servicing of community groups, is relatively speaking much more ambitious, sharply defined and institutionally embedded. The RCN's membership grew rapidly and has stayed high in a way that has not been true of Muintir's (save at times in the past). Taken on a purely spatial basis the RCN's task, with six counties (as against 26) to serve, might appear less onerous than Muintir's. As a supra-local alliance the RCN has been able to use its 12 sub-regional rural support networks (Acheson et al. 2004, p. 63), its professional staff, regular publications and annual conferences to help build a sense that very different community groups are all part of the same movement.

Of course, Muintir in the past had some of the features that we can now associate with the RCN, but for various reasons these weakened or disappeared as time passed. Uneasy with its rural self-image, Muintir has even sought to distance itself from its rural origins in recent years by sub-titling itself 'Irish Communities in Action' and 'The Irish Community Development Movement'. Strapped for professional resources as it is, Muintir, though recently active on the question of post office closures, has not been able to expand its advocacy efforts in anything like the way achieved by RCN. In any event others have now emerged, notably Irish Rural Link, to perform the advocacy role Muintir has tentatively sought to build for itself. Something else that has not helped Muintir is the situation where whole community groups (such as community councils) in the RoI have been declining in number relative to communities of interest (Crickley and Devlin 1990, p. 54).

RCN, by virtue of being a member of a policy community, has been at the heart of a rural development regime in a way that the Muintir leadership can only dream of. Indeed partnership with the state has always been a double-edged sword for Muintir in recent times; state funding has helped keep it alive (by funding Muintir's Community Alert groups) but (its absence) has also restricted its ability to develop the sort of community development (based on community councils) that is the basis of its distinctive approach. The RCN has contrastingly benefited from the 'peace process' and the availability of a relative abundance of funding for peace and reconciliation. What is also evident is that the RCN, for all its undoubted commitment to partnership, owes much of its effectiveness to being able to strike a dynamic and productive balance between partnership and selective opposition to state policy.

RCN may have achieved much, but has it been able to exert any real control over the structural forces that are driving rural decline in NI? Agriculture, for instance, is still contracting in terms of the number of farms, many services are under attack and the position of certain groups (such as the young, the old, farm women and ethnic minorities) remains precarious. All this might be read as evidence of how much the power of the forces driving rural decline lie beyond the countervailing power of even a well-organized and influential supra-local community alliance like the RCN. It is also possible to say, of course, that things would be appreciably worse were it not for the presence of the RCN.

### References

Acheson, N., B. Harvey, J. Kearney and A. Williamson (2004) *Two Paths, One Purpose: Voluntary Action in Ireland, North and South.* Dublin: Institute of Public Administration.

Brody, H. (1973) *Inishkillane: Change and Decline in the West of Ireland.* London: Allen Lane.

Burger, T. (1987) *Max Weber's Theory of Concept Formation: History, Laws and Ideal Types.* Durham: Duke University Press.

Crawley, M. (2002) *Protestant Communities in Border Areas.* Cookstown: Rural Community Network (NI).

Crickley, A. and M. Devlin (1990) 'Community Work in the Eighties – An Overview', in *Community Work in Ireland: Trends in the 80s, Options for the 90s.* Dublin: Combat Poverty Agency.

Cronin, M. (1997) *The Blueshirts and Irish Politics.* Dublin: Four Courts Press.

Crowley, E. (2006) *Land Matters: Power Struggles in Rural Ireland.* Dublin: The Lilliput Press.

Etzioni, A. (1995) *The Spirit of Community: Rights, Responsibilities and the Communitarian Agenda.* London: Fontana Press.

Fitzduff, N. (2003) 'Forward', in J. Greer and M. Murray (eds), *Rural Planning and Development in Northern Ireland.* Dublin: Institute of Public Administration, pp. xvii–xxiii.

Fitzduff, N. (2005) 'Forward', in M. McEldowney, M. Murray, B. Murtagh and K. Sterrett (eds), *Planning in Ireland and Beyond: Multidisciplinary Essays in Honour of John V. Greer.* Belfast: School of Environmental Planning, Queen's University Belfast, pp. xiii–xiv.

Friedmann, J. (1992) *Empowerment: The Politics of Alternative Development.* Oxford: Blackwell.

Gaventa, J., J. Morrissey and W. R. Edwards (1995) 'Empowering People: Goals and Realities', *Forum for Applied Research and Public Policy*, 10: 116–21.

Hanna, M. G. and B. Robinson (1994) *Strategies for Community Empowerment: Direct Action and Transformative Approaches to Social Change Practice.* Lewiston, New York: The Edwin Mellen Press.

Hoppen, K. T. (1984) *Elections, Politics, and Society in Ireland 1832–1885.* Oxford: Clarendon Press.

ISG (1987) *Action Research Projects Involved in the Second European Programme to Combat Poverty: Short Descriptions.* Köln: ISG Sozialforschung und Gesellschaftspolitik.

Kissane, B. (2002) *Explaining Irish Democracy.* Dublin: University College Dublin Press.

Kitching, G. (1989) *Development and Underdevelopment in Historical Perspective: Populism, Nationalism and Industiralization.* London: Routledge.

Klandermans, B. (2004) 'Why Social Movements Come into Being and Why People Join Them', in J. R. Blau (ed.), *The Blackwell Companion to Sociology.* Oxford: Blackwell, pp. 268–81.

Larkin, E. (1984) 'Foreword', to W. L. Feingold, *The Revolt of the Tenantry: The Transformation of Local Government in Ireland 1872–1886.* Boston: Northeastern University Press, pp. xi–xvii.

Lee, J. J. (1980) 'Patterns of Rural Unrest in Nineteenth Century Ireland: A Preliminary Survey', in L. M. Cullen and F. Furet (eds), *Ireland and France, 17th–20th Centuries: Towards a Comparative Study of Rural History.* Paris: Editions de l'Ecole des Hautes Etudes en Sciences Sociales, pp. 223–37.

Lipton, M. (1977) *Why Poor People Stay Poor: A Study of Urban Bias in World Development.* London: Temple Smith.

Mair, P. (1987) *The Changing Irish Party System: Organisation, Ideology and Electoral Competition.* New York: St. Martin's Press.

McCall, C. and L. O'Dowd (2008) 'Hanging Flower Baskets, Blowing in the Wind? Third-sector Groups, Cross-border Partnerships, and the EU Peace Programs in Ireland', *Nationalism and Ethnic Politics*, 14: 29–54.

Midgley, J. (1995) *Social Development: The Developmental Perspective in Social Welfare.* London: Sage.

Morriss, P. (2002) *Power: A Philosophical Analysis* (2nd edn). Manchester: Manchester University Press.

Murray, M. R. and J. V. Greer (1999a) *State-Community Interaction in the Changing Arena of Rural Development in Northern Ireland.* Coleraine: Association for Voluntary Action Research in Ireland.

Murray, M. R. and J. V. Greer (1999b) 'The Changing Governance of Rural Development: State-Community Interaction in Northern Ireland', *Policy Studies*, 20, 1: 37–50.

Murray, M. and B. Murtagh (2004) *Equity, Diversity and Interdependence: Reconnecting Governance and People through Authentic Dialogue.* Aldershot: Ashgate.

Murtagh, B. (2003) 'Dealing with the Consequences of a Divided Society in Troubled Communities', in J. V. Greer and M. Murray (eds), *Rural Planning and Development in Northern Ireland.* Dublin: Institute of Public Administration, pp. 168–90.

Nairn, T. (1998) 'The Curse of Rurality: Limits of Modernisation Theory', in J. A. Hall (ed.), *The State of the Nation: Ernest Gellner and the Theory of Nationalism.* Cambridge: Cambridge University Press, pp. 107–34.

Ó Cinnéide, S. and Walsh, J. (1990) 'Multiplication and Divisions: Trends in Community Development in Ireland since the 1960s', *Community Development Journal*, 24: 326–36.

O'Connor, E. (1988) *Syndicalism in Ireland.* Cork: Cork University Press.

O'Dowd, L. (2005) 'Republicanism, Nationalism and Unionism: Changing Contexts, Cultures and Ideologies', in J. Cleary and C. Connolly (eds), *The Cambridge Companion to Modern Irish Culture.* Cambridge: Cambridge University Press, pp. 78–95.

O'Leary, D. (2000) *Vocationalism and Social Catholicism in Twentieth-century Ireland: The Search for a Christian Social Order.* Dublin: Irish Academic Press.

Putnam, R. D. (2000) *Bowling Alone: The Collapse and Revival of American Community.* New York: Simon & Schuster.

RCN (2003) *'You Feel You'd Have No Say': Border Protestants and Community Development.* Cookstown: Rural Community Network.

Robson, T. (2000) *The State and Community Action.* London: Pluto Press.

Rolston, B. (1997) 'Overview of Community Development in Northern Ireland', *Inishowen Development Journal,* Summer: 15–19.

Rynne, S. (1960) *Father John Hayes: Founder of Muintir na Tíre.* Dublin: Clonmore and Reynolds.

Tierney, M. (2004) *The Story of Muintir na Tíre 1931–2001 – the First Seventy Years.* Tipperary: Muintir na Tíre Publications.

Varley, T. (2003) 'Populism, the Europeanised State and Collective Action in Rural Ireland', in M. Blanc (ed.), *Innovations, Institutions and Rural Change.* Luxembourg: Office for Official Publications of the European Communities, pp. 127–68.

Varley, T. (2006) 'Negotiating Power and Powerlessness: Community Interests and Partnership in Rural Ireland', in S. Healy, B. Reynolds and M. Collins (eds), *Social Policy in Ireland: Principles, Practice and Problems* (2nd edn). Dublin: The Liffey Press, pp. 403–18.

Wilson, T. M. and H. Donnan (2006) *The Anthropology of Ireland.* Oxford: Berg.

Chapter 20

# The Road to Sustainable Transport: Community Groups, Rural Transport Programmes and Policies in Ireland

Henrike Rau and Colleen Hennessy

## Introduction

Today many rural dwellers in the Republic of Ireland (RoI) depend on the private car to access services, employment, education, healthcare and recreation and thus shoulder a disproportionate share of the burden of insufficient public transport (see O'Shea in this volume; Roberts et al. 1999; McDonagh 2006).[1] But car dependency does not only affect rural Ireland: those who live in Irish cities, towns and their sub-urbanized hinterlands also experience accessibility problems and reduced 'walkability' resulting from sprawl and a lack of transport alternatives (Leyden 2003; Wickham 2006). Recent census figures illustrate the geographical and demographic expansion of urban centres such as Dublin, Cork and Galway which are now extending well into their (semi-) rural hinterlands where public transport is often unavailable (CSO 2007; see also McDonald and Nix 2005). Car dependency thus represents a key challenge for national transport policy in general, and rural public transport provision in particular.

In Northern Ireland (NI), car dependency is less pronounced, partly because of the greater availability of public and community transport coinciding with planning and land use policies aimed at increasing population density (see Murray and Murtagh 2007). Nevertheless, rural dwellers in NI are also more likely than their urban counterparts to experience access problems and 'mobility deprivation' (see Nutley and Thomas 1992, 1995). This suggests that in the RoI and NI (rural) transport policy is inextricably linked to questions of equity and social inclusion. Transport policy thus needs to have regard for the social consequences of transport and mobility decisions as well as fiscal and technical concerns (see Lohan and Wickham 1999; Lucas, Grosvenor and Simpson 2001; Pickup and Guiliano 2005; Kenyon 2006; McDonagh 2006).

---

1   Note that UK towns with populations under 10,000 are classed as rural in relation to transport policy while Irish towns/villages with populations under 5,000 are considered rural and therefore eligible to avail of the Rural Transport Programme.

Local transport services provided by community and voluntary organizations have been the focus of recent rural transport strategies in Ireland north and south, resulting in two distinct approaches with regard to policy background and 'degree of fit' with policies targeting regional balance and sustainable rural development. On the one hand, the introduction of services under the Irish Government's Rural Transport Programme (RTP), formerly the Rural Transport Initiative (RTI), has significantly improved accessibility and reduced social and geographical isolation for target groups in some parts of rural Ireland. This initiative was a response to repeated calls by rural-based community and voluntary organizations for improved transport infrastructure and non-conventional, demand-based transport solutions as an essential prerequisite for their work. However, the lack of mainstream public transport in many other locations continues to pose serious challenges, in particular for vulnerable sections of the rural population such as women, older people, those in low-income households without access to a car, people with disabilities and children and young people (Pobal/Department of Transport 2006; Fitzpatrick Associates/DoT 2006).

Since the late 1990s changes in transport policy, legislation and practice in Northern Ireland – in particular the establishment of the Rural Transport Fund (RTF) in 1998 – have led to improved rural bus services and better conditions for mostly demand-responsive community transport, though increasing demand means that further improvements are now necessary (DRDNI 2007; see also Nutley 2001 for a detailed account). Many parts of rural NI are now served by community transport initiatives which provide both stand-alone and feeder services that complement the existing public transport network. Overall, community transport services are now an important lifeline for many rural communities in NI and the RoI, in particular those experiencing decline.

Despite the growth in community transport (CT), the role of community and voluntary groups in providing inclusive and sustainable rural transport remains under-explored, particularly regarding their relationship with the state and statutory bodies and their (lack of) involvement in policy making. Historically community transport in the RoI has received little attention and even today its importance in tackling (rural) social exclusion is seldom recognized by policy makers and the public. The situation in NI differs somewhat in that there is greater recognition of the CT sector, some of which is attributable to the PR and lobbying work of the Community Transport Association Northern Ireland (see CTANI 2005).

More importantly, however, the community and voluntary sector's (henceforth CVS) involvement in transport policy both in NI and the RoI has been very limited, though recently the CVS has been given a greater role in both the development *and* implementation of rural transport policy at a local level:

> The RTI was introduced as a response to the growing level of community interest and involvement in finding local solutions to transport difficulties in rural Ireland. The lessons emerging from this action research programme will feed into the development of rural transport policy (Pobal/DoT 2006).

Nevertheless, the community transport sector in the RoI continues to play an 'execution only' role most of the time by offering transport services to vulnerable target groups such as people with disabilities, often without sound financial backup. This partly resembles experiences in Northern Ireland up until the late 1990s (see Nutley 2001).

Using existing statistics, excerpts from official documents and agency reports and fieldnotes relating to research conducted in 2006/7 in County Galway, this chapter will compare the policy background and outcomes of CT provision in rural Ireland north and south, paying particular attention to the involvement of the CVS in rural transport policy design and implementation. We will argue that the provision of rural transport, while often positive for targeted rural service users, could potentially undermine the work of some CVS groups because of its resource-intensive nature and its ever-increasing remit. Using the RTP and recent CT initiatives in Northern Ireland as case studies, we will show how different dimensions of rural transport – social, economic and organisational – can mesh in novel and sometimes unpredictable ways, bringing both positive and negative consequences for the state, the CVS and its service users.

It will be shown that based on experiences in Northern Ireland, the success of CT initiatives depends on continuous state funding, strong local government, including the establishment of regional transport authorities, and most importantly, an adequate 'mainstream' rural transport system to tie in with demand-responsive community transport. Community transport tends to 'outsource' rural transport services to private operators and the CVS while the responsibility for transport policy, legislation and the evaluation of transport programmes remains with Government departments and state agencies. This both reflects and perpetuates the CVS's relative powerlessness vis-à-vis other interested parties (state, business, transport unions) and raises issues about its long-term sustainability, the risk of potential cooption by the state and the moderation of more radical approaches to rural development.

## Sustainability, social inclusion and rural transport

International research has shown that public transport can play an important role in tackling social exclusion, in particular in remote rural areas (Lucas et al. 2001; Hine and Mitchell 2003; Cass, Shove and Urry 2005). More importantly, existing studies highlight the state's role as key transport policy maker and legislator and the importance of strong local government and regional planning bodies in improving accessibility and addressing unmet mobility needs. Experiences in NI indicate that a good bus network and a legislative framework and transport licensing system conducive to the introduction of community transport services can help address issues of (mobility) deprivation, especially in rural areas (Nutley 2001; Frawley 2007). This implies that a clearly defined regulatory framework and comprehensive transport policy provision by the state, in particular for rural transport, continue to be an important pre-condition for successful (demand-responsive) CT.

In the RoI the state remains largely responsible for transport policy and legislation but has gradually shifted responsibility for the provision of transport infrastructure and services to the private sector, for example through public-private partnerships (see Killen 2007 for an overview). This trend towards greater involvement of the private sector has transformed both transport infrastructure provision (as in the prioritization of road construction) and policy priorities (seen in the focus on large-scale critical infrastructure and (inter-)urban transport under *Transport 21*). Rural transport, on the other hand, has not been given the same priority, and policy in this area remains patchy and partly reliant on outdated legislation, such as the 1932 Road Transport Act which regulates bus licensing arrangements. Moreover, the Irish Government's reluctance to prioritize integrated public transport as a necessary infrastructural development and to allocate public funds accordingly has meant that problems regarding accessibility and social exclusion continue to impact on both urban and rural dwellers. The introduction of the RTI/RTP, while welcome and important, must thus be seen as an initial step only in addressing the problem of rural transport which ultimately requires a coherent transport strategy as well as a detailed roadmap for implementation.

The RTI was launched in 2001 and started off in 2002 as a small-scale pilot project aimed at addressing the lack of public transport in rural Ireland:

> The overall aim of the RTI has been 'to encourage innovative community-based initiatives to provide transport services in rural areas, with a view to addressing the issue of social exclusion in rural Ireland, which is caused by lack of access to transport (Fitzpatrick Associates/DoT 2006, p. iii).

A gradual shift towards a more permanent, large-scale roll-out of the programme followed which introduced public transport services in more remote rural areas. There are now 34 mostly demand-responsive projects ranging from island minibus schemes to demand-based hackney services (see also Pobal's RTI information pack 2005 for some selected case studies at http://www.pobal.ie/media/Publications/RTI/Casestudies.pdf). Placing the pilot RTI on a permanent financial footing in 2006 as the Rural Transport Programme (RTP) signalled the Irish State's recognition of the role of transport services in tackling social exclusion in rural Ireland. That said, the relative marginality, in financial terms, of the RTP – €9m in 2007 and rising to about €18m in subsequent years (Fitzpatrick Associates 2006, p. i) – and its continued focus on target groups – older people and people with disabilities – suggests a certain reluctance to recognize the wider role of public transport for the sustainable development of the country as a whole. This contrasts with the situation in Northern Ireland, with its existing public transport links (e.g. Translink services) that are now being complemented with community-run services aimed at previously neglected, mobility-deprived areas.

Despite the proposed widening of the remits of the RTP, the programme still presents itself as a project-based solution to a systemic problem; funding is handled on a case-by-case basis. Community-level research suggests that both geographic and interest-based communities still have critical transport needs so it is unclear what criteria are used to fund and evaluate both individual projects and the RTP as a whole and how successful the programme is in addressing transport problems in rural Ireland. At present there is no actual data that shows the efficacy of the RTP in tackling 'mobility deprivation' nationally, though case study evidence suggests that many RTP-sponsored services significantly improve the lives of users and their families (see http://www.pobal.ie/media/Publications/RTI/Casestudies.pdf). As part of the RTI review in 2006, transport consultants Fitzpatrick Associates analysed the survey material from each county's audit to predict the continued unmet national need for rural transport. Their work suggested that 380,000 rural dwellers had unmet rural transport needs in 2005 and that this figure could grow to over 450,000 by 2021. For the key target groups initially identified by the RTI (older people, people with disabilities, low income households and young people), the numbers with unmet needs could grow from 200,000 in 2005 to over 240,000 by 2021 (Fitzpatrick Associates 2006, p. 27). While it is unclear whether these unmet transport needs are unique to rural areas,[2] these figures reveal that there is a large contingent of rural dwellers who would benefit from the mainstreaming of the RTP to include *all* potential users of rural transport services.

*State funding and the role of local government*

Public funding for transport services remains a contested issue both in NI and the RoI, leading to suggestions that 'the desire to operate public transport with minimal or no public subvention appears, in a European context at least, to be a distinctly UK and Irish phenomenon' (Fitzpatrick Associates 2006, p. 12).[3] The persistence of this 'slim state' transport policy paradigm in the RoI contrasts with social democratic policies elsewhere in Europe that link public transport to positive 'externalities' (e.g. better health, environmental protection, access to employment opportunities) that merit substantial state subvention. This said, many state-subsidized transport systems in Europe are also changing (for better or worse), partly as a result of EU efforts to increase competition in the transport sector.

Lack of funding remains a major barrier to public transport provision in the RoI. Up to the late 1990s under-investment also curbed public transport development

---

2   National evidence suggests that 'transport poverty' also exists in sub-urban and disadvantaged areas of cities where more and more Irish people are residing (Lohan and Wickham 1999; Wickham 2006).

3   Similarly, greater CVS involvement and the farming out of public services to the community or private sector appear to be particularly pronounced in the UK and Ireland compared to other countries. Fitzpatrick Associates argue that 'the desire to operate public transport with minimal or no subvention appears, in a European context at least, to be a distinctly UK and Irish phenomenon. In many other European countries, good public transport is perceived to deliver "positive externalities", and this is seen as meriting considerable levels of Government subvention' (2006, p. 12).

in NI. In 2002, however, the then Minister for Regional Development in NI, Peter Robinson, indicated that the *Regional Transportation Strategy for Northern Ireland 2002–2012* (RTS) would mark a clear departure from 'decades of under-investment and an *ad hoc* approach to transport planning' (DRDNI 2002, p. iii). This new way of thinking was also reflected in the RTS document:

> [...] Northern Ireland has suffered from decades of underinvestment in its roads and public transport. The United Kingdom generally has fallen behind best European practice in transportation investment and, in its turn, Northern Ireland compares unfavourably to levels of transportation investment per capita in England, Scotland and Wales. [...] There is now acceptance that investment in roads and transport is a high priority for public expenditure (DRDNI 2002, p. 1).

In fact, legislation passed prior to the RTS had already recognized the social inclusion benefits of public transportation. Two major reports published in 1998 by the then Department of Environment, Transport & Regions – *A New Deal for Transport* and *Moving Forward: The Northern Ireland Transport Policy Statement* – identified integrated and sustainable CT as a means of reducing social exclusion and resulted in the establishment of the Rural Transport Fund (RTF) in 1998. According to Nutley (2001), the RTF in NI was unprecedented because it positively discriminated in favour of rural areas and attempted to integrate CT operations with conventional public transport to improve accessibility and complement the work of other local development agencies.

Factors other than state subvention are also likely to influence public transport provision. According to Fitzpatrick Associates (2006), transport policy and planning in the RoI is shaped by centralization and weak local government (McDonald and Nix 2005; McDonagh 2006; Rau and McDonagh forthcoming; see also O'Broin and Waters 2007). McDonagh (2006) argues that local government needs to be given greater responsibility for and involvement in transport provision and that the Government's top-down transport policy solutions undermine rural and community development policies and encourage car dependency. Over-reliance on the car is further exacerbated by serious deficiencies in the existing public transport network, including infrequent services in more remote rural areas and the comparatively poor quality and availability of public transport options (McDonagh 2006, pp. 361–3). This contrasts with the situation in Northern Ireland where recent moves towards a three-tier system – central Government (policy, legislation and regulation), public transport authority (design and management of public transport services, development of local public transport policy) and transport operators (service delivery) – has meant that local authorities have now much greater input into transport planning and are able to buy in commercial transport services and/or fund CT to address accessibility and social exclusion issues (DRDNI 2007).

The first national study on rural transport needs in the RoI – *Rural Transport: A National Study from a Community Perspective* (2000) – was commissioned by the Area Development Management (ADM; now Pobal) and resulted from strategic plans submitted by partnership and community groups. Following this study, the Interdepartmental Working Group on Rural Transport directed the County Development Boards to conduct rural transport audits and needs assessments in each county to inform the Board's ten-year-strategy and the development of national strategies, including the RTI. Given that local government in the RoI has normally no authority to plan, fund or regulate public transport, its input into the RTP represents an important, yet somewhat uncertain, step towards greater involvement in national transport policy and planning.

*Public transport and social exclusion: Perceptions, policy and practice*

The relationship between sustainable public transport and social inclusion in the RoI remains poorly understood and existing research has largely concentrated on Dublin and other (sub) urban areas (Lohan and Wickham 1999; Wickham 2006). To-date information and data about rural transport needs have mostly been gathered by communities themselves to document development needs and lobby for resources. The ADM study highlighted the urgent need for a long-term strategy in the rural transport sector and identified the lack of transport as 'a major barrier to social and economic development' (ADM 2000, Preface). The ensuing consultation process confirmed that all parties recognized the 'cost-benefit problem' inherent to rural transport and the need for state funding for projects that could not finance themselves. However, the report also acknowledged the need for equal access for rural and urban dwellers and all groups in society as a valuable (yet difficult to quantify) trade-off to financial cost because equal access to opportunities is seen as a pre-requisite to social inclusion (p. 77). A similar perspective is reflected in the concept of 'rural proofing', that is, attempts at reducing accessibility gaps between urban and rural areas, which has informed recent transport strategy in Northern Ireland (DRDNI 2002, p. 2).

At this stage it is important to note that Irish public transport planning in general and the RTP in particular tend to connect public transport to socially and economically disadvantaged areas and groups. This clearly affects user profiles and shapes public perceptions of more sustainable transport alternatives. For example, the majority of organizations and limited companies managing the RTP originally provided social services for specific target groups, such as the elderly and people with disabilities. Their involvement in rural transport services thus appears to have attached a 'charity connotation' which stigmatizes public transport users, at least in the minds of some, with the labelling of RTI services as the 'old persons bus' being just one example of this (Fitzpatrick Associates 2004, p. 41). Moreover, transport is seen as an additional service for the CVS's specific target groups which 'naturally' falls within the remit of their work, which partly explains why the consequences for groups and organizations involved in the RTP have hitherto remained under-explored.

As a result, CT projects in rural areas are often viewed as marginal rather than mainstream, thereby consolidating the distinction between the socially included and excluded and exacerbating perceptions of peripherality and neglect (Frawley 2007). This runs contrary to Preston and Raje's (2007) recommendation that 'problems of the immobile socially excluded should not be analysed in isolation from the mobile included. Accessibility planning should not be limited to analysing social exclusion' (p. 10). Moreover, anecdotal evidence suggests that in many locations outside Dublin, public transport is mostly used by people who do not have access to a car (low-income families, immigrants, students, occasionally tourists) and that this has impacted significantly on public perceptions of the status of public transport vis-à-vis the car. As a result, the growing demand for rural (and urban) public transport which cuts across all sections of the community and which puts considerable pressure on community transport providers to expand their services is largely ignored.

The apparent need for additional rural transport services to tackle social exclusion has resulted in calls for increased RTP funding. Groups that represent transport-poor areas such as the North Galway Rural Transport Steering Group (NGRTSG) have lobbied Pobal for additional RTP projects. NGRTSG, with the assistance of Galway Rural Development and Clare Accessible Transport, commissioned a North Galway Transport Feasibility Study to inform its planning process. The study was completed in 2006 and demonstrated that the lack of public transport in the area presented itself as an obstacle to 'a reasonable quality of life for many people' (Grimes 2006, p. 20). The lack of transport services in North Galway contributed to social exclusion in the following ways:

- Isolation within the home reinforces social isolation for those in disadvantaged areas;
- Difficulties in accessing services, many of which are located outside the area;
- Excessive cost in availing of employment opportunities which can lead to forsaking skilled careers;
- Inhibiting business and enterprise development and the location of new businesses and subsequently the potential creation of employment; and
- Forsaking formative sporting, leisure and educational opportunities (Grimes 2006, p. 20).

It is important to note here that this level of research, although encouraged and required by Pobal, seems to actually compound the short-term, disjointed nature of RTP project planning that occurs all over Ireland.

*Travelling North and South: (Missed) Opportunities in developing cross-border initiatives*

Community transport initiatives in Northern Ireland and the Republic of Ireland differ significantly in terms of prevalence, quality and connectivity, all of which raise concerns for cross-border community development programmes that depend on these transport services. NI community transport legislation introduced during the 1990s improved opportunities for setting up transport initiatives, reduced bureaucracy (e.g. by simplifying the licensing process) and was intended to bring community transport efforts in NI into line with those in other parts of the UK (Nutley 2001, p. 52). Previously there was no provision of minibus permits (and associated CT funding) and local authorities had no power to subsidize transport services. The Transport (Amendment) (NI) Order 1990 added a section to the Transport Act (NI) 1967 which enabled district councils, education agencies and health boards to grant 'Section B' permits to local groups for the running of minibus services. Nutley (2001) points out that prior to 1990, the biggest barrier (outside of funding) for CT in NI was the complicated PSV license which increased costs and deterred volunteer drivers. 'Section B' permits solved most of these problems.

The Republic of Ireland, on the other hand, has currently no coherent legislative framework for community and non-conventional transport schemes, and licensing is both centralized and cumbersome. Navigating the fragmented and often confusing transport regulation system takes a huge amount of time and makes it difficult to determine authority and responsibility in relation to public transport provision. This is highlighted by local attempts to plan for and put in place auxiliary services for the RTP, such as bus shelters. The Community and Enterprise Unit in Galway County Council is the lead agent on the Rural Bus Shelter Scheme, as determined in the County Development Strategy, and has repeatedly tried to pilot the scheme in Portumna, Co. Galway (the location of the SE Galway Local Bus RTI). However, despite much work locally, progress has been slow because there are neither a commercial or statutory body nor relevant policy measures that cover the installation and maintenance of vital auxiliary facilities such as good-quality bus shelters, adequate lighting, ticket vending machines and display units for timetables (Frawley 2007). Again, this suggests that a coherent national policy for community and non-conventional transport services is most critical for the sustainability of community transport in the RoI.

Finally, the potential for North-South collaboration appears to be of relevance here. Some collaboration has occurred in relation to the integration of transport services. In 2005 an EU-funded cross-border advice and information project (CTAIS) was established by Community Transport Northern Ireland (CTANI) and the Community Transport Association of Ireland (CTAI) to provide support, guidance and training for local transport operators. Nutley (2001) acknowledges the key role played by cross-border umbrella group Community Connections in the successful coordination of projects. However, differences in legislation such as the absence of an equivalent to the 'Section B' permits in the RoI have hampered cross-border integration of services (CTAIS 2006). Even today a NI minibus under permit is technically not allowed to operate south of the border which suggests that very little work has taken place to ensure legislative consistency (e.g. with regard to permits) and facilitate cross-border policy learning. This is regrettable considering that the shift towards community-run rural transport occurred approximately around the same time and would have offered ample opportunity for an all-island policy debate. Moreover, a cross-border approach to addressing issues of rural isolation and social exclusion would have benefited vulnerable groups in both NI and the RoI. The mid-term report for the EU-funded *Positive Ageing Cross Border Project*, published in October 2007, revealed that poor transport networks continue to be a significant barrier to positive ageing both north and south of the border and called for immediate NI/RoI cooperation to tackle rural isolation (*RTE News at 6 o'clock*, Monday, 1 October 2007).

### Rural transport and development: The changing relationship between state and community and voluntary sector (CVS)

The relationship between the Irish state and the community and voluntary sector has been subject to considerable debate in Ireland in relation to both rural governance in general (Taylor 2005; Ó Riain 2006; The Wheel 2007) and local area partnerships in particular (Sabel 1995; Varley and Curtin 2006). Interestingly, public transport provision – an area traditionally associated with (sub-) urban settings – now constitutes a key arena for multi-level rural governance. Community involvement in rural transport in the RoI is also a response to the serious crisis arising from (unintended) inconsistencies in Irish transport policy, including tensions between the need for stricter traffic law enforcement to improve road safety (as with the clampdown on drink driving) and the persistent lack of alternatives to the car (as with demand-responsive night busses to and from rural pubs). More importantly, the growing involvement of non-governmental actors in local transport-related partnership agreements raises important questions about the implications for both the State and non-state actors such as community and voluntary groups.

The Irish Government's decision to hand over some of the responsibility for rural transport to community and voluntary organizations added a labour-intensive and costly task to their already extensive range of services. This proved particularly difficult for CVS groups with little or no prior experience and with few local connections to agencies and other development groups:

> The development of the RTI has not been without its challenges and difficulties. [...it] has placed a considerable burden of management, administration and governance on RTI groups, with the people involved having little prior experience of such matters in most cases...the burden of governance has been especially onerous for groups with no links to existing development groups (Fitzpatrick Associates 2006, p. 63).

More importantly, the introduction of project-based funding under the RTP to encourage CT initiatives indicated a 'paradigm shift' in transport policy (Vigar 2002) and signalled the Irish state's partial withdrawal from the *provision* of rural transport. At the same time, responsibility for (rural) transport policy remained with the state (Department of Transport) and various national transport agencies

(NRA, RPA). Despite (or perhaps because of) this 'division of labour', a clearly articulated policy framework for (rural) public transport is proving elusive and, as a result, community transport in the RoI continues to operate in a policy environment characterized by serious gaps and outdated legislation.[4] This makes it extremely difficult for many CVS groups to coordinate their efforts so as to complement existing rural transport, thereby preventing wider integration that could potentially alleviate some of the pressure on both commercial and voluntary rural transport providers (see Fitzpatrick Associates 2004). Most crucially, however, current partnership arrangements in relation to rural transport both maintain and reflect the *uneven distribution of policy making powers* among different parties, including the CVS. This partly reflects the situation in other transport-related policy arenas, such as centralized planning for the provision of road infrastructure, but also suggests some unique patterns attributable to the specific structures of decision-making and knowledge transfer in the context of rural transport policy in Ireland.

The development agency, Pobal, plays a particular role here in that it functions as intermediary between the Department of Transport and individual community groups and organizations, entrusted with the administration and management of the RTP. Undoubtedly, Pobal's past involvement in rural transport has proved vital for the success of RTI/RTP measures to date, even though resources and skills will need to be extended for Pobal to meet the demands of any future RTP expansion (Fitzpatrick Associates 2006). This is crucial as two of the four key recommendations made in the Farrell Grant Sparks report (2000) commissioned by Pobal – the development of a national rural public transport policy and the prioritization of Government funding for rural transport services – remain largely unmet. This contrasts with Northern Ireland's more integrated approach to public transport which places CT initiatives within a comparatively coherent policy framework aimed at tackling rural isolation, inaccessibility and exclusion. And while there are some parallels between NI and the RoI regarding limited funds and the uncertainty of funding in the long term, community and voluntary organizations in Northern Ireland (such as CTANI) have achieved a considerable amount of influence in transport decision-making.

Despite delays in devising a coherent rural transport policy framework to underpin the RTI/RTP, many participating CVS groups availed of funding to provide much-needed transport for their target groups. Existing case studies show that many (though not all) CVS groups in receipt of RTI/RTP funding were able to make their services more accessible and feed back their experiences and recommendations to Pobal and to the Department of Transport, thereby strengthening their position as community representatives (Fitzpatrick Associates 2004; Pobal/DoT 2005, 2006). Nevertheless, the impacts of transport provision on the *development principles and*

---

4 For example, the Fitzpatrick report (2006) *Progressing Rural Public Transport in Ireland* compared conditions in the RoI with international practice and identified key problems related to Ireland's approach to transport policy, particularly public transport and rural transport strategy.

*practices* of these groups remain poorly understood. For example, participation in the RTP partnership process has occasionally highlighted the lack of political influence and lobbying power of the CVS (see Meade 2005 and Lee 2006 for a general discussion).

There is also some evidence that reliance on funding from the Department of Transport has partly re-shaped the work of some CVS groups by softening more critical approaches to rural transport and community development and increasing the likelihood of co-option, at least to some extent (see Lee 2003 and Daly 2007 for a general discussion). Certainly some groups have found themselves in the role of sole transport provider which added to their already extensive list of commitments, increased their dependency on continuous state funding and limited their capacity to plan strategically. Moreover, small groups in particular have found the level of administration required to participate in the RTI/RTP a major source of frustration that has restricted their capacity to develop services, thereby endangering their long-term sustainability (Fitzpatrick Associates 2004, pp. 42–3). These experiences partly resemble observations by community development practitioners in Northern Ireland:

> There is some concern that Government services are being 'downloaded' onto small organizations that do not have adequate resources to cope...Moreover, unstable policy environments, fierce competition with private and statutory providers, and the burden of administrating...service[s...] are problematic. Perhaps most importantly for the future of community development in Northern Ireland, public service delivery may compromise the independence of community and voluntary organizations, damage their campaigning and lobbying role, and enable Government to withdraw from providing services they should provide (Lewis 2006, p. 12).

The CVS's reluctance to threaten to withdraw capital investment or labour power, together with its status as a group without electoral mandate to make policy, has been shown to further weaken its bargaining power.[5] Transport is a very expensive and technical sector and the CVS offers neither finance nor technical expertise to the Irish Government and hence has no real power in determining how transport policy is made in relation to the RTP, especially if a change in strategy or investment is required. If the CVS is going to be involved in the rolling out of rural transport services on a wider scale – a possibility mooted in the Fitzpatrick report (2006) – will it be able to significantly influence national transport policy or plan for the services it is asked to provide?

Given the current lack of central transport planning, CVS groups frequently need to deal with different state agencies – including the Departments of Social

---

5    It is important to note that other groups are equally affected by the uneven power distribution in Irish partnership agreements, such as Environmental NGOs which hitherto have not even been part of the social partnership process.

and Family Affairs, Education and Transport, An Garda Síochána, HSE and local authorities – on a case-by-case basis in order to plan and run the RTP. For this reason, route licensing presents itself as a significant barrier for many groups (including local authorities who are in the same position as the CVS in this regard) and hampers their efforts to respond to unmet rural transport needs. Problems also arise for CVS groups trying to establish schemes in areas not covered by the RTP, such as in towns over 5,000 people which do not qualify for rural transport funding (see Frawley 2007).

Finally, the CVS's relationship with the state is also critical on a *local level*, and current plans to reform local government in Ireland are also very likely to change the way (public) transport services are planned and delivered. International experiences show that such transformative processes can both assist and prevent collaboration between CVS groups and local authorities. Banks and Orton's (2005) study of community development workers in local authorities in England suggests that local government reform poses challenges for both the local authorities themselves, which are often the nexus between communities and the state, as well as the community development sector. Their work highlights five high-tension areas in the relationship between CVS and local authorities that include a) complex accountabilities, b) the visibility of work, c) working to both national and local priorities, d) locally based community development work versus strategic policy-level work and e) tensions between the desire to improve representative democracy by supporting councillors and the need to develop participatory democracy by engaging local people in decision-making processes (pp. 102–5). All this suggests that the transformation of local government structures can also bring about very significant changes in the relationship between local authorities and the CVS, and that community transport in rural Ireland is likely to be affected by future reforms.

Overall, it can be shown that the RTP has increased the capacity of many (though not all) CVS groups to deliver vital transport services for their target groups. Evaluation studies (Fitzpatrick Associates 2004, 2006) highlight the many successes of the RTP but also identify a number of obstacles, including the burden of administration, the persistence of unmet transport needs and the need for a coherent rural transport policy.

## Towards greater sustainability? Arguments for and against the CVS's involvement in transport policy-making

Should then the CVS continue or even extend its involvement in the planning and provision of rural transport? And what would this mean for the long-term sustainability of community and voluntary organizations in rural Ireland? So far this chapter has shown that while there are many positives arising from improved public transport for rural dwellers in Ireland, the decision to do so could also pose problems, in particular with regard to the long-term sustainability of the

CVS's skill and resource base. More importantly, the relationship between the CVS and the Irish state must be the focal point of any debate in favour or against the continued participation of community groups in the running of the RTP. With (rural) service users with unmet transport needs pitted against the interests of the state, the CVS finds itself hemmed in and faces a conflict between community development principles and practice. Community development is by nature a process aimed at supporting communities to build their capacity to influence their own opportunities through local solutions. National initiatives and top-down policy making, both of which are evident in the context of the RTP, leave little room for these capacity-building processes to flourish, thereby not only compromising community development principles but also threatening the effectiveness of local services.

Given the existing commitment of many community and voluntary organizations to provide transport for their own members as well as their service users, many of which now depend on the continuation of the RTP, it seems almost impossible for the CVS to review its decision at this stage. Moreover, the CVS's weak position in the context of the social partnership talks mentioned previously implies that challenging the *status quo* or arguing for greater involvement of the state in the provision of regular rural transport could prove difficult. The experiences of RTI groups so far point to a real risk of 'programme analysis' (Lee 2006), where the primary focus is the challenge of sustaining programme *structures*, not reviewing and reshaping programme *actions* to tackle poverty and social exclusion where necessary. To put the RTP on a sustainable footing in the long term, it thus seems necessary for the Irish state to devolve decision-making powers regarding (rural) transport to the local level and provide additional resources. The latter seems particularly important to avoid unsustainable competition for resources among community and voluntary groups that could potentially undermine the role of the CVS as a champion of community development.

### Conclusions

This chapter has argued that the farming out of rural transport to the CVS and private operators represents an (almost uniquely) Irish approach to service provision based on minimal involvement of the state in delivering and maintaining public goods. Undoubtedly, the RTP has provided real opportunities for learning and capacity building for those involved in the management and operation of rural transport services and has (in-)directly contributed to local employment in rural areas affected by economic and demographic decline. However, focusing on rural transport provision has also clearly reduced the ability of some CVS groups, especially those without prior knowledge in the area of rural transport, to attend to wider community development issues such as more balanced regional development and the policy influence of local groups. A long-term assessment of the future viability and sustainability of rural community transport strategies such as the RTP that have the potential to undermine the capacity of community and voluntary organizations to pursue their development goals thus seems to be of paramount importance.

There is now considerable debate in Ireland about the role of community development, as the Irish state contracts out more and more essential social services to this sector 'on the cheap'. This contract points to an inherent conflict, often identified within the voluntary sector, where the state provides resources to organizations whose stated (or original) aim is to change the state or provide support to communities in this process (Lee 2006, p. 17). This type of cooption can be problematic and funding lines can come under threat because of conflict between activists and funding bodies over the way resources are used, including research necessary for the development and articulation of social change proposals (Lee 2006). There is also a (perceived) shift from traditional government structures towards (local) participatory governance networks. While this does not necessarily reduce the influence of the state, it essentially moves responsibility for the delivery of essential public services such as transport to non-state actors, including private businesses and the CVS (Larragy and Bartley 2007, Rau and McDonagh forthcoming). As a result, the relationship between the CVS and the Irish state has undergone fundamental changes, producing both positive and negative outcomes for those involved.

The recent rejuvenation of community transport in Northern Ireland suggests that adequate funding and a strategic rural transport policy framework are necessary prerequisites for successful community-run schemes (Nutley 2001). Rural transport policy in the Republic of Ireland, however, remains ambiguous, in particular with regard to the future role of the CVS in the provision of rural transport. This chapter has shown that the lack of an adequate rural transport and mobility policy framework that integrates the views of the CVS and other interested parties remains a major barrier to sustainable transport provision. Subsequently the Government's commitment to financing the RTP must also coincide with joined-up, long-term policy provision to ensure sustainability and promote integration.

Overall, the Irish Government's renewed financial commitment to the RTP recognizes that access and mobility are critical factors in promoting social inclusion in rural Ireland. This said, it also raises serious questions about the state's commitment to an integrated public transport system that presents an accessible alternative to the private vehicle. The community and voluntary sector has proven useful for connecting the most excluded populations to mainstream transport links in isolated areas and for providing targeted services for their members. However, recent figures suggest that the number of people with unmet transport needs remains high and that measures for improving public transport in (rural) Ireland must go beyond the current 'target group approach' which has hitherto characterized the RTP. Recent proposals to widen the coverage of the RTP to include *all* potential rural transport service users (Fitzpatrick Associates 2006, esp. pp. v–xi) would thus necessitate a detailed review of the role of the CVS in providing rural transport services that takes into account the issues raised in this chapter.

## References

Banks, S. and A. Orton (2005) '"The Grit in the Oyster": Community Development Workers in a Modernizing Local Authority', *Community Development Journal*, 42, 1: 97–113.

Cass, N., E. Shove and J. Urry (2005) 'Social Exclusion, Mobility and Access', *The Sociological Review*, 53, 3: 539–55.

Central Statistics Office – CSO (2007) *Census 2006: Principal Demographic Results*. Dublin: Stationery Office.

Community Transport Advice and Information Service (CTAIS) (2006) 'CTAIS secures Funding for Cross Border Services Study', *CTAIS Newsletter*, No. 4. http://www.communitytransport.ie/CTAIS%20NL4%20final.pdf (accessed 5/12/2007).

Community Transport Association Northern Ireland (CTANI) (2005) *CTA Communication Strategy for Northern Ireland*. http://www.communitytransport-ni.com/publications.php (accessed 6/12/2007).

Daly, S. (2007) 'Mapping Civil Society in the Republic of Ireland', *Community Development Journal*, Advance Access, 31/07/2007, doi:10.1093/cdj/bsl051. http://cdj.oxfordjournals.org/papbyrecent.dtl (accessed 19/12/2007).

Department of Regional Development Northern Ireland (DRDNI) (2002) *Regional Transportation Strategy for Northern Ireland 2002–2012*. http://www.drdni.gov.uk/rts (accessed 29/11/2007).

Department of Regional Development Northern Ireland (DRDNI) (2007) *Future of Public Transport Services in Northern Ireland: Change Programme*. http://www.drdni.gov.uk/ptr_changeprogv4.pdf (accessed 9/08/2008).

Department of Transport (IRL) (2005) *Transport 21: Connecting Communities, Promoting Prosperity*. http://www.transport21.ie/ (12/12/2007).

Department of Transport (IRL) (2007) *Revised Guidelines for the Consideration of Passenger Road Licence Applications under the Road Transport Act, 1932*. http://www.transport.ie/viewitem.asp?Id=9682&Lang=ENG&loc=1701 (accessed 30/11/2007).

Donaghy, K.P., S. Poppelreuter and G. Rudinger (2005) *Social Aspects of Sustainable Transport: Transatlantic Perspectives*. London: Ashgate.

Farrell Grant Sparks (2000) *Rural Transport: A National Study from a Community Perspective*. Dublin: Farrell Grant Sparks.

Fitzpatrick Associates/ADM (2004) *External Evaluation of Rural Transport Initiative*. http://www.pobal.ie/media/Publications/RTI/RTI%20external%20evaluation.pdf (accessed 12/12/2007).

Fitzpatrick Associates/Department of Transport (2006) *Progressing Rural Public Transport in Ireland – A Discussion Paper*, report commissioned by the Department of Transport. http://transport.ie/upload/general/7903-0.pdf (accessed 15/06/2007).

Frawley, C. (2007) *Transport Policy and the Community and Voluntary Sector in Ireland: Community Development's Role in Public Service Provision*.

Unpublished dissertation, MA programme in Community Development, NUI, Galway.

Grimes, M. (2006) *North Galway Transport Feasibility Study*. Galway: Galway Rural Development.

Hine, J. and F. Mitchell (2003) *Transport Disadvantage and Social Exclusion: Exclusionary Mechanisms in Transport in Urban Scotland*. London: Ashgate.

Interdepartmental Working Group on Rural Transport (2001) *Report of the Interdepartmental Working Group on Rural Transport*. http://www.transport. ie/upload/general/2646.pdf (accessed 15/06/2007).

Kenyon, S. (2006) 'Reshaping Patterns of Mobility and Exclusion? The Impact of Virtual Mobility upon Accessibility, Mobility and Social Exclusion', in M. Sheller and J. Urry (eds), *Mobile Technologies of the City*. London: Routledge, pp. 102–20.

Killen, J. E. (2007) 'Transport', in B. Bartley and R. Kitchin (eds), *Understanding Contemporary Ireland*. London: Pluto Press, pp. 100–111.

Larragy, J. and B. Bartley (2007) 'Transformations in Governance', in B. Bartley and R. Kitchin (eds), *Understanding Contemporary Ireland*. London: Pluto Press, pp. 197–207.

Lee, A. (2003) 'Community Development in Ireland', *Community Development Journal*, 38, 1: 48–58.

Lee, A. (2006) *Community Development: Current Issues and Challenges*. Dublin: Combat Poverty Agency.

Lewis, H. (2006) *New Trends in Community Development*. L/Derry: INCORE.

Leyden, K. (2003) 'Social Capital and the Built Environment: The Importance of Walkable Neighborhoods', *American Journal of Public Health*, 93, 9: 1546–51.

Lohan, M. and J. Wickham (1999) 'The Transport Rich and the Transport Poor: Car Dependency and Social Class in Four European Cities'. Paper presented at Conference 'Urbanism and Suburbanism at the End of the Century', November 26th–27th 1999, NUI, Maynooth, Ireland. http://www.tcd.ie/ERC/pastprojects/carsdownloads/Transport%20Rich.pdf (accessed 3/10/2007).

Lucas, K., T. Grosvenor and R. Simpson (2001) *Transport, the Environment and Social Exclusion*. York: Josef Rowntree Foundation.

Mac Connell, S. (2006) 'Non-farm Families Opting Out of Rural Activities', *Irish Times*, 24/01/2006.

McDonagh, J. (2006) 'Transport Policy Instruments and Transport-related Exclusion in rural Republic of Ireland', *Journal of Transport Geography*, 14: 355–66.

McDonagh, J. (2007) 'Rural Development', in B. Bartley and R. Kitchin (eds), *Understanding Contemporary Ireland*. London: Pluto Press, pp. 88–99.

McDonald, F. and J. Nix (2005) *Chaos at the Crossroads*. Dublin: Gandon.

Meade, R. (2005) 'We Hate it Here, Please Let Us Stay! Irish Social Partnership and the Community/Voluntary Sector's Conflicted Experiences of Recognition', *Critical Social Policy*, 25, 3: 349–73.

Murray, M. R. and B. Murtagh (2007) 'Strategic Spatial Planning in Northern Ireland', in B. Bartley and R. Kitchin (eds), *Understanding Contemporary Ireland.* London: Pluto Press, pp. 112–23.

Nutley, S. D. (2001) 'Community Transport in Northern Ireland – A Policy Review', *Irish Geography*, 34, 1: 50–68.

Nutley, S. D. and C. Thomas (1992) 'Mobility in Rural Ulster: Travel Patterns, Car Ownership and Local Services', *Irish Geography*, 25: 67–82.

Nutley, S. D. and C. Thomas (1995) 'Spatial Mobility and Social Change: The Mobile and the Immobile', *Sociologia Ruralis*, 35: 24–39.

O'Broin, D. and E. Waters (2007) *Governing Below the Centre: Local Governance in Ireland.* Dublin: TASC at New Island.

Ó Riain, S. (2006) 'Social Partnership as a Mode of Governance: Introduction to the Special Issue', *The Economic and Social Review*, 37, 3: 311–18.

Pickup, L. and G. Giuliano (2005) 'Transport and Social Exclusion in Europe and the US', in K. P. Donaghy, S. Poppelreuter and G. Rudinger (eds), *Social Aspects of Sustainable Transport: Transatlantic Perspectives.* Aldershot: Ashgate, pp. 38–49.

Pobal/Department of Transport (2005) *RTI Information Pack.* http://www.pobal. ie/media/ Publications/RTI/Casestudies.pdf (accessed 2/08/2007).

Pobal/Department of Transport (2006) *Rural Transport Initiative: A Driving Force Behind Inclusion*, September 2006 leaflet. Dublin: Pobal/ Department of Transport.

Preston, J. and F. Rajé (2007) 'Accessibility, Mobility and Transport-related Social Exclusion', *Journal of Transport Geography*, 15, 3: 151–60.

Rau, H. and J. McDonagh (forthcoming) 'Governance and Transport Policy in Ireland', in M. Millar and M. Adshead (eds), *Governance and Public Policy in Ireland.* Dublin: Irish Academic Press.

Sabel, C. (1995) *Ireland: Local Partnerships and Social Innovation.* Paris: OECD.

Taylor, G. (2005) *Negotiated Governance and Public Policy in Ireland.* Manchester: Manchester University Press.

Varley, A. and C. Curtin (2006) 'The Politics of Empowerment: Power, Populism and Partnership in Rural Ireland', *The Economic and Social Review*, 37, 3: 423–46.

Vigar, G. (2002) *The Politics of Mobility: Transport, the Environment and Public Policy.* London: Spon Press.

The Wheel (2007) *Supporting and Enabling Voluntary Activity: A Manifesto on the Cross-cutting Needs of Community and Voluntary Organisations.* Dublin: The Wheel.

Wickham, J. (2006) *Gridlock: Dublin's Transport Crisis and the Future of the City.* Dublin: TASC at New Island.

# CONCLUSION

Chapter 21

# Sustainability and Getting the Balance Right in Rural Ireland

John McDonagh, Tony Varley and Sally Shortall

Very reasonably Eden (2000) can describe sustainable development as a 'slippery concept'; the diverse interpretation of 'needs' it permits, she contends, forms a key attraction for policy-makers and lobby groups, as it allows sustainable development 'to mean what one would like it to mean' (p. 111). Ever present in contemporary political rhetoric, the compilation of as many as 70 different usages is testament to sustainable development's diverse and contested connotations. Nonetheless debates about the prospects for achieving sustainable development are particularly relevant at this moment in time. Growing world populations, escalating demands on natural resources, energy and food, provide the backdrop to the major global challenge of reducing poverty and inequality without damaging and degrading our environment. In an increasingly inter-connected world, the relentless consumption of goods and services, and the rising demand for greater levels of production to satisfy it, present a testing challenge to the Brundtland call for 'development that meets the needs of the present without compromising the ability of future generations to meet their own needs' (WCED 1987).

We have seen how the ideal of achieving 'sustainable' rural development has become a key dimension of EU, national, regional, and local policy. There is a broad acceptance of the view that the concept of sustainability envelops the notion of a 'living countryside', and implies in particular the striking of some sort of viable balance between the economy, the environment and society. Achieving such balance is by no means an easy task as economic, social and environmental objectives can frequently be at odds with one another. As Holling (2000) reminds us, sustainable development 'is not an ecological problem, nor an economic one, nor a social one. It is a combination of all three. And yet actions to integrate all three typically have short-changed one or more' (p. 1).

## Balance or imbalance?

Each of the contributors to this book has attempted, in one way or another, to engage with the politics of sustainability by considering how oppositions between balanced and imbalanced development are working themselves out in a variety of guises north and south of the Irish border. In the process many interpretations of and engagements with sustainability have been brought to the discussion that reflect variously on the opposition between 'survival *via* balance' and 'survival *via* imbalance'. How to transform forms of imbalanced development into balanced development has emerged as a leading challenge and one that ultimately gives rise to a politics of sustainability. In these politics the state (and the EU) as well as various organized interests (some based outside the countryside) have emerged as major actors. We have seen how consensus and co-operation, conflict and competition have featured variously in this politics of sustainability, and how such politics can produce 'winners' and 'losers'.

In Scannell and Turner's contribution, and also in that of Flynn's, there is considerable focus on the EU's use of an array of legal measures to redress the perceived imbalance that leaves the environment at risk of further degradation. What is striking about the discussions in these chapters that touch on the autonomy and capacity of the EU and the state is how crucial yet how often ineffective each of these actors can be. Witness the Nitrates Directive that remained on the shelf for over ten years in the South, and the ongoing failure in both jurisdictions to impose a 'balance' in rural housing policy that would be more favourable to the environment.

The emphasis in Mullally and Motherway's discussion falls on state/EU efforts that strive through institutional change to create some of the conditions for achieving sustainable development. We have seen how many of these efforts have fallen short in view of the way the environment continues to be subordinated to economic 'necessities'; and how, compared to social partnership bodies at the national level, the local partnerships for sustainable development have remained the poor relation.

Scott's contribution addresses the problem of spatial imbalance at the regional level. By no means has the ever more powerful position of Dublin as an urban region been ignored in regional planning strategies, though what is evident as well is that, for reasons tied up with market forces and the state, the measures used to address this spatial imbalance have not worked.

A major imbalance explored in Tovey's account of environmental management is the controlling influence of scientific knowledge and how this can lead to the exclusion of lay actors and knowledge. The danger when science dominates in this way is that the sorts of knowledge that rural people can usefully bring to bear on environmental management is often overlooked in favour of more elitist forms of knowledge.

The opposition between balance and imbalance is no less apparent in the second section of the book. Feehan and O'Connor describe attempts at rebalancing Irish agriculture through the resort to pluriactivity and multifunctionality. They describe how the move toward a post-productivitist society presents farmers with a particularly difficult balancing act to play as are urged to be competitive producers across a range of commodities while remaining environmentally sound in their farming practices.

When it comes to achieving balanced development the forestry and fishing industries have faced their own peculiar challenges. Only in recent years has the historic neglect of forestry in Ireland begun to be redressed – just under 10 per cent of its land is under forestry, the lowest in the EU. What is also clear is that planting programmes have led to some conflict with local communities and environmental interests. Having extensively explored the features of Irish forestry that can make for imbalances, Tomlinson and Fennessy can still conclude that the expansionist policy being pursued (and its ability to embrace change) is broadly on the right track 'economically, socially and environmentally'.

Imbalance in the form of declining fish stocks and fishing communities are very much at the core of Meredith and McGinley's discussion. Their account shows how smaller fishers have lost out and how mismanagement on the state's part has been a major impediment to the achievement of survival via balance in the sea fisheries.

The power of the large supermarkets to dominate aquaculture is a leading theme in Phyne's chapter. To survive in the new context of production, small and large-scale aquacultural interests have no choice but to accommodate themselves to a range of market, environmental and regulatory requirements.

Grimes and Roper take the expectation that the IT revolution would almost of itself reduce disparities between urban and rural areas as unrealistic. In certain respects, a more deeply embedded imbalance is what we see emerging as the IT revolution rolls on, with little to indicate any great change where the countryside is concerned in the near future.

The need to move from conflict to consensus and to rebalance environmental concerns and economic development is the central theme of McAreavey, McDonagh and Heneghan's chapter on tourism. Notwithstanding the presence of multiple stakeholders and interest groups all with legitimate (if competing) claims of their own, it is contended that any workable balance can only be achieved through adaptation, collaboration and consultation with stakeholders, however complex the politics of this may prove to be in practice.

The book's fourth section considers the impacts of different forms of social differentiation on sustainability. Haase's work on the demography of rural decline clearly identifies the imbalance between rural and urban labour markets that result in a continuing flow of people from the countryside in search of work. As market forces of their own cannot come to the rescue here (they are often responsible for the malaise), the state is seen to have a crucial role to play in countering the forces that make for imbalanced development.

McGrath's chapter on rural youth echoes many of Haase's observations regarding rural labour markets and one again highlights the pervasiveness and intractability of rural imbalances. Rural youth in Ireland often feel excluded from decision-making, left behind in terms of social facilities and in a sense 'forced' to move in order to avail of acceptable employment opportunities.

The notion of cumulative decline provides O'Shea's discussion of rural ageing and public policy with its starting point. To counter the tendencies producing decline, O'Shea makes a strong case for 'social entrepreneurship' among the elderly as well as for a host of changes in public policy.

Shortall and Byrne's chapter focuses on a different form of imbalance, namely that which arises from the patriarchal domination of women in rural Ireland. Such domination has a long history and has tended to persist. In spite of the state's role in gender proofing policy and positive discrimination that has made a difference, gender equality is still a long way off.

Linguistic imbalance, reflecting and the growing dominance of English and the continuing decline of Irish, are addressed in Mac Donnacha and Ó Giollagáin's paper where the nature of this decline, its extent and attempts at countering it are all explored. Once again, considerable significance is assigned to state intervention if the demise of Irish as a minority language is to be avoided.

Some linkages between sustainability and civil society supply the theme for the final section of the book. The background to Barry and Doran's chapter on environmental collective action is the way the dominant model of liberal capitalism has caused the often severe environmental degradation, a phenomenon carried to new heights during the 'Celtic Tiger' period in Ireland. What the environmental movement needs to do to resist such imbalanced and often highly destructive development is explored in some depth.

Can supra-local alliances of community groups counteract some of the imbalances that produce rural decline is the focus of Varley's discussion. More specifically, he considers the conditions in which a partnership with the state approach adopted by his two cases – Muintir na Tíre and the Rural Community Network – can make some significant difference.

Rau and Hennessy, in the final chapter, consider the record and prospects for community transport in Ireland. Imbalance here presents itself in the way rural areas have been, and continue to be, poorly served in terms of a range of services (including transport). Much success has been achieved by community transport initiatives, based on collaborative partnerships, in meeting the needs of many rural residents, though it is doubtful if the high cost of this to community groups can be borne in the longer term.

**Final remarks**

At the heart of the politics of sustainability, as conceived in this volume, is to be found some opposition between balance and imbalance. The context in which this opposition presents itself, in view of the diversity of development arenas and the tendency for economic, political and social conditions to shift over time, is always likely to be complex and fluid. Typically how this opposition works itself out will involve a range of state and civil society actors. Many contributors have drawn attention to the centrality of the state's role and to that of the EU in stimulating, regulating and co-ordinating a broad range of development initiatives. These different processes may attract consensus and co-operation as well as conflict and competition, as different organized interests try to influence the state and the EU so as to advance their version of rural sustainable development. The optimistic view here is that everyone can win. In our discussion, however, just as much significance has been given to the pessimistic possibility that the pursuit of sustainability is likely to produce losers as well as winners.

# Index